Natural Products: Therapeutic Properties and Beyond II

Natural Products: Therapeutic Properties and Beyond II

Editors

Ana Paula Duarte
Ângelo Luís
Eugenia Gallardo

MDPI • Basel • Beijing • Wuhan • Barcelona • Belgrade • Manchester • Tokyo • Cluj • Tianjin

Editors
Ana Paula Duarte
Universidade da Beira Interior
Portugal

Ângelo Luís
Universidade da Beira Interior
Portugal

Eugenia Gallardo
Universidade da Beira Interior
Portugal

Editorial Office
MDPI
St. Alban-Anlage 66
4052 Basel, Switzerland

This is a reprint of articles from the Special Issue published online in the open access journal *Molecules* (ISSN 1420-3049) (available at: https://www.mdpi.com/journal/molecules/special_issues/NaturalProducts_Therapeutic).

For citation purposes, cite each article independently as indicated on the article page online and as indicated below:

LastName, A.A.; LastName, B.B.; LastName, C.C. Article Title. *Journal Name* **Year**, *Volume Number*, Page Range.

ISBN 978-3-0365-5641-3 (Hbk)
ISBN 978-3-0365-5642-0 (PDF)

Cover image courtesy of Ana Paula Duarte

Contents

About the Editors . . . . **vii**

Ana Paula Duarte, Ângelo Luís and Eugenia Gallardo
Natural Products: Therapeutic Properties and Beyond II
Reprinted from: *Molecules* **2022**, *27*, 6140, doi:10.3390/molecules27196140 . . . . **1**

Weaam Alhallaf and Lewis B. Perkins
The Anti-Inflammatory Properties of Chaga Extracts Obtained by Different Extraction Methods against LPS-Induced RAW 264.7
Reprinted from: *Molecules* **2022**, *27*, 4207, doi:10.3390/molecules27134207 . . . . **5**

Zheling Feng, Zhujun Fang, Cheng Chen, Chi Teng Vong, Jiali Chen, Ruohan Lou, Maggie Pui Man Hoi, Lishe Gan and Ligen Lin
Anti-Hyperglycemic Effects of Refined Fractions from *Cyclocarya paliurus* Leaves on Streptozotocin-Induced Diabetic Mice
Reprinted from: *Molecules* **2021**, *26*, 6886, doi:10.3390/molecules26226886 . . . . **23**

Muhammad Waleed Baig, Madiha Ahmed, Nosheen Akhtar, Mohammad K. Okla, Bakht Nasir, Ihsan-Ul Haq, Jihan Al-Ghamdi, Wahidah H. Al-Qahtani and Hamada AbdElgawad
Caralluma tuberculata N.E.Br Manifests Extraction Medium Reliant Disparity in Phytochemical and Pharmacological Analysis
Reprinted from: *Molecules* **2021**, *26*, 7530, doi:10.3390/molecules26247530 . . . . **47**

Adila Nazli, Muhammad Zafar Irshad Khan, Madiha Ahmed, Nosheen Akhtar, Mohammad K. Okla, Abdulrahman Al-Hashimi, Wahidah H. Al-Qahtani, Hamada Abdelgawad and Ihsan-ul- Haq
HPLC-DAD Based Polyphenolic Profiling and Evaluation of Pharmacological Attributes of *Putranjiva roxburghii* Wall.
Reprinted from: *Molecules* **2022**, *27*, 68, doi:10.3390/molecules27010068 . . . . **65**

Beenish Khanzada, Nosheen Akhtar, Mohammad K. Okla, Saud A. Alamri, Abdulrahman Al-Hashimi, Muhammad Waleed Baig, Samina Rubnawaz, Hamada AbdElgawad, Abdurahman H. Hirad, Ihsan-Ul Haq and Bushra Mirza
Profiling of Antifungal Activities and In Silico Studies of Natural Polyphenols from Some Plants
Reprinted from: *Molecules* **2021**, *26*, 7164, doi:10.3390/molecules26237164 . . . . **85**

Carmen Arlotta, Valeria Toscano, Claudia Genovese, Pietro Calderaro, Giuseppe Diego Puglia and Salvatore Antonino Raccuia
Nutraceutical Content and Genetic Diversity Share a Common Pattern in New Pomegranate Genotypes
Reprinted from: *Molecules* **2022**, *27*, 389, doi:10.3390/molecules27020389 . . . . **101**

Areej Al-Qahtani, Jamaan Ajarem, Mohammad K. Okla, Samina Rubnawaz, Saud A. Alamri, Wahidah H. Al-Qahtani, Ahmad R. Al-Himaidi, Hamada Abd Elgawad, Nosheen Akhtar, Saleh N. Maodaa and Mostafa A. Abdel-Maksoud
Protective Effects of Green Tea Supplementation against Lead-Induced Neurotoxicity in Mice
Reprinted from: *Molecules* **2022**, *27*, 993, doi:10.3390/molecules27030993 . . . . **119**

Joana Gonçalves, Miguel Castilho, Tiago Rosado, Ângelo Luís, José Restolho, Nicolás Fernández, Eugenia Gallardo and Ana Paula Duarte
In Vitro Study of the Bioavailability and Bioaccessibility of the Main Compounds Present in Ayahuasca Beverages
Reprinted from: *Molecules* **2021**, *26*, 5555, doi:10.3390/molecules26185555 . . . . **133**

Ana Ruas, Angelica Graça, Joana Marto, Lídia Gonçalves, Ana Oliveira, Alexandra Nogueira da Silva, Madalena Pimentel, Artur Mendes Moura, Ana Teresa Serra, Ana Cristina Figueiredo and Helena M. Ribeiro
Chemical Characterization and Bioactivity of Commercial Essential Oils and Hydrolates Obtained from Portuguese Forest Logging and Thinning
Reprinted from: *Molecules* **2022**, *27*, 3572, doi:10.3390/molecules27113572 . . . . **149**

Mónica Zuzarte and Lígia Salgueiro
Essential Oils in Respiratory Mycosis: A Review
Reprinted from: *Molecules* **2022**, *27*, 4140, doi:10.3390/molecules27134140 . . . . **167**

Adriana Monserrath Orellana-Paucar and María Gabriela Machado-Orellana
Pharmacological Profile, Bioactivities, and Safety of Turmeric Oil
Reprinted from: *Molecules* **2022**, *27*, 5055, doi:10.3390/molecules27165055 . . . . **191**

About the Editors

Ana Paula Duarte

Ana Paula Duarte is a Professor of Pharmacognosy in the Faculty of Health Sciences of the University of Beira Interior (UBI), and a researcher at the Health Sciences Research Centre of UBI, of which she was coordinator from 2015 to 2019. She is the Coordinator of Pharmaceutical Sciences PhD Program. She has a Pharmaceutical Sciences degree, (PharmD, 6-year program) (1986) and a PhD in Wood Chemistry (1996). She has been a professor at UBI since 1987 and a full professor since 2009. She published 106 papers in international peer-reviewed journals (orcid: 0000-0003-3333-5977), 9 book chapters, 1 national patent, and has over 1900 citations and an h-index of 24 (Scopus). Her research interests focus on extracting the polyphenols of plants and evaluating their biological activities, namely those related to antioxidant activity. Her research interests include the impact and mechanism of action of natural polyphenols in oxidative-stress-related diseases, as well as the bioaccessibility and bioavailability of these compounds.

Ângelo Luís

Ângelo Luís completed his PhD in Biochemistry at the University of Beira Interior in 2014. In 2016, he began a post-doctoral fellowship in Food Chemistry-Mathematical Modeling with the University of Beira Interior and Santander-Totta. Since December 2019, he has worked as an Assistant Researcher and an integrated member of the Health Sciences Research Centre of the University of Beira Interior in the Natural Products and Microbial Research group. He has published 50 SCI papers (h-index=18), 4 book chapters, and 2 national patents. His main research interests are the study of the bioactive properties of plant extracts and essential oils and the development of new biodegradable/edible/bioactive food packaging materials.

Eugenia Gallardo

Eugenia Gallardo has been an Assistant Professor of Toxicology at the Faculty of Health Sciences of the University of Beira Interior since 2009, and she is a researcher at the Health Sciences Research Centre of the University of Beira Interior. She has a degree in Pharmaceutical Sciences (2000), a Master's of Advanced Studies in Toxicology (2002), and a European PhD in Toxicology (2006) obtained at the University of Santiago de Compostela (Spain). In 2011, she became the Head of the Laboratory of Fármaco-Toxicologia, UBImedical (Portugal). She published 121 SCI papers, 16 book chapters and guest-edited 1 book. Her main areas of interest are bioanalysis, toxicology, and drug monitoring, particularly with regard to the development of analytical methods for analysis of environmental substances, pesticides, drugs and their abuse, with special focus on biofluids and other bioactive compounds in non-biological specimens (plants and extracts).

molecules

MDPI

Editorial

Natural Products: Therapeutic Properties and Beyond II

Ana Paula Duarte [1,2,*], Ângelo Luís [1,2,*] and Eugenia Gallardo [1,2,*]

1 Centro de Investigação em Ciências da Saúde, Universidade da Beira Interior (CICS-UBI), Av. Infante D. Henrique, 6201-556 Covilhã, Portugal
2 Laboratório de Fármaco-Toxicologia, Ubimedical, Universidade da Beira Interior, Estrada Municipal 506, 6200-284 Covilhã, Portugal
* Correspondence: apduarte@fcsaude.ubi.pt (A.P.D.); angelo.luis@ubi.pt (Â.L.); egallardo@fcsaude.ubi.pt (E.G.); Tel.: +35-127-532-9002 (A.P.D. & Â.L. & E.G.)

1. Introduction

Historically, natural products have contributed to drug discovery as a source of active molecules due to their great diversity and structural complexity. Thus, they have contributed to the development of drugs for applications in different therapeutic areas. In recent decades, there has been a paradigm shift in drug discovery strategies that has allowed for identifying new natural products that exhibit activities on therapeutic targets. Newman and Cragg studied the origin of 1330 new drugs that had been approved between 1981 and 2010 and found that 64% of them were somewhat related to natural compounds [1]. In a recent review by these same authors, it was noted that, within all of the drugs newly approved by the Food and Drug Administration between January 1981 and September 2019, compounds derived from natural products ranked second [2]. Besides the importance of the discovery of new molecules based on natural compounds, the concern today is focused on the therapeutic potential of secondary metabolites classified as drugs of abuse, such as derivatives of cannabis [3] and psilocybin [4], or even on the use of plants used ancestrally in medicine as well [5,6]. On the other hand, with the development of computational techniques, a decision has been made to study the possibilities of analyzing the pharmacological potential of natural products or their derivatives and converting these molecules into low toxicity active products. However, apart from the use of naturally occurring compounds in the field of health, they have been studied and are increasingly used in solutions, for instance in the agrochemical and food industries.

After the success of the Special Issue "Natural Products: Therapeutic Properties and Beyond I", this second edition aims to categorize the state of the art concerning scientific research on natural products, including their applications as compounds with added value to human health. This issue intends to be used as a text for academia or as a reference tool for researchers, particularly for those working in the fields of medicinal chemistry, toxicology, phytochemistry, and natural product chemistry, and for health and industry professionals.

2. Contributions

This Special Issue gathers nine research papers and two reviews covering developments in the understanding of photochemical profiles, bioactivities, and safety of several natural products with applications in different fields. Concerning therapeutic properties, the anti-inflammatory, antioxidant, antifungal, antibacterial, and antidiabetic potentials of different natural compounds are addressed.

Inflammation is a physiological immune response of the body to injury, characterized by fever, swelling, and pain, and is usually implicated in the pathogenesis of a variety of diseases including, asthma, heart disease, cancer, and diabetes. In response to the increasing interest in the health-promoting effects of chaga (*Inonotus obliquus*), Alhallaf and Perkins [7] have shown that chaga collected in Maine, USA, exhibits significant anti-inflammatory properties against LPS-activated 264.7 RAW macrophages. Their results suggest that

Citation: Duarte, A.P.; Luís, Â.; Gallardo, E. Natural Products: Therapeutic Properties and Beyond II. *Molecules* **2022**, *27*, 6140. https://doi.org/10.3390/molecules27196140

Received: 5 September 2022
Accepted: 16 September 2022
Published: 20 September 2022

extracts produced from accelerated solvent extraction and traditional aqueous chaga tea extraction of chaga powder are superior to other conventional methods (maceration, soxhlet, and reflux) when using NO production and the expression of TNF-α, IL-6, and IL-1β as measurements of anti-inflammatory potential.

The incidence of diabetes mellitus is increasing at an alarming rate worldwide, and the search for new drugs is currently a challenge. Feng et al. [8] studied the anti-hyperglycemic effects of refined fractions obtained from *Cyclocarya paliurus* leaves. The leaves of this plant have been widely used in ethnic medicine or herbal teas to treat diabetes in China. In recent years, the extracts of *C. paliurus* leaves were shown to reduce blood glucose and to improve insulin sensitivity on different diabetic models. These authors evaluated the chemical profiles of ethanol and the water extracts of *C. paliurus* using liquid chromatography coupled to tandem mass spectrometry and have also studied their anti-hyperglycemic effects on STZ-induced mice using glucose tolerance tests, insulin tolerance tests, and homeostasis model assessments of basal insulin resistance. The authors concluded that polysaccharides and flavonoids, but not triterpenoids, were responsible for the anti-diabetic effects of *C. paliurus* leaves.

Baig et al. [9] analyzed the aerial parts of *Caralluma tuberculata* N. E. Br. phytochemically. It is an edible plant traditionally used in the treatment of several diseases such as dysentery, jaundice, constipation, stomach pain, freckles, and pimples. The authors observed the dependence of phytochemicals and bioactivities on the polarity of the extraction solvent. Significant responses were observed in vitro by the test extracts, namely antifungal activities, antileishmanial activities, brine shrimp cytotoxicity, THP-1 leukemia cell line cytotoxicity, and protein kinase inhibition. These findings revealed the therapeutic potential of *C. tuberculate*.

Nazli et al. [10] explored the therapeutic potential of *Putranjiva roxburghii* through different bioactivity studies. This plant is found in Southeast Asia, and traditionally, its uses include the treatment of rheumatism, muscle twisting, fever, arthralgia, pain, and inflammation. The authors studied different parts of the plant (leaf, steam, and fruits), and apart from carrying out a phytochemical analysis, the authors stated that crude extracts are a potential reservoir of phytoconstituents, namely the considerable antioxidant, antibacterial, antidiabetic, and cytotoxic compounds that can be considered as potential candidates for the treatment of different ailments.

An interesting study that combines in silico and experimental sets was presented by Khanzada et al. [11]. These authors explored and identified the mechanism of action of the antifungal agents of edible plants. Eight plants were selected: *Cinnamomum zeylanicum*, *Cinnamomum tamala*, *Amomum subulatum*, *Trigonella foenumgraecum*, *Mentha piperita*, *Coriandrum sativum*, *Lactuca sativa*, and *Brassica oleraceae var. italica*. These common plants were selected due to their minimal/nontoxicity, significant antioxidant and antimicrobial properties, and their frequent use in routine diet. The active phytochemicals of plants were quantified using high-performance liquid chromatography in combination with a diode array detector. In silico studies were performed to understand the underlying antifungal mechanism of detected polyphenols. The findings revealed that *C. zeylanicum*, *C. tamala*, and *A. subulatum* represented good sources of such antifungal polyphenols, and the authors suggested that the high phenolic content of the plant extracts was responsible for their antioxidant and fungal inhibitory activities. They further suggested that the polyphenols detected could be used alone or in synergy with fungal antibiotics to reduce their toxic effects and to increase antifungal efficacy.

The study conducted by Arlotta et al. [12] aimed to evaluate the nutraceutical and genetic diversity of novel pomegranate genotypes (G1–G5) in comparison with leading commercial pomegranate varieties (Wonderful, Primosole, Dente di Cavallo, and Valenciana). The results showed that pomegranate juice is an excellent source of minerals that are essential for human health. Consuming one fruit per day may cover the daily requirement of many minerals, especially potassium, with most present in WD, DC, and G5 among the genotypes investigated. The G1–G3 and G5 genotypes presented total phenolic contents and antioxidant activities comparable with those of the commercial genotypes, representing a valid alternative to the most common cultivars for nutraceutical purposes.

Lead is a chemical toxicant that can cause severe damage to the blood and many body organs such as the liver, the kidneys, the brain, the spleen, and the lungs. It is one of the most important toxic heavy elements in the environment that can penetrate the blood–brain barrier, resulting in lead poisoning, which can cause non-traumatic brain injury. Al-Qahtani et al. [13] demonstrated that the oral administration of green tea extracts improved lead-associated pathological changes in the biochemical and neurobehavioral responses of mice treated for lead poisoning in a significant manner.

Concerning the kinetics of natural products in the human body, Gonçalves et al. [14] evaluated in vitro the bioavailability and bioaccessibility of the main compounds present in ayahuasca beverages. Ayahuasca is a psychoactive beverage traditionally used for divine cults and medicine, but it is also known because of its recreational use. There are no studies concerning the fate of the active compounds of ayahuasca formulations after ingestion, namely concerning their absorption to general circulation for distribution. In this study, the authors observed that compounds such as *N*,*N*-dimethyltryptamine, harmine, harmaline, harmol, harmalol, and tetrahydroharmine were released from the matrix during the in vitro digestion process, becoming bioaccessible. Similarly, some of these compounds (except harmalol and harmol) were absorbed after incubation with the cell monolayer, becoming bioavailable.

Currently, consumers have a growing interest in substances with natural origins, such as essential oils, which have been widely used for several purposes. There has been a growing interest from different industries such as pharmaceuticals, cosmetics, and food, in using essential oils, mainly due to their biological properties, such as antifungal, antibacterial, and antioxidant activities. Ruas et al. [15] studied nationally produced (mainland Portugal and the Azores archipelago) essential oils and hydrolates obtained from forest logging and the thinning of *Eucalyptus globulus*, *Pinus pinaster*, *Pinus pinea*, and *Cryptomeria japonica*. Some of the essential oils and hydrolates showed relevant antioxidant and antimicrobial properties. Thus, it can be concluded that essential oils could be used as natural antioxidants or cosmetic preservatives, for example. Moreover, such products address the demand for sustainable and responsibly sourced odors accepted by consumers.

The review of Zuzarte et al. [16] aimed to collect and systematize relevant information on the antifungal effects of various essential oils and volatile compounds against the main types of respiratory mycoses that impact health systems (namely *Aspergillus fumigatus*, *Candida auris*, and *Cryptococcus neoformans*). The authors collected additional information on the main mechanisms of action underlying the antifungal effects of essential oils and current limitations in clinical applications. The authors stated that essential oils were rich in phenolic compounds and appeared to be very effective in respiratory mycosis, but their application in clinical practice required more comprehensive *in vivo* studies and human trials to assess efficacy and tolerability.

Orellana-Paucar et al. presented another interesting review about turmeric oil [17]. The aims of this review were to discuss its pharmacokinetics, pointing to a potential application of its active molecules in therapy. Its therapeutic potential includes antioxidant, anti-inflammatory, analgesic, antinociceptive, neuroprotective, cardiovascular, antidiabetic, nephroprotective, anticancer, antibacterial, antifungal, antiparasitic, and insecticidal properties. Most research studies reported interesting pharmacological effects without any associated toxicity. However, more studies are required to evaluate the possible clinical application of the active components of this essential oil to determine the pharmacological profile of the isolated active compounds and their bioavailability, efficacy, and safety to maximize their therapeutic benefits.

Overall, this Special Issue of *Molecules* brings together great contributions and continues to advance the discovery of natural products and the development of new applications in different areas of health sciences.

Author Contributions: A.P.D., Â.L. and E.G. conceived, designed, and wrote this editorial. All authors have read and agreed to the published version of the manuscript.

Funding: This work was supported by projects UIDB/00709/2020 and UIDP/00709/2020 (CICS-UBI) carried out with national funds from the Foundation for Science and Technology (FCT) and co-financed by community funds. Ângelo Luís acknowledges the contract for scientific employment in the scientific area of microbiology financed by FCT.

Acknowledgments: Thanks are due to all the authors and peer reviewers for their valuable contributions to this Special Issue.

Conflicts of Interest: The authors declare no conflict of interest.

References

1. Newman, D.J.; Cragg, G.M. Natural products as sources of new drugs over the 30 years from 1981 to 2010. *J. Nat. Prod.* **2012**, *75*, 311–335. [CrossRef] [PubMed]
2. Newman, D.J.; Cragg, G.M. Natural Products as Sources of New Drugs over the Nearly Four Decades from 01/1981 to 09/2019. *J. Nat. Prod.* **2020**, *83*, 770–803. [CrossRef]
3. Gonçalves, J.; Rosado, T.; Soares, S.; Simão, A.Y.; Caramelo, D.; Luís, Â.; Fernández, N.; Barroso, M.; Gallardo, E.; Duarte, A.P. Cannabis and Its Secondary Metabolites: Their Use as Therapeutic Drugs, Toxicological Aspects, and Analytical Determination. *Medicines* **2019**, *6*, 31. [CrossRef]
4. Vargas, A.S.; Luís, Â.; Barroso, M.; Gallardo, E.; Pereira, L. Psilocybin as a New Approach to Treat Depression and Anxiety in the Context of Life-Threatening Diseases-A Systematic Review and Meta-Analysis of Clinical Trials. *Biomedicines* **2020**, *8*, 331. [CrossRef] [PubMed]
5. Gonçalves, J.; Luís, Â.; Gallardo, E.; Duarte, A.P. Psychoactive Substances of Natural Origin: Toxicological Aspects, Therapeutic Properties and Analysis in Biological Samples. *Molecules* **2021**, *26*, 1397. [CrossRef] [PubMed]
6. Cruz, A.; Domingos, S.; Gallardo, E.; Martinho, A. A unique natural selective kappa-opioid receptor agonist, salvinorin A, and its roles in human therapeutics. *Phytochemistry* **2017**, *137*, 9–14. [CrossRef] [PubMed]
7. Alhallaf, W.; Perkins, L.B. The Anti-Inflammatory Properties of Chaga Extracts Obtained by Different Extraction Methods against LPS-Induced RAW 264.7. *Molecules* **2022**, *27*, 4207. [CrossRef] [PubMed]
8. Feng, Z.; Fang, Z.; Chen, C.; Vong, C.T.; Chen, J.; Lou, R.; Hoi, M.P.; Gan, L.; Lin, L. Anti-Hyperglycemic Effects of Refined Fractions from Cyclocarya paliurus Leaves on Streptozotocin-Induced Diabetic Mice. *Molecules* **2021**, *26*, 6886. [CrossRef] [PubMed]
9. Baig, M.W.; Ahmed, M.; Akhtar, N.; Okla, M.K.; Nasir, B.; Haq, I.-U.; Al-Ghamdi, J.; Al-Qahtani, W.H.; AbdElgawad, H. Caralluma tuberculata N.E.Br Manifests Extraction Medium Reliant Disparity in Phytochemical and Pharmacological Analysis. *Molecules* **2021**, *26*, 7530. [CrossRef] [PubMed]
10. Nazli, A.; Irshad Khan, M.Z.; Ahmed, M.; Akhtar, N.; Okla, M.K.; Al-Hashimi, A.; Al-Qahtani, W.H.; Abdelgawad, H.; Haq, I.-U. HPLC-DAD Based Polyphenolic Profiling and Evaluation of Pharmacological Attributes of Putranjiva roxburghii Wall. *Molecules* **2022**, *27*, 68. [CrossRef] [PubMed]
11. Khanzada, B.; Akhtar, N.; Okla, M.K.; Alamri, S.A.; Al-Hashimi, A.; Baig, M.W.; Rubnawaz, S.; AbdElgawad, H.; Hirad, A.H.; Haq, I.-U.; et al. Profiling of Antifungal Activities and In Silico Studies of Natural Polyphenols from Some Plants. *Molecules* **2021**, *26*, 7164. [CrossRef] [PubMed]
12. Arlotta, C.; Toscano, V.; Genovese, C.; Calderaro, P.; Puglia, G.D.; Raccuia, S.A. Nutraceutical Content and Genetic Diversity Share a Common Pattern in New Pomegranate Genotypes. *Molecules* **2022**, *27*, 389. [CrossRef] [PubMed]
13. Al-Qahtani, A.; Ajarem, J.; Okla, M.K.; Rubnawaz, S.; Alamri, S.A.; Al-Qahtani, W.H.; Al-Himaidi, A.R.; Elgawad, H.A.; Akhtar, N.; Maodaa, S.N.; et al. Protective Effects of Green Tea Supplementation against Lead-Induced Neurotoxicity in Mice. *Molecules* **2022**, *27*, 993. [CrossRef] [PubMed]
14. Gonçalves, J.; Castilho, M.; Rosado, T.; Luís, Â.; Restolho, J.; Fernández, N.; Gallardo, E.; Duarte, A.P. In vitro study of the bioavailability and bioaccessibility of the main compounds present in ayahuasca beverages. *Molecules* **2021**, *26*, 5555. [CrossRef] [PubMed]
15. Ruas, A.; Graça, A.; Marto, J.; Gonçalves, L.; Oliveira, A.; da Silva, A.N.; Pimentel, M.; Moura, A.M.; Serra, A.T.; Figueiredo, A.C.; et al. Chemical Characterization and Bioactivity of Commercial Essential Oils and Hydrolates Obtained from Portuguese Forest Logging and Thinning. *Molecules* **2022**, *27*, 3572. [CrossRef] [PubMed]
16. Zuzarte, M.; Salgueiro, L. Essential Oils in Respiratory Mycosis: A Review. *Molecules* **2022**, *27*, 4140. [CrossRef] [PubMed]
17. Orellana-Paucar, A.M.; Machado-Orellana, M.G. Pharmacological Profile, Bioactivities, and Safety of Turmeric Oil. *Molecules* **2022**, *27*, 5055. [CrossRef] [PubMed]

molecules

MDPI

Article

The Anti-Inflammatory Properties of Chaga Extracts Obtained by Different Extraction Methods against LPS-Induced RAW 264.7

Weaam Alhallaf [1,2] and Lewis B. Perkins [1,*]

[1] School of Food & Agriculture, Rogers Hall, University of Maine, Orono, ME 04469, USA; weaam.al@maine.edu

[2] College of Education for Women, University of Baghdad, Baghdad 17001, Iraq

* Correspondence: bperkins@maine.edu

Abstract: Chaga, a sclerotia formed by the *Inonotus obliquus* fungus, has been widely recognized for a number of medicinal properties. Although numerous scientific investigations have been published describing various biological activities of chaga from different geographical locations, little work has focused on chaga harvested in the USA or extraction techniques to maximize anti-inflammatory properties. The aim of this study was to investigate the anti-inflammatory properties of chaga collected in Maine (USA) extracted using traditional aqueous (hot water steeping) methods against lipopolysaccharide (LPS)-induced RAW 264.7 macrophages. Chaga extracts obtained from both conventional (ethanol/water) extraction methods and an accelerated solvent extraction method (ASE) at optimized conditions were compared to aqueous extracts (tea) obtained from chaga in the powder form (P) and powder form in tea bags (B) based on their effect on both nitric oxide (NO) production and pro-inflammatory cytokine expression, in particular, the expression of TNF-α, interleukin-6 (IL-6), and interleukin-β (IL-1β). Phenolic acid extracts from chaga and individual phenolic acid standards were also investigated for their effect on the same parameters. Results indicated that various chaga extracts have significant anti-inflammatory activity on LPS-stimulated RAW 264.7 cells. The inhibitory effect was through a decrease in the production of NO and the downregulation of TNF-α, IL-6, and IL-1β in RAW 264.7 macrophages. ASE1 (novel, optimized ethanol/water extraction) and P6 (six-minute steeping of powder in 100 °C water) extracts showed the highest inhibitory activity on NO production and on the expression of the inflammatory cytokines, compared to extracts obtained by conventional extraction methods.

Keywords: natural product extracts; nutraceuticals; inflammation treatment; healthful fungal products

Citation: Alhallaf, W.; Perkins, L.B. The Anti-Inflammatory Properties of Chaga Extracts Obtained by Different Extraction Methods against LPS-Induced RAW 264.7. *Molecules* **2022**, *27*, 4207. https://doi.org/10.3390/molecules27134207

Academic Editors: Ana Paula Duarte, Ângelo Luís and Eugenia Gallardo

Received: 31 May 2022
Accepted: 23 June 2022
Published: 30 June 2022

1. Introduction

Inflammation is a physiological immune response of the body to injury, characterized by fever, swelling, and pain, and is usually implicated in the pathogenesis of a variety of diseases including, asthma, heart disease, cancer, and diabetes [1] During the inflammatory process, large amounts of pro-inflammatory mediators such as nitric oxide (NO), prostaglandin E2 (PGE2), tumor necrosis factor-α (TNF-α), interleukin-6 (IL-6), and interleukin-1β (IL-1β) are generated, mostly for the primary protection of the host [2,3]. However, excess uncontrolled production of these inflammatory products can lead to oxidative stress [3,4]. Recently, there has been considerable interest in finding anti-inflammatory agents with low toxicity, from natural sources [1]. Phenolic compounds are among the many natural substances extracted from plant and fungal sources attracting attention for anti-inflammatory and other biological activities.

Previous studies have shown that chaga (*Inonotus obliquus*) contains various bioactive components with a range of chemical characteristics and polarities [5–7]. Accordingly, extraction of bioactive compounds from chaga requires various solvents targeted to specific chemical profiles. For example, petroleum ether and chloroform are used to extract lanostane-type triterpenoids from chaga; whereas water and aqueous alcohol are suitable solvents targeted for polysaccharides, melanin pigments, and phenolic compounds [8,9]. Aqueous preparations of chaga have been used since the 12th century in Eastern Europe to treat a variety of ailments without toxic effect. The traditional preparation method for chaga tea is steep or boil the sclerotia (small pieces or powder) in water for a few hours or a few days. The resultant extracts from these processes can be immediately consumed or stored at a suitable temperature [10]. Currently, "quick brewing" is a more common way to consume chaga, most often as a tea. Chaga in powder form, usually in tea bags, is steeped in hot water (80–100 °C) for a several minutes, strained, and then consumed as a tea. Investigations of the chemical composition and biological properties of aqueous extracts from chaga have shown therapeutic effects against diabetes via multiple pathways, including antioxidative [11]. Research has found that orally administered aqueous extract of chaga can ameliorate acute inflammation [12]. However, no investigation has been carried out to examine the chemical composition of chaga tea in powder and tea-bag form.

Alcoholic extracts from chaga are characterized by high phenolic content. It has been reported that phenolics are the main chemical compounds involved in chaga's biological effects—including its antioxidant, anti-cancerous, antimicrobial, and anti-inflammatory activities [13–19]. A variety of conventional and advanced extraction methods have been used to isolate phenolic compounds from chaga [20,21]. Recent work [22] found that accelerated solvent extraction (ASE) increases the total phenolic content and enhances the DPPH-scavenging activity of chaga extracts, as compared to conventional extraction methods. In addition, the concentrations of individual phenolic acids significantly increased compared to concentrations in other extracts. Previous studies have demonstrated that alcoholic extracts from chaga possess significant anti-inflammatory effects in vivo and in vitro. However, no studies have investigated the effect of the extraction method on the anti-inflammatory properties of chaga extracts.

The focus of the work presented in this manuscript is the investigation of the effect of chaga extracts obtained from a number of consumer-based and laboratory extraction techniques on anti-inflammation in LPS-stimulated 264.7 cells. Extraction methods include powdered (P3, P6, P10) and bagged powdered (B3, B6, B10) chaga steeped in 100 °C water for 3, 6, and 10 min. Chemical analysis was performed on these aqueous-based extracts. Traditional extraction in 70/30 ethanol/water (*v*/*v*) using maceration (ME), soxhlet (SE), reflux (RE); and accelerated solvent extraction (ASE) at 130 (1), 150 (2) and 170 °C (3) were also tested on the aforementioned anti-inflammatory cell model. This is the first research focused on birch chaga collected in Maine, USA and also examines the biologically active chemical features of chaga tea steeping in bag and powder form.

2. Results

2.1. Chemical Composition

The chemical yield and the bioactive content of the crude extracts from the powdered form of chaga were significantly higher ($p < 0.05$) than those of the bagged form at all extraction temperatures (Table 1). For example, P6 yielded 30.21 ± 0.01% crude polysaccharide containing 17.56 ± 0.01% carbohydrate, 10.23 ± 1.02% protein, 4.12 ± 0.44% uronic acid, and 17.61 ± 0.05% (*w*/*w*) phenolic content, while B6 yielded 17.33 ± 0.03% crude polysaccharide containing 11.56 ± 0.01% carbohydrate, 7.11 ± 0.5% protein, 2.57 ± 0.71% uronic acid, and 8.33 ± 0.05% (*w*/*w*) phenolic content. The results also indicated that brewing time had no significant effect on the yield and the chemical content of the crude extracts of both bagged and powdered form. The B3 extract resulted in a carbohydrate content of 10.26 ± 0.08% (*w*/*w*), which was comparable to the carbohydrate content of 11.56 ± 0.01% and 11.81 ± 0.04% (*w*/*w*) obtained from B6 and B10 extracts, respectively. Similarly, P3, P6, and P10 extracts had carbohydrate content of 17.02 ± 0.01%, 17.56 ± 0.01%, and 18.31 ± 0.05% (*w*/*w*), respectively. The same trends were observed for the yield, protein, uronic acid, and total phenolic content of crude polysaccharide extracts of chaga tea obtained from different brewing times.

Table 1. Major chemical content of the crude polysaccharide extracts from chaga sclerotia. The data represent the mean ± SD of triplicate experiments.

Sample	Yield%	Carbohydrate%	Protein%	Uronic Acid%	TPC%
P3	30.66 ± 0.05 [a]	17.02 ± 0.01 [a]	10.53 ±1.66 [a]	4.11 ± 0.47 [a]	16.77 ± 0.13 [ab]
P6	30.21 ± 0.01 [a]	17.56 ± 0.01 [a]	10.23 ±1.02 [a]	4.12 ± 0.44 [a]	17.61 ± 0.05 [a]
P10	31.33 ± 0.04 [a]	18.31 ± 0.05 [a]	11.02 ±1.28 [a]	4.11± 0.26 [a]	17.04 ± 0.03 [a]
B3	15.53 ± 0.01 [c]	10.26 ± 0.08 [b]	7.14 ± 1.83 [b]	3.01 ± 0.53 [b]	6.49 ± 0.04 [d]
B6	17.33 ± 0.03 [b]	11.56 ± 0.01 [b]	7.11 ± 0.5 [b]	2.57 ± 0.71 [b]	8.33 ± 0.05 [c]
B10	17.33 ± 0.02 [b]	11.81 ± 0.04 [b]	7.23 ± 2.01 [b]	2.84 ± 1.22 [b]	8.33 ± 0.06 [c]

P: powdered chaga; B: bagged powdered chaga steeped at 100 °C for 3, 6, or 10 min. Different letters (a,b,c,d) in the same column denote statistically different means ($p < 0.05$) according to ANOVA and Tukey's test% (w/w).

2.2. Cell Viability

RAW 264.7 cells were treated with various chaga extracts or pure phenolic standards: 50, 100, and 150 μg/mL chaga extracts; or 25, 50, and 100 μM vanillic acid (VA), caffeic acid (CA), or syringic acid (SA); or 5, 10, and 20 μM protocatechuic acid (PA) or protocatechuic aldehyde (PCA) to assess their effects on cell viability using the MTT (3-(4,5-dimethylthiazolyl-2)-2,5-diphenyltetrazolium bromide) assay. This assay measures the activity of the mitochondrial succinate-tetrazolium reductase of living cells and its ability to cleave the tetrazolium salts to formazan crystals, resulting in a color change that can be monitored spectrophotometrically [23]. The data were expressed as percent cell viability compared to control (0.05% DMSO). The results revealed that chaga extracts and the pure standards caused no cytotoxicity at the examined concentrations in RAW 264.7 cells (Figure 1) and subsequent experiments were performed at these concentrations.

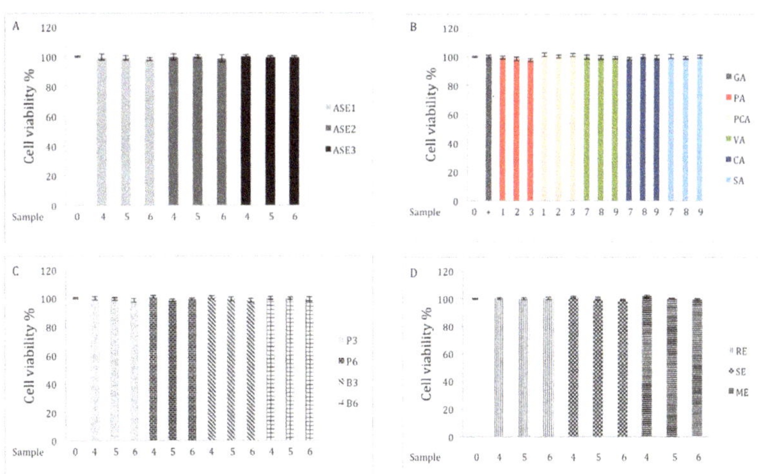

Figure 1. Effects of different extracts and phenolics on cell viability of RAW264.7 cells. Extracts were obtained by (**A**) ASE conditions, (**B**) pure phenolic acid standards, (**C**) crude polysaccharide extracts, and (**D**) different extraction methods. Cells were stimulated with 1 μg/mL of LPS plus varying concentrations of samples (0 = media, GA = 2 μM; 1 = 5 μM; 2 = 10 μM; 3 = 20 μM; 4 = 50 μg/mL; 5 = 100 μg/mL; 6 = 150 μg/mL; 7= 25 μM; 8 = 50 μM; and 9 = 100 μM).

2.3. Inhibition of NO Production in LPS-Stimulated RAW 264.7 Macrophages

Inhibitory effects of various chaga extracts and pure phenolic acid standards on NO production on LPS-induced RAW 264.7 cells were examined. Stimulation of the cells with

1 μg/mL LPS increased the NO levels to 44.53 ± 0.23 μM as compared to 4.8 ± 0.12 μM in the negative control group (Figure 2). All ASE extracts significantly ($p < 0.05$) reduced the level of NO production at all tested concentrations (Figure 2A). It is of note that at 150 μg/mL, ASE1, ASE2, and ASE3 reduced the concentration of NO released from RAW 264.7 cells by 66.82%, 61.61%, and 43.34%, respectively, compared with the LPS group. We further investigated the inhibitory effect of the main phenolic acids found in chaga extracts on the inhibition of NO production. Results showed that treating the LPS-induced cells with various concentrations of PA, PCA, and CA significantly ($p < 0.05$) reduced the production of NO, while treatments with SA and VA did not exhibit any effect on the induced cells (Figure 2B). At the highest examined concentrations, the levels of NO released from the induced cells decreased by 54.35%, 58.66%, and 41.60% after treatments with PA, PCA, and CA, respectively. However, the VA and SA treatments did alter the level of NO production at all concentrations.

All crude polysaccharide extracts separated from chaga tea in the powder form showed a significant inhibitory effect ($p < 0.05$) on NO production (Figure 2C). However, crude polysaccharide extracts separated from the bag form of chaga tea exhibited a significant inhibitory effect ($p < 0.05$) on NO production only at the highest examined concentration of the extracts compared to the LPS group. For example, at 150 μg/mL, P6 and B6 extracts reduced the nitrile concentration in the supernatants by 67.76% and 37.31%, respectively, compared to the LPS group.

To investigate the effect of the conventional extraction methods on the inhibition of NO production, RAW264.7 cells were incubated with extracts obtained by different extraction methods in the presence of 1 μg/mL of LPS. ME had no effect on the inhibition of NO production, but both RE and SE extracts showed significant inhibition ($p < 0.05$) on the production of NO, compared to the LPS group (Figure 2D). At 150 μg/mL, RE and SE extracts reduced the nitrile concentration in the supernatants by 51.42% and 52.61%, respectively, compared to the LPS group.

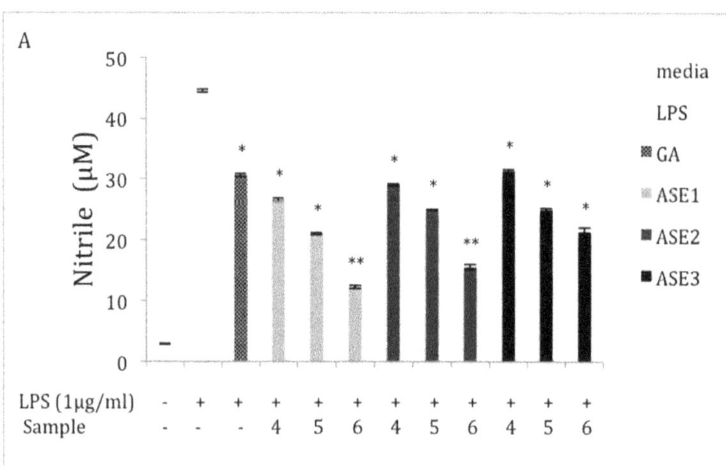

Figure 2. *Cont.*

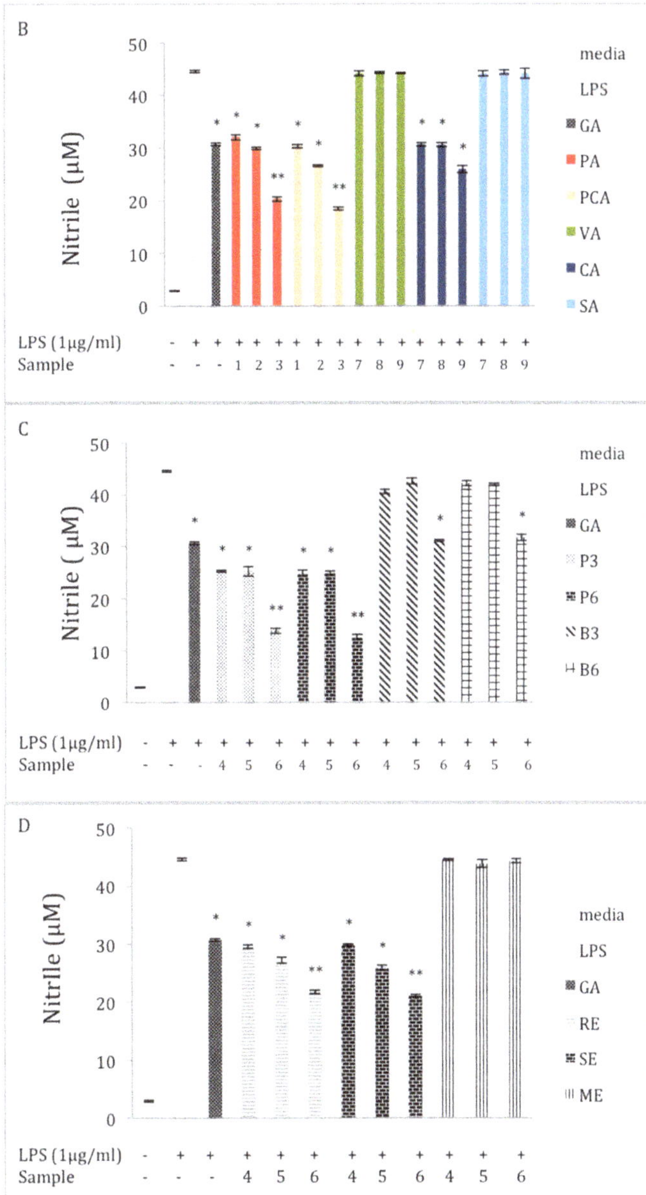

Figure 2. Effect of extracts and phenolics on production of nitric oxide (NO) in macrophage RAW 264.7 cells (**A**) ASE extracts; (**B**) pure phenolic standards; (**C**) crude polysaccharide extracts; and (**D**) different extraction methods. Cells were cultured in the absence (−) or presence (+) of LPS (1 µg/mL) with various concentrations of different samples (- denotes no sample) for 24 h (0 = media; GA = 2 µM; 1 = 5 µM; 2 = 10 µM; 3 = 20 µM; 4 = 50 µg/mL; 5 = 100 µg/mL; and 6= 150 µg/mL; 7 = 25 µM; 8 = 50 µM; and 9 = 100 µM). NO production was measured by the Griess reagent and is represented as mean ± standard error (SE) in the bars. Statistical significance $p < 0.05$ (*) and $p < 0.01$ (**) was determined using one-way analysis of variance for independent means, followed by Tukey's HSD test.

2.4. Inhibition of TNF-α Production in LPS-Stimulated RAW 264.7 Macrophages

To evaluate the anti-inflammatory effect of chaga, the inhibitory effects of various extracts and pure phenolic standards on the expression of the pro-inflammatory cytokines TNF-α, 1L-6, and IL-1β in LPS-stimulated RAW 264.7 cells were investigated. We found an increase in the expression of TNF-α in the LPS stimulation group compared to the control group (Figure 3). Twenty-four hours of incubation with ASE extracts significantly inhibited the level of TNF-α ($p < 0.05$) in the LPS-induced cells, compared with the LPS group. At a concentration of 150 μg/mL, ASE1, ASE2, and ASE3 extracts reduced the level of TNF-α released from RAW 264.7 cells by 42.90%, 35.94%, and 31.15%, respectively, as compared to the LPS group (Figure 3A). The effect of the pure phenolic standards on the expression of TNF-α is presented in Figure 3B. All phenolic acids except VA and SA significantly ($p < 0.05$) decreased the secretion of TNF-α at all examined concentrations, compared to the LPS treatment. At the highest concentrations, PA, PCA, and CA reduced the expression of TNF-α by 44.3%, 46.7%, and 32.5%, respectively, compared to the LPS treatment. The results from the crude polysaccharide extracts showed that extracts from the powder form of chaga tea had significant inhibitory effects ($p < 0.05$) on the level of TNF-α, compared with the LPS group. At a concentration of 150 μg/mL, P3 and P6 extracts reduced the level of TNF-α in the supernatants by 37.2% and 37.5%, respectively (Figure 3C).

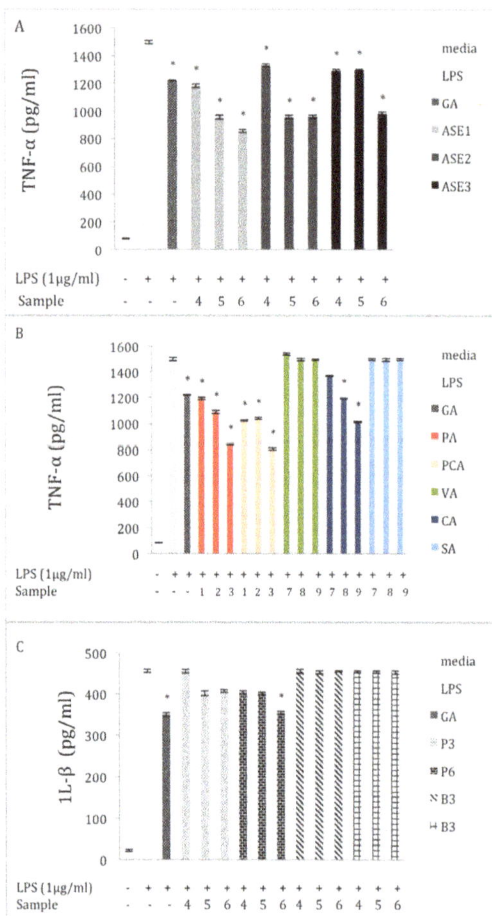

Figure 3. *Cont.*

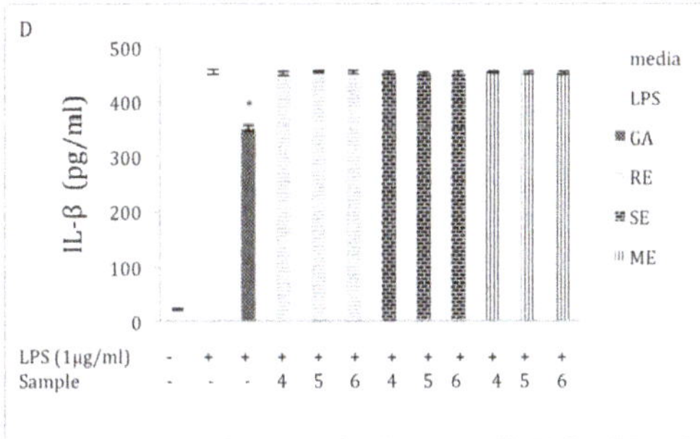

Figure 3. Effect of extracts and phenolics on tumor necrosis factor-α (TNF-α) expression in macrophage RAW 264.7 cells (**A**) ASE extracts; (**B**) pure phenolic standards; (**C**) crude polysaccharide extracts; and (**D**) different extraction methods. Cells were cultured in the absence or presence of LPS (1 μg/mL) with various concentrations of different samples for 24 h (0 = media; GA = 2 μM; 1 = 5 μM; 2 = 10 μM; 3 = 20 μM; 4 = 50 μg/mL; 5 = 100 μg/mL; and 6 = 150 μg/mL; 7 = 25 μM; 8 = 50 μM; and 9 = 100 μM). TNF-α production was determined through an ELISA. The data represent the mean ± SE of triplicate experiments. Statistical significance $p < 0.05$ (*) was determined using one-way analysis of variance for independent means, followed by Tukey's HSD test.

The level of NO inhibited by chaga extracts depended on extraction method. Further examination found that extraction method affected the attenuation level of TNF-α. At a concentration of 150 μg/mL, extracts made by RE and SE had significant effects ($p < 0.05$) on the level of TNF-α, compared to LPS treatment. For example, TNF-α expression was inhibited by 22.3%, and 22.1% after treatment with 150 μg/mL of RE, and SE, respectively. The results also showed that ME extract had no inhibitory effect on the level of TNF-α at any concentration (Figure 3D).

2.5. *Inhibition of IL-6 Production in LPS-Stimulated RAW 264.7 Macrophages*

The expression of IL-6 cytokine increased to (755 ± 0.42 pg/mL) in macrophage cells after stimulation with LPS. However, when various chaga extracts obtained by ASE were added at 50, 100, and 150 μg/mL, these increases were significantly ($p < 0.05$) reduced. For example, at 150 μg/mL, ASE1, ASE2, and ASE3 extracts reduced the level of IL-6 released from RAW 264.7 cells by 57.3%, 49.4%, and 50.4%, respectively, compared with the LPS group (Figure 4A). The inhibition activity of phenolic acid standards on the expression of IL-6 in LPS-induced 264.7 RAW cells is illustrated in Figure 4B. The same trend of TNF-α expression inhibition for all levels of IL-6 was observed for all assayed phenolic acids except VA and SA ($p < 0.05$). The levels of IL-6 at all examined concentrations were suppressed compared to the LPS treatment. At the highest concentrations, PA, PCA, and CA reduced the expression of IL-6 by 53.7%, 56.6%, and 39.5%, respectively, compared to the LPS treatment. The effect of the polysaccharide extracts on the expression of IL-6 was similar to their effect on TNF-α; the expression of IL-6 was significantly reduced ($p < 0.05$) after treating the induced cells with P3 and P6 extracts at different concentrations. For example, the IL-6 level was reduced by 56.8% and 57.1% after the LPS-induced cells were treated with 150 μg/mL of P3 and concentration (Figure 4C). Figure 4D shows that extracts obtained by SE and RE had a significant inhibitory effect ($p < 0.05$) on the expression of IL-6 cytokine compared with the LPS group. At a concentration of 150 μg/mL, RE and SE inhibited the cytokine level by 26.6% and 26.9%, respectively, compared with the LPS group. No inhibitory effect of ME extract on the level of IL-6 was observed at any concentration.

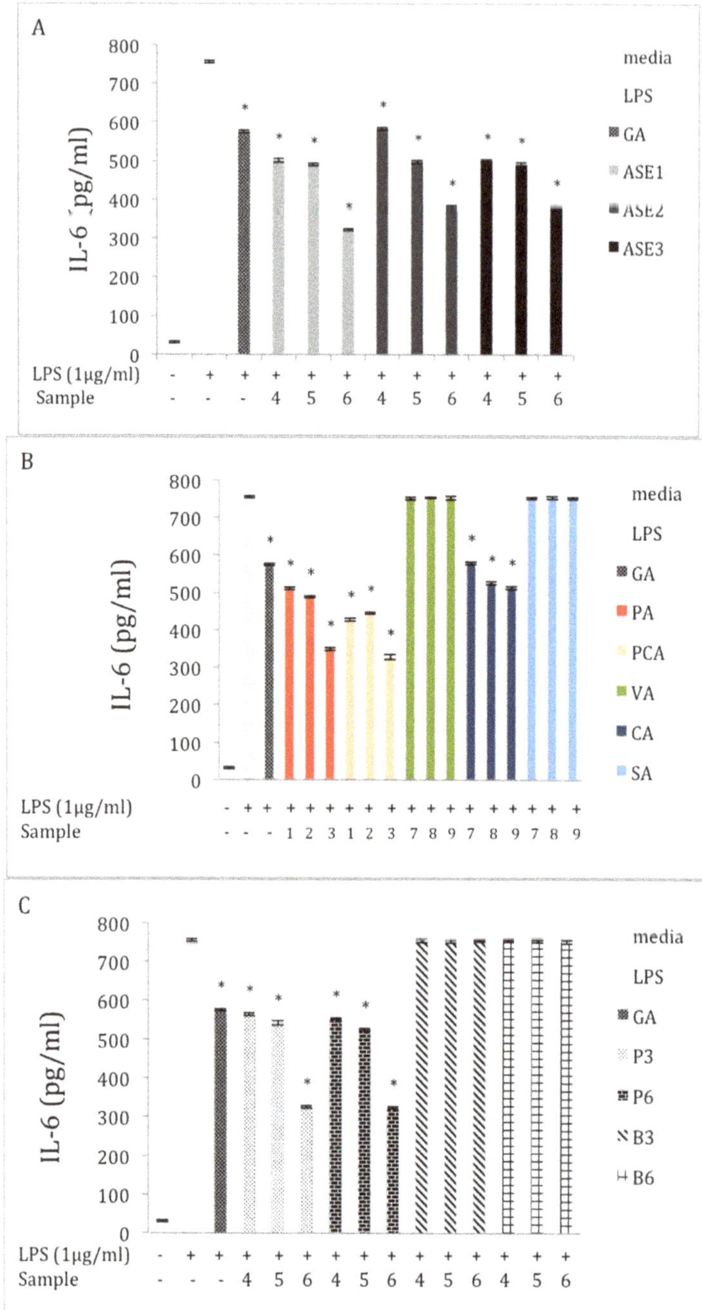

Figure 4. *Cont.*

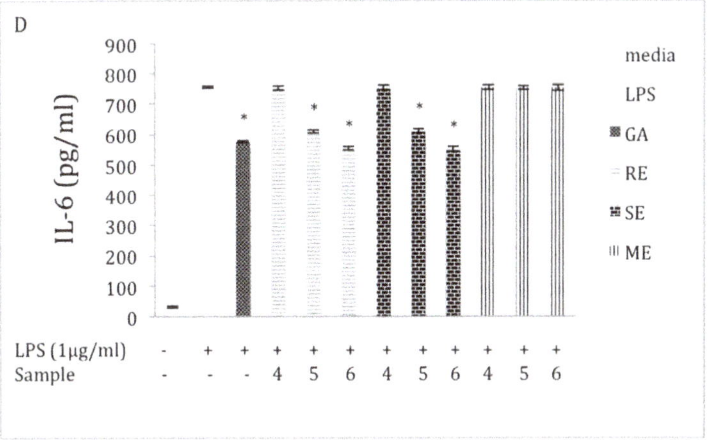

Figure 4. Effect of extracts and phenolics on IL-6 expression in macrophage RAW 264.7 cells (**A**) extracts obtained by ASE; (**B**) pure phenolic acid standards; (**C**) crude polysaccharide extracts; and (**D**) different extraction methods. Cells were cultured in the absence or presence of LPS (1 µg /mL) with various concentrations of different samples for 24 h (0 = media; GA = 2 µM; 1 = 5 µM; 2 = 10 µM; 3 = 20 µM; 4 = 50 µg/mL; 5 = 100 µg/mL; and 6 = 150 µg/mL; 7 = 25 µM; 8 = 50 µM; and 9 = 100 µM). IL-6 production was determined through an ELISA. The data represent the mean ± SD of triplicate experiments. Statistical significance $p < 0.05$ (*) was determined using one-way analysis of variance for independent means, followed by Tukey's HSD test.

2.6. Inhibition of IL-β Production in LPS-Stimulated RAW 264.7 Macrophages

Our data indicated an increase in the expression of IL-β in the LPS-stimulation group compared to the control group (Figure 5). At 150 µg/mL, all chaga extracts obtained by ASE displayed a significant inhibitory effect ($p < 0.05$) on the level of IL-1β as compared to the LPS group (Figure 5A). For example, ASE1 reduced the level of IL-1β by 22.6% compared to the LPS group. Similarly, the level of IL-1β decreased significantly ($p < 0.05$) after treating the LPS-induced cells with the highest concentrations of PA and PCA, respectively (Figure 5B). For example, the concentration of IL-1β in the supernatants was decreased by 22.6% and 21.5%, after incubation of the LPS-induced cells with highest concentration of PA and PCA, respectively. The effect of the polysaccharide extracts on the expression of IL-β is presented in Figure 5C. Polysaccharide extracts from the chaga obtained from the tea-bag form did not affect the level of inflammatory cytokine. However, at 150 µg/mL, P6 significantly ($p < 0.05$) reduced the level of IL-1β, by 21.5%, compared to the LPS treatment. Figure 5D presents the effect of the extraction method on the inhibition activity of the extracts against IL-1β in the LPS-induced macrophages. None of the extracts obtained by the described conventional methods affected the expression of the IL-1β at any concentration compared to the LPS group.

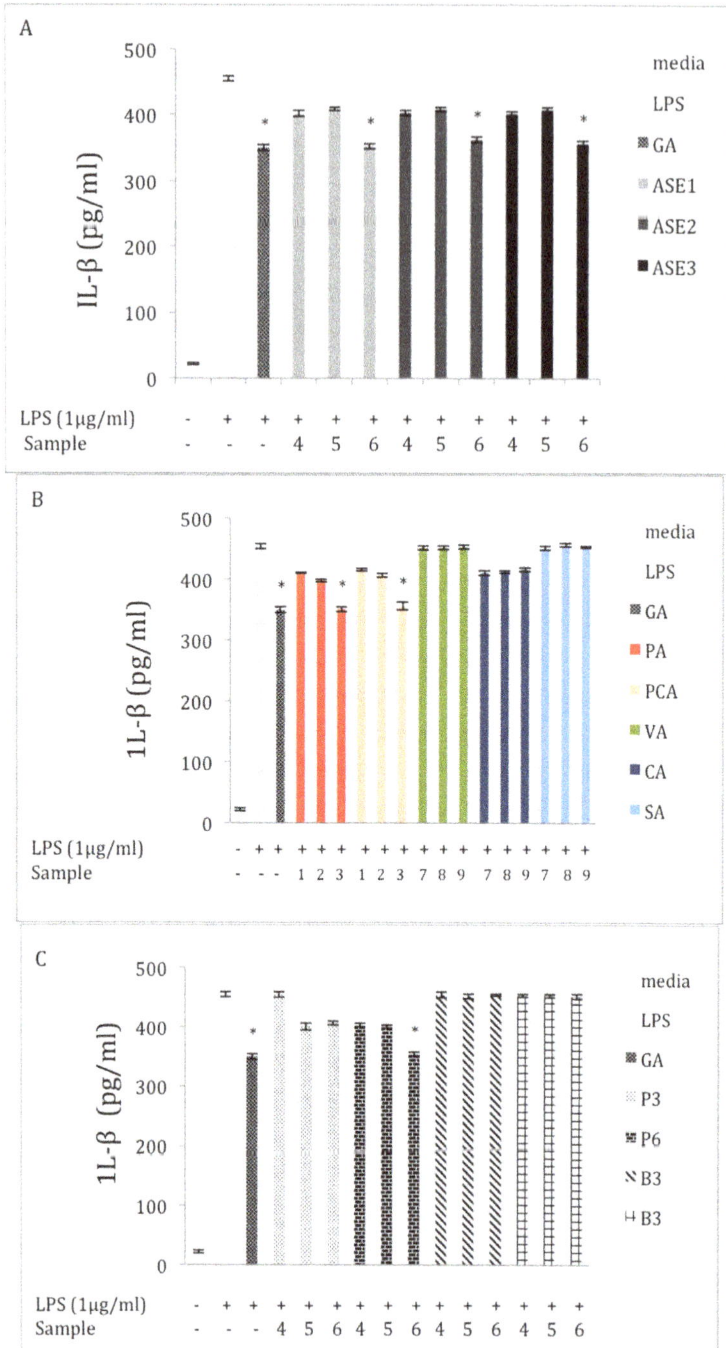

Figure 5. *Cont.*

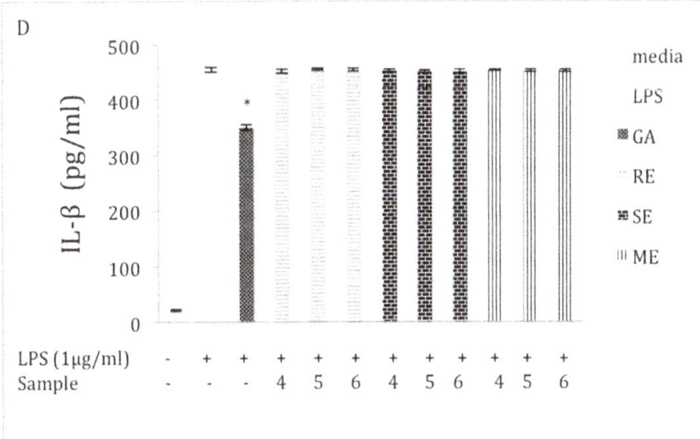

Figure 5. Effect of extracts and phenolics on IL-β expression in macrophage RAW 264.7 cells (**A**) extracts obtained by ASE; (**B**) pure phenolic acid standards; (**C**) crude polysaccharide extracts; and (**D**) different extraction methods. Cells were cultured in the absence or presence of LPS (1 μg /mL) with various concentrations of different samples for 24 h (0 = media; GA = 2 μM; 1 = 5 μM; 2 = 10 μM; 3 = 20 μM; 4 = 50 μg/mL; 5 = 100 μg/mL; and 6 = 150 μg/mL; 7 = 25 μM; 8 = 50 μM; and 9 = 100 μM). IL-β production was determined through an ELISA. The data represent the mean ± SD of triplicate experiments. Statistical significance $p < 0.05$ (*) was determined using one-way analysis of variance for independent means, followed by Tukey's HSD test.

3. Discussion

Inflammation is a physiological immune response of body tissues against physical, chemical, and biological stimuli such as tissue injury, chemical toxins, or pathogens [3]. Lipopolysaccharide, an endotoxin, is an integral outer membrane component of gram-negative bacteria, and the most potent trigger for microbial initiators of inflammatory response [24]. Macrophages play a vital role in the immune system and are associated with inflammatory diseases as initiators of pathogen- or tissue-induced inflammation. Macrophages activated by LPS treatment produce a wide variety of inflammatory markers, including NO TNF-α, IL-6, and IL-1β, primarily for the protection of the host. However, excess and uncontrolled production of inflammatory products often leads to excessive inflammatory response and oxidative stress [4,25,26]. Anti-inflammatory agents produce an anti-inflammatory effect through regulating cytokines and these inflammatory mediators. Therefore, monitoring the expression of these mediators is vital for understanding the inflammatory process and provides a measure to evaluate the effects of anti-inflammatory agents [1].

Nitric oxide is a multi-functional mediator that plays an important role in cellular signaling and a variety of physiological functions in many cells and tissues, including the brain, the vasculature, and the immune system [27]. Evidence indicates that the overproduction of NO is a significant contributor to inflammatory processes and may provide an indicator of the degree of inflammation [28,29]. Therefore, the inhibition of NO overproduction is a useful measure for assessing the anti-inflammatory effects of drugs [30].

Inflammation is characterized by the production of a wide variety of free radicals, nitrogen reactive species, and cytokines—such as TNF-α, IL-6, and IL-1β—which act as modulators throughout the inflammation process [31]. TNF-α stimulates the production of other cytokines such as IL-6 and IL-1β. IL-6 is a multifunctional cytokine with pro- and anti-inflammatory properties that plays a central role in immune and inflammatory responses [32]. IL-1β is also a multifunctional cytokine that has been implicated in pain, fever, inflammation, and autoimmune conditions [33]. High levels of these cytokines elicit number of physiological effects including septic shock, inflammation, and cytotoxicity [2,34]. Thus, inhibiting the expression of cytokines in macrophage cells is very important during the anti-inflammatory response.

For up to 1000 years, chaga has been recognized for its medicinal properties and used widely for a plethora of illnesses in the world's northern hemisphere. Numerous scientific reports, mostly with Asian and eastern European roots, have investigated the chemical composition and the biological activities of chaga from various geographical locations. However, no previous study has utilized chaga from the U.S. The results of this study show that different extracts of chaga sclerotia collected from Maine have significant anti-inflammatory activity in LPS-stimulated RAW 264.7 cells. The inhibitory effect occurred through a decrease in the production of the NO and a downregulation of TNF-α, IL-6, and IL-1β in RAW 264.7 macrophages, with no effect on cell viability at a concentration range of (50–150 μg/mL). It is especially interesting that the results detailed in this study also show that phenolic extracts obtained from different extraction methods have different anti-inflammatory properties. All extracts produced from the accelerated solvent extraction method and most conventional methods significantly reduce the level of NO, while the ME extract had no effect on the NO production at any concentration. The inhibitory effect of chaga extracts can likely be attributed to the content of phenolic compounds and resulting free-radical scavenging activity. Phenolic compounds have been widely recognized as natural molecules with potential antioxidant activity. It is well known that oxidative stress can activate a variety of inflammatory mediators that contribute to the inflammation process. Such oxidative stress can be inhibited by compounds with high antioxidant activity such as phenolic compounds. Previous studies have also reported that alcohol extracts of chaga are rich in polyphenolic compounds that possess strong antioxidant activity and can protect cells against oxidative damage. Such extracts have been reported to attenuate inflammation reactions and decrease the production of inflammatory mediators in 264.7 macrophages.

Earlier research focused on small phenolic compounds as a main constituent of alcoholic extracts of chaga and noted significant contribution to the antioxidant activity. However, no reports have examined the anti-inflammatory effect of these constituents using RAW 264.7 cells. In our previous work [22] we increased the extraction of phenolic acids from chaga by optimizing an ASE extraction method. In the current study, we investigated the NO-inhibitory activities of small phenolic compounds on stimulated RAW 264.7 cells. Results show phenolic acids influence different inflammatory mediators; some of the compounds tested did not affect the production of NO or the expression of the inflammatory cytokines, while others had a significant effect. For example, both VA and SA did not alter the production of NO and the expression of TNF-α, IL-6, and IL-1β at any of the examined concentrations. However, PA, PCA, and CA significantly reduced the production of NO and attenuated the expression of TNF-α, IL-6, and IL-1β at all the examined concentrations. This is in agreement with previous studies, which demonstrated anti-inflammatory properties for PA, PCA, and CA [35]. However, other research suggests that phenolic compounds with only one phenol ring, such as the tested compounds, generate a lesser anti-inflammatory effect, through inhibition of cytokine production; it has been hypothesized that other mechanisms may be involved in the anti-inflammatory action of phenolics [36]. Our work suggests that some of the small phenolic compounds present in chaga play a vital role in anti-inflammatory activity.

Previous evidence demonstrates that polysaccharides from many sources have a variety of therapeutic effects, including antioxidant and anti-inflammatory activities. Polysaccharides in chaga from Russia, China, and South Korea have been reported to act as immune-modulators and possess anti-inflammation properties from in vivo and in vitro studies. In this study, we examined the chemical structure of crude polysaccharides extracted from chaga tea, in both powder and bagged form, collected from Maine, USA. We evaluated the anti-inflammation effects of these extracts using LPS-stimulated RAW 264.7 cells. The results indicate that more components of chaga could be extracted from the powder form than from bagged form. For example, P6 yielded 30% polysaccharide containing 17.56% carbohydrate, 10.23% protein, 4.12% uronic acid, and 17.61% phenolic content;

while B6 yielded 17.33% polysaccharide containing 11.81% carbohydrate, 7.23% protein, 2.84% uronic acid, and 8.33% phenolic content. Our data are in agreement with previous reports suggesting that more bioactive constituents can be extracted from different raw sources in powder forms than other forms. We attribute the higher extraction efficiency to the greater surface area of powdered form, which allows better penetration of the solvents to target analytes and thus higher extraction efficiency. The results show that the crude polysaccharides of chaga extracts in both forms have high phenolic content, indicating that these phenolic compounds are bound to macromolecules in chaga such as polysaccharides and melanin. Chaga contains high amounts of the water-soluble macromolecule pigments known as melanin. The dark color of the extracts suggests that a relatively high amount of melanin is present. Other reports also suggested that crude polysaccharide extract from chaga has melanin and melanin-associated phenolic compounds.

Our work demonstrates that crude polysaccharides obtained from both the powder and the bagged form significantly inhibit LPS-induced NO production in RAW 264.7 cells. We observed a higher NO inhibitory effect of polysaccharides obtained from chaga tea in the powdered form, in comparison to those in bagged form at the same concentration. Polysaccharides obtained from the powdered form significantly inhibited the production of TNF-α and IL-6; however, none of the crude polysaccharide extracts obtained in either form altered the expression of IL-1β. However, since the extract is crude, it contains high values of phenolic compounds and melanin, which may contribute to the anti-inflammatory effect of the extracts.

4. Materials and Methods

4.1. Fungal Material

Chaga sclerotia of various sizes and ages were collected from yellow birch (*Betula alleghaniensis*) in Maine forests. Removing chaga from tree surfaces with a hatchet or hammer, samples were collected by volunteers throughout the year and transported to the University of Maine Analytical Food Chemistry Laboratory where the entire sclerotia were immediately frozen and lyophilized (Model 7754511, Labconco Corporation, Kansas City, MO, USA) and ground using an electric food processor (Nutribullet, model-NBR-1201M, Los Angeles, CA, USA). The resulting powder was passed through a 20-mesh (0.84 mm) sieve and only particles with a diameter smaller than 0.84 mm (20-mesh) were collected. All processed samples were stored at 20 °C. A standard sample for all experimental work was created by combining and thoroughly mixing equal weights of individual processed powdered chaga. All sample extractions and assays were performed in triplicate.

4.2. Reagents

Folin–Ciocalteu (FC) reagent, 1,1-diphenyl-2-picrylhydrazyl (DPPH), 3,4-dihydroxybenzoic acid, caffeic acid, syringic acid, 3,4-dihydroxybenzaldehyde, bovine serum albumin (BSA), galacturonic acid, Griess reagent, 3-(4,5-dimethylthiazol-2-yl)-2,5-diphenyl-tetrazolium bromide (MTT), dimethyl sulfoxide (DMSO), and *Escherichia coli* LPS were purchased from Sigma–Aldrich (St. Louis, MO, USA). Ethanol, sodium carbonate, vanillic acid, diatomaceous earth, and Ottawa sand were purchased from Fisher Scientific (Fair Lawn, NJ, USA). Ultrapure water was obtained from a Millipore water system (EMD Millipore, Billerica, MA, USA). The murine macrophage (RAW 264.7) cell line, Dulbecco's modified media (DMEM), heat-inactivated fetal bovine serum (FBS), and penicillin-streptomycin were obtained from Gibco Life Technologies. For the enzyme-linked immunosorbent assay (ELISA) the TNF-α, IL-6, and IL-1β ELISA kits were obtained from e-Bioscience, Inc. (Cincinnati, OH, USA). Cytokine ELISA kits were obtained from R&D Systems (Minneapolis, MN, USA). All reagents and solvents were HPLC or analytical grade.

4.3. Preparation of Polysaccharide Extracts

Chaga samples (1.5 g) placed in teabags (B) or in powder form (P) were infused (steeped) in 200 mL of boiled distilled water at 100 °C (for 3, 6, and 10 min). The infusions were filtered through sterilized gauze. Four volumes of cold 95% ethanol were added to the aqueous extract after concentrating to 30% of the original volume with rotary evaporator under reduced pressure at 60 °C. The extracts were kept at 4 °C overnight to isolate the crude polysaccharides. The precipitate was recovered by centrifugation at 20 min, 2000× *g* (Rotavapor R3000, Buchi, Switzerland), washed with acetone to remove adherent sugar residue and other small molecules, and dialyzed for two days with distilled water (cut-off Mw 8000 Da). The retained portion was concentrated; deproteinated with Sevag reagent ($CHCl_3$:BuOH, 4:1, *v*/*v*) for 30 min under magnetic force stirring and the procedure was repeated two times. Finally, the extracts were centrifuged to remove insoluble material and the supernatant was lyophilized in the freeze–dry apparatus for 48 h to produce the crude polysaccharide extracts from the powder form (P3, P6, and P10) and from the bagged form (B3, B6, and B10), depending on the brewing time.

4.4. *Preparation of Phenolic Extracts*

4.4.1. Green Extraction, Accelerated Solvent Extraction (ASE)

ASE was performed with a Dionex (Sunnyvale, CA, USA) ASE 200 instrument with solvent controller, following the procedure described by Alhallaf et al. [22]. Briefly, a dried, ground sample of chaga (1 g) was placed in a stainless-steel extraction cell, preheated for 2 min, and extracted with 70% (*v*/*v*) aqueous ethanol or 66% (*v*/*v*) aqueous ethanol. The extractions, shown in Table 2, were performed at three temperature ranges (130 °C, 150 °C, and 170 °C) for 30 min (two cycles for every sample) at a pressure of 1500 psi. Once the extraction was complete, the suspension obtained was centrifuged (10 min, 2000× *g*) and the solvent was removed using a rotary evaporator (Rotovap R3000, Buchi, Switzerland). The resulting powders were stored at −20 °C for further experiments.

Table 2. Extraction conditions of chaga using accelerated solvent extraction ASE.

Extraction Conditions		Extract
Temperature °C	ETOH%	
170	66	ASE1
150	70	ASE2
130	70	ASE3

4.4.2. Conventional Solvent Extraction (CSE)

Maceration Extraction (ME)

Chaga powder (1 g) was macerated with 25 mL of 70% (*v*/*v*) aqueous ethanol for 48 h at room temperature. After filtration through a Whatman no. 1 filter paper, the solvent was removed using a rotary evaporator (Rotovap R3000, Buchi, Switzerland). The resulting residue was then dissolved and filtered in accordance with the procedure defined in Section 4.4.1. Extraction was carried out in triplicate.

Reflux Extraction (RE)

Chaga powder (1 g) was mixed with 25 mL of 70% (*v*/*v*) aqueous ethanol in a round-bottom flask. The extraction mixture was then refluxed in a water bath at 70 °C for 3 h. The resulting residue was dissolved and filtered in accordance with the procedure defined in Section 4.4.1. Extraction was carried out in triplicate.

Soxhlet Extraction (SE)

Chaga (1 g) was continuously extracted with 500 mL of 70% (*v*/*v*) aqueous ethanol for 48 h at 70 °C in a soxhlet apparatus. The resulting residue was then dissolved and filtered in accordance with the procedure defined in Section 4.4.1. Extraction was carried out in triplicate.

4.5. Determination of Total Phenolic Content (TPC)

The total phenolic content (TPC) of the extracts was determined by the Folin–Ciocalteu method described by [37]. Briefly, 20 μL of supernatant was mixed with 90 μL of a 10-fold diluted Folin–Ciocalteu reagent in a 96-well microplate. After standing for 5 min at room temperature, 90 μL of 6% sodium carbonate (Na_2CO_3) solution was added and the mixture was incubated at room temperature for 90 min. The absorbance was measured at 750 nm in a spectrophotometric microplate reader (Bio-Tek ELx808, Winooski, VT, USA). The absorbance of the extract was compared with a gallic acid standard curve for estimating the concentration of TPC in the sample. The TPC was expressed as milligrams of gallic acid equivalent per gram of dry weight chaga (mg GAE/g DW).

4.6. Determination of Total Neutral Carbohydrate Contents

The carbohydrate content of the polysaccharide extracts was determined using a slightly modified phenol-sulphuric acid method reported by Masuko et al. [38]. Briefly, 1 mL of sample solution, 0.05 mL of 80% phenol, and 5 mL of concentrated sulfuric acid were mixed and shaken. Following this, the mixture was held at room temperature for 10 min and the absorbance was measured at 490 nm. The total carbohydrate content was calculated using D-glucose as a standard.

4.6.1. Determination of Uronic Acid Content

The uronic acid content of the polysaccharide extracts were measured according to the Blumenkrantz method [39] using D-galacturonic acid as a standard. Briefly, 0.2 mL of sample solution and 1.2 mL of sulphuric acid/tetraborate solution were mixed and shaken. The mixture was kept at 100 °C for 5 min, 20 mL of m-hydroxydiphenyl reagent was added and the absorbance was measured within 5 min at 520 nm.

4.6.2. Determination of Protein Content

The total protein content of the polysaccharide extracts was determined by the Bradford method [40] using bovine serum albumin as a standard. Ten μL of sample solution and 200 μL of Bradford reagent were mixed and the mixture was incubated at room temperature for 5 min before reading the absorbance at 595 nm.

4.7. Cell Culture

The RAW264.7 cell line, derived from murine macrophages, was purchased from the American Type Culture Collection (Rockville, MD, USA). The cells were maintained at 37 °C in a humidified atmosphere of 5% (*v*/*v*) CO2 in Dulbecco's Modified Eagle's Medium (DMEM) supplemented with glutamine (1 mM), 10% fetal bovine serum (FBS; ATCC; Manassas, VA, USA), penicillin (50 U/mL), and streptomycin (50 μg/m). Medium was changed every two days. In all experiments, cells were grown to 70–80% confluence and subjected to no more than 20 cell passages.

4.7.1. Measurement of Cell Viability

Cell viability was assessed by the MTT 3-(4,5-dimethylthiazolyl-2)-2,5-diphenyltetrazolium bromide) assay. The assay is based on the ability of mitochondria in viable cells to reduce the yellow tetrazolium salt MTT to purple formazan crystals. The method was performed according to the manufacturer's procedure [23] with some modifications. The cells were cultured in 96-well plates at a density of 1×10^4 cells/well for 24 h. The cells were then treated with the samples at different concentrations (50, 100, and 150 μg/mL (chaga extract) or 25, 50, and 100 μM (vanillic acid, caffeic acid, and syringic acid) or 5, 10, and 20 μM (protocatechuic acid and protocatechuic aldehyde) for 24 h in a humidified 5% CO_2 atmosphere at 37 °C. After the incubation period, the media was removed and 100 μL of fresh medium and 10 μL of MTT solution were added to each well, and the plate was incubated for 2 h at 37 °C. Finally, the cell culture medium was discarded, and the formazan blue formed in the cell was re-suspended in 200 μL solubilization solution. The quantity of formazan (an indicator of cell viability) is measured by recording changes in absorbance at 540 nm using a spectrophotometric microplate reader (Bio-Tek ELx808, VT, USA). Of note, extracts and standards were dissolved in 0.05% DMSO. Cells treated with 0.05% DMSO was used as control and cells were treated with 2μM gallic acid was used as positive control. All experiments were performed in triplicate. % Cell viability was calculated using the following equation:

$$\% \text{ Cell viability } = \frac{\text{Absorbance of the extract}}{\text{Absorbance of the media}} \times 100 \quad (1)$$

4.7.2. Measurement of NO Production

Inhibitory effects of chaga extracts and the pure phenolic acid standards on the production of NO in RAW 264.7 cells were evaluated using a method modified from the previously reported work of Sun et al [41]. RAW 264.7 cells in 10% FBS-DMEM (without phenol red) were seeded (at 1×10^5 cells/well) in 12 well plates. Cells were incubated for 24 h at 37 °C. Cells were then treated with varying concentrations of samples (50, 100, and 150 μg/mL (chaga extract) or 25, 50, and 100 μM (vanillic acid, caffeic acid, and syringic acid) or 5, 10, and 20 μM (protocatechuic acid and protocatechuic aldehyde)) for 2 h. The cells were then treated with 1 μg/mL LPS (Sigma–Aldrich) for 24 h at 37 °C. After 24 h, 100 μL of cell culture medium was mixed with 100 μL of Griess reagent, incubated at room temperature for 15 min and the absorbance was measured at 540 nm in an ELISA microplate reader (Bio-Tek ELx808, VT, USA). The values were compared with a sodium nitrite standard curve (5–100 μM).

4.7.3. Cytokine Measurement

To assess the anti-inflammatory effect of chaga extracts and the pure phenolic acid standards on the expression of TNF-α, IL-6, and IL-1β were quantified using ELISA kits (e-Bioscience, Inc., Cincinnati, OH, USA). The assays were performed according to instructions provided by the manufacturer.

4.8. Statistical Analysis

Statistical analyses were performed with SPSS v25 (SPSS, Chicago, IL, USA). All results were expressed as the mean ± the standard error of triplicate analysis. Statistical significance ($p < 0.05$) was determined using one-way analysis of variance for independent means, followed by Tukey's HSD test.

5. Conclusions

In response to increasing interest in on the health-promoting effects of chaga (*Inonotus obliquus*), we have shown that chaga collected in Maine, USA, exhibits significant anti-inflammatory properties against LPS-activated 264.7 RAW macrophages. Results suggest that extracts produced from accelerated solvent extraction and traditional aqueous chaga tea extraction of chaga powder are superior to other conventional methods (maceration, soxhlet, and reflux) when using NO production and the expression of TNF-α, IL-6, and IL-1β as measurements of anti-inflammatory potential. Although not quantified in any of extracts, we tested individual phenolic acids reported in similar extracts from numerous other studies on the RAW264.7 cell model and noted strong anti-inflammatory responses from caffeic acid (CA), protocatechuic acid (PA) and protocatechuic aldehyde (PCA), but not vanillic (VA) or syringic acids (SA). Also of interest is that powdered chaga extracted without incorporation into a traditional teabag resulted in higher total phenolic and carbohydrate content and produced a higher anti-inflammatory response on the cell model, with P6, the powdered form steeped at 100 °C for six minutes, exhibiting the highest response. Carbohydrates found in the aqueous tea extracts may also contribute to the anti-inflammatory response. Ongoing research is focused on the further development of methodology to maximize phenolic and carbohydrate content and includes identification and quantitation of individual phenolic acids and carbohydrates in chaga extracts in support of the development of promising new anti-inflammatory supplements.

Author Contributions: Conceptualization, all authors; methodology, W.A.; resources, L.B.P.; data curation, W.A.; writing—original draft preparation, W.A.; writing—review and editing, all authors; visualization, all authors; supervision, L.B.P.; project administration, L.B.P.; funding acquisition, L.B.P. All authors have read and agreed to the published version of the manuscript.

Funding: This project was supported by the USDA National Institute of Food and Agriculture, Hatch Project Number ME0-21817 through the Maine Agricultural and Forest Experiment Station.

Institutional Review Board Statement: Not applicable.

Informed Consent Statement: Not applicable.

Data Availability Statement: The data presented in this study are available on request from the corresponding author.

Conflicts of Interest: The authors declare no conflict of interest.

References

1. Taofiq, O.; Martins, A.; Barreiro, M.F.; Ferreira, I.C. Anti-inflammatory potential of mushroom extracts and isolated metabolites. *Trends Food Sci. Technol.* **2016**, *50*, 193–210. [CrossRef]
2. Garlanda, C.; Di Liberto, D.; Vecchi, A.; La Manna, M.P.; Buracchi, C.; Caccamo, N.; Salerno, A.; Dieli, F.; Mantovani, A. Damping excessive inflammation and tissue damage in *Mycobacterium tuberculosis* infection by Toll IL-1 receptor 8/single Ig IL-1-related receptor, a negative regulator of IL-1/TLR signaling. *J. Immunol.* **2007**, *179*, 3119–3125. [CrossRef]
3. Jeong, J.B.; Jeong, H.J. Rheosmin, a naturally occurring phenolic compound inhibits LPS-induced iNOS and COX-2 expression in RAW264.7 cells by blocking NF-κB activation pathway. *Food Chem. Toxicol.* **2010**, *48*, 2148–2153. [CrossRef]
4. Dai, B.; Wei, D.; Zheng, N.-N.; Chi, Z.-H.; Xin, N.; Ma, T.-X.; Zheng, L.-Y.; Sumi, R.; Sun, L. Coccomyxa Gloeobotrydiformis polysaccharide inhibits lipopolysaccharide-induced inflammation in RAW 264.7 Macrophages. *Cell. Physiol. Biochem.* **2018**, *51*, 2523–2535. [CrossRef]
5. Duru, K.C.; Kovaleva, E.G.; Danilova, I.G.; van der Bijl, P. The pharmacological potential and possible molecular mechanisms of action of *Inonotus obliquus* from preclinical studies. *Phytotherapy Res.* **2019**, *33*, 1966–1980. [CrossRef] [PubMed]
6. Burmasova, M.A.; Sysoeva, M.A. Chemical composition and biological activity of the BuOH Fraction from chaga melanin. *Pharm. Chem. J.* **2017**, *51*, 292–294. [CrossRef]
7. Shashkina, M.Y.; Shashkin, P.N.; Sergeev, A.V. Chemical and medicobiological properties of chaga (review). *Pharm. Chem. J.* **2006**, *40*, 560–568. [CrossRef]
8. Mazurkiewicz, W. Analysis of aqueous extract of *Inonotus obliquus*. *Acta Pol. Pharm. Drug Res.* **2007**, *63*, 497–501.
9. Zheng, W.; Zhang, M.; Zhao, Y.; Miao, K.; Pan, S.; Cao, F.; Dai, Y. Analysis of antioxidant metabolites by solvent extraction from sclerotia of Inonotus obliquus (Chaga). *Phytochem. Anal.* **2011**, *22*, 95–102. [CrossRef]
10. Géry, A.; Dubreule, C.; André, V.; Rioult, J.-P.; Bouchart, V.; Heutte, N.; De Pécoulas, P.E.; Krivomaz, T.; Garon, D. Chaga (Inonotus obliquus), a future potential medicinal fungus in oncology? A chemical study and a comparison of the cytotoxicity against human lung adenocarcinoma cells (A549) and human bronchial epithelial cells (BEAS-2B). *Integr. Cancer Ther.* **2018**, *17*, 832–843. [CrossRef]
11. Diao, B.-Z.; Jin, W.-R.; Yu, X.-J. Protective Effect of polysaccharides from *Inonotus obliquus* on Streptozotocin-induced diabetic symptoms and their potential mechanisms in rats. *Evid.-Based Complement. Altern. Med.* **2014**, *2014*, 841496. [CrossRef]
12. Mishra, S.K.; Kang, J.-H.; Kim, D.-K.; Oh, S.H.; Kim, M.K. Orally administered aqueous extract of Inonotus obliquus ameliorates acute inflammation in dextran sulfate sodium (DSS)-induced colitis in mice. *J. Ethnopharmacol.* **2012**, *143*, 524–532. [CrossRef] [PubMed]
13. Eid, J.I.; Al-Tuwariijiri, M.M.; Prasad, C. Chaga mushroom (*Inonotus obliquus*) inhibits growth of both lung adeno-carcinoma (A549) cells and *Aspergillus fumigtatus*. *Curr. Top. Nutraceutical Res.* **2018**, *16*, 289–296. [CrossRef]
14. Glamočlija, J.; Ciric, A.; Nikolic, M.; Fernandes, Â.; Barros, L.; Calhelha, R.C.; Ferreira, I.; Sokovic, M.; van Griensven, L.J. Chemical characterization and biological activity of Chaga (*Inonotus obliquus*), a medicinal "mushroom". *J. Ethnopharmacol.* **2015**, *162*, 323–332. [CrossRef]
15. Van, Q.; Nayak, B.; Reimer, M.; Jones, P.; Fulcher, R.; Rempel, C. Anti-inflammatory effect of Inonotus obliquus, Polygala senega L., and Viburnum trilobum in a cell screening assay. *J. Ethnopharmacol.* **2009**, *125*, 487–493. [CrossRef]
16. Nakajima, Y.; Sato, Y.; Konishi, T. Antioxidant Small Phenolic Ingredients in *Inonotus obliquus* (persoon) Pilat (Chaga). *Chem. Pharm. Bull.* **2007**, *55*, 1222–1226. [CrossRef]
17. Nakajima, Y.; Nishida, H.; Matsugo, S.; Konishi, T. Cancer cell cytotoxicity of extracts and small phenolic compounds from chaga [*Inonotus obliquus* (persoon) Pilat]. *J. Med. Food* **2009**, *12*, 501–507. [CrossRef]
18. Park, Y.K.; Lee, H.B.; Jeon, E.-J.; Jung, H.S.; Kang, M.-H. Chaga mushroom extract inhibits oxidative DNA damage in human lymphocytes as assessed by comet assay. *BioFactors* **2004**, *21*, 109–112. [CrossRef]
19. Park, Y.-M.; Won, J.-H.; Kim, Y.-H.; Choi, J.-W.; Park, H.-J.; Lee, K.-T. In vivo and in vitro anti-inflammatory and anti-nociceptive effects of the methanol extract of *Inonotus obliquus*. *J. Ethnopharmacol.* **2005**, *101*, 120–128. [CrossRef]

20. Hwang, A.Y.; Yang, S.C.; Kim, J.; Lim, T.; Cho, H.; Hwang, K.T. Effects of non-traditional extraction methods on extracting bioactive compounds from chaga mushroom (*Inonotus obliquus*) compared with hot water extraction. *LWT* **2019**, *110*, 80–84. [CrossRef]
21. Seo, H.-K.; Lee, S.-C. Antioxidant activity of subcritical water extracts from chaga mushroom (*Inonotus obliquus*). *Sep. Sci. Technol.* **2010**, *45*, 198–203. [CrossRef]
22. Alhallaf, W.; Bishop, K.; Perkins, L.B. Optimization of accelerated solvent extraction of phenolic compounds from chaga using response surface methodology. *Food Anal. Methods* **2022**, *in press*. [CrossRef]
23. Mosmann, T. Rapid colorimetric assay for cellular growth and survival: Application to proliferation and cytotoxicity assays. *J. Immunol. Methods* **1983**, *65*, 55–63. [CrossRef]
24. Dobrovolskaia, M.A.; Vogel, S.N. Toll receptors, CD14, and macrophage activation and deactivation by LPS. *Microbes Infect.* **2002**, *4*, 903–914. [CrossRef]
25. Han, S.; Sung, K.-H.; Yim, D.; Lee, S.; Cho, K.; Lee, C.-K.; Ha, N.-J.; Kim, K. Activation of murine macrophage cell line RAW 264.7 by Korean propolis. *Arch. Pharmacal Res.* **2002**, *25*, 895–902. [CrossRef]
26. Park, H.-J.; Kim, I.-T.; Won, J.-H.; Jeong, S.-H.; Park, E.-Y.; Nam, J.-H.; Choi, J.; Lee, K.-T. Anti-inflammatory activities of ent-16αH,17-hydroxy-kauran-19-oic acid isolated from the roots of Siegesbeckia pubescens are due to the inhibition of iNOS and COX-2 expression in RAW 264.7 macrophages via NF-κB inactivation. *Eur. J. Pharmacol.* **2007**, *558*, 185–193. [CrossRef]
27. Paige, J.S.; Jaffrey, S. Pharmacologic manipulation of nitric oxide signaling: Targeting NOS dimerization and protein-protein interactions. *Curr. Top. Med. Chem.* **2007**, *7*, 97–114. [CrossRef]
28. Alderton, W.; Cooper, C.; Knowles, R.G. Nitric oxide synthases: Structure, function and inhibition. *Biochem. J.* **2001**, *357*, 593–615. [CrossRef]
29. Gilmore, T.D. The Rel/NF-κB signal transduction pathway: Introduction. *Oncogene* **1999**, *18*, 6842–6844. [CrossRef]
30. Kim, H.S.; Ye, S.-K.; Cho, I.H.; Jung, J.E.; Kim, D.-H.; Choi, S.; Kim, Y.S.; Park, C.-G.; Kim, T.-Y.; Lee, J.W.; et al. 8-hydroxydeoxyguanosine suppresses NO production and COX-2 activity via Rac1/STATs signaling in LPS-induced brain microglia. *Free Radic. Biol. Med.* **2006**, *41*, 1392–1403. [CrossRef]
31. Adams, D.O.; Hamilton, T.A. The cell biology of macrophage activation. *Annu. Rev. Immunol.* **1984**, *2*, 283–318. [CrossRef]
32. Beutler, B.; Cerami, A. The Biology of Cachectin/TNF—A Primary Mediator of the Host Response. *Annu. Rev. Immunol.* **1989**, *7*, 625–655. [CrossRef]
33. Dinarello, C.A. The IL-1 family and inflammatory diseases. *Clin. Exp. Rheumatol.* **2002**, *20*, S1–S13.
34. Sánchez-Miranda, E.; Lemus-Bautista, J.; Pérez, S.; Pérez-Ramos, J. Effect of Kramecyne on the Inflammatory Response in Lipopolysaccharide-Stimulated Peritoneal Macrophages. *Evid.-Based Complement. Altern. Med.* **2013**, *2013*, 762020. [CrossRef]
35. Juman, S.; Yasui, N.; Ikeda, K.; Ueda, A.; Sakanaka, M.; Negishi, H.; Miki, T. Caffeic Acid phenethyl Ester suppresses the production of pro-inflammatory cytokines in hypertrophic adipocytes through lipopolysaccharide-stimulated macrophages. *Biol. Pharm. Bull.* **2012**, *35*, 1941–1946. [CrossRef]
36. Miles, E.A.; Zoubouli, P.; Calder, P. Differential anti-inflammatory effects of phenolic compounds from extra virgin olive oil identified in human whole blood cultures. *Nutrition* **2005**, *21*, 389–394. [CrossRef]
37. Jaramillo-Flores, M.E.; González-Cruz, L.; Cornejo-Mazón, M.; Alvarez, L.D.; Gutierrez, G.; Hernández-Sánchez, H. Effect of thermal treatment on the antioxidant activity and content of carotenoids and phenolic compounds of cactus pear cladodes (*Opuntia ficus-indica*). *Food Sci. Technol. Int.* **2003**, *9*, 271–278. [CrossRef]
38. Masuko, T.; Minami, A.; Iwasaki, N.; Majima, T.; Nishimura, S.-I.; Lee, Y.C. Carbohydrate analysis by a phenol–sulfuric acid method in microplate format. *Anal. Biochem.* **2005**, *339*, 69–72. [CrossRef]
39. Allen, J.; Brock, S.A. Tailoring the message. *Minn. Med.* **2000**, *83*, 45–48.
40. Emami Bistgani, Z.; Siadat, S.A.; Bakhshandeh, A.; Ghasemi Pirbalouti, A.; Hashemi, M. Interactive effects of drought stress and chitosan application on physiological characteristics and essential oil yield of Thymus daenensis Celak. *Crop J.* **2017**, *5*, 407–415. [CrossRef]
41. Sun, J.; Zhang, X.; Broderick, M.; Fein, H. Measurement of Nitric Oxide Production in Biological Systems by Using Griess Reaction Assay. *Sensors* **2003**, *3*, 276–284. [CrossRef]

molecules

MDPI

Article

Anti-Hyperglycemic Effects of Refined Fractions from *Cyclocarya paliurus* Leaves on Streptozotocin-Induced Diabetic Mice

Zheling Feng [1,†], Zhujun Fang [2,†], Cheng Chen [1], Chi Teng Vong [1], Jiali Chen [1], Ruohan Lou [1], Maggie Pui Man Hoi [1], Lishe Gan [3,*] and Ligen Lin [1,*]

1 State Key Laboratory of Quality Research in Chinese Medicine, Institute of Chinese Medical Sciences, University of Macau, Avenida da Universidade, Taipa, Macao 999078, China; yb77508@um.edu.mo (Z.F.); mb85812@um.edu.mo (C.C.); gigict.vong@gmail.com (C.T.V.); yc07516@um.edu.mo (J.C.); mc05836@um.edu.mo (R.L.); MagHoi@um.edu.mo (M.P.M.H.)

2 Department of Clinical Pharmacy, Zhejiang Provincial Key Laboratory for Drug Evaluation and Clinical Research, The First Affiliated Hospital, College of Medicine, Zhejiang University, Hangzhou 310000, China; 1320127@zju.edu.cn

3 School of Biotechnology and Health Sciences, Wuyi University, Jiangmen 529020, China

* Correspondence: ganlishe@163.com (L.G.); ligenl@um.edu.mo (L.L.)

† These authors contributed equally to this work.

Citation: Feng, Z.; Fang, Z.; Chen, C.; Vong, C.T.; Chen, J.; Lou, R.; Hoi, M.P.M.; Gan, L.; Lin, L. Anti-Hyperglycemic Effects of Refined Fractions from *Cyclocarya paliurus* Leaves on Streptozotocin-Induced Diabetic Mice. *Molecules* **2021**, *26*, 6886. https://doi.org/10.3390/molecules26226886

Academic Editors: Ana Paula Duarte, Ângelo Luís and Eugenia Gallardo

Received: 9 October 2021
Accepted: 12 November 2021
Published: 15 November 2021

Abstract: To identify the chemical components responsible for the anti-hyperglycemic effect of *Cyclocarya paliurus* (Batal.) Iljinsk (Juglandaceae) leaves, an ethanol extract (CPE) and a water extract (CPW) of *C. paliurus* leaves, as well as their total flavonoids (CPF), triterpenoids (CPT) and crude polysaccharides (CPP), were prepared and assessed on streptozotocin (STZ)-induced diabetic mice. After being orally administrated once a day for 24 days, CPF (300 mg/kg), CPP (180 mg/kg), or CPF+CPP (300 mg/kg CPF + 180 mg/kg CPP) treatment reversed STZ-induced body weight and muscle mass losses. The glucose tolerance tests and insulin tolerance tests suggested that CPF, CPP, and CPF+CPP showed anti-hyperglycemic effect in STZ-induced diabetic mice. Furthermore, CPF enhances glucose-stimulated insulin secretion in MIN6 cells and insulin-stimulated glucose uptake in C2C12 myotubes. CPF and CPP suppressed inflammatory cytokine levels in STZ-induced diabetic mice. Additionally, CPF and CPP improved STZ-induced diabetic nephropathy assessed by H&E staining, blood urea nitrogen content, and urine creatinine level. The molecular networking and Emperor analysis results indicated that CPF showed potential anti-hyperglycemic effects, and HPLC–MS/MS analysis indicated that CPF contains 3 phenolic acids and 9 flavonoids. In contrast, CPT (650 mg/kg) and CPC (300 mg/kg CPF + 180 mg/kg CPP + 650 mg/kg CPT) did not show anti-hyperglycemic effect. Taken together, polysaccharides and flavonoids are responsible for the anti-hyperglycemic effect of *C. paliurus* leaves, and the clinical application of *C. paliurus* need to be refined.

Keywords: *Cyclocarya paliurus*; flavonoids; polysaccharides; triterpenoids; anti-hyperglycemia

1. Introduction

Diabetes mellitus is increasing at an alarming rate worldwide, especially in developing countries [1,2]. Type 2 diabetes is characterized by hyperglycemia and insulin deficiency [3–5]. Long-term hyperglycemia causes damage, dysfunction, and failure of various organs, especially the eyes, kidneys, nerves, heart, and blood vessels [6]. Hyperglycemia-associated muscle mass and functional loss appear in the very early stage of diabetes [7]. A recent study showed high blood glucose decelerates WWP1-associated ubiquitous degradation of the transcription factor KLF15, resulting in muscle atrophy [8]. Diabetic nephropathy is a major complication of diabetes; poorly controlled diabetes causes damage to blood vessel clusters in the kidneys, leading to kidney damage and high blood pressure [9]. Since the clinical application of anti-diabetic drugs, including sulfonylureas, metformin, and thiazolidinediones, always have unpleasant side effects such as nausea, weight gain, headache, and dizziness [10,11], more and more interest has been directed towards natural products for the discovery of anti-hyperglycemic agents.

Cyclocarya paliurus (Batal.) Iljinsk (Juglandaceae) is native to eastern and central China, with the Chinese name "Qing Qian Liu" or "sweet tea tree" [12]. The leaves of *C. paliurus* have been widely used as ethnic medicine or herbal tea to treat diabetes in China. In recent years, the extracts of *C. paliurus* leaves were revealed to reduce blood glucose and improve insulin sensitivity on different diabetic models [13–15]. Polysaccharides from *C. paliurus* leaves were reported to display anti-diabetic activity in alloxan-induced mice [16]. Flavonoids isolated from *C. paliurus* leaves showed potential anti-diabetic activity in high-fat-diet-fed and streptozotocin (STZ)-stimulated mice; and the major constituents, including quercetin-3-*O*-glucuronide and kaempferol-3-*O*-glucuronide, are responsible for the anti-hyperglycemic activity [17]. Several triterpenoids isolated from *C. paliurus* leaves were found to enhance insulin-stimulated glucose uptake in both C2C12 myotubes and 3T3-L1 adipocytes [18]. However, the anti-hyperglycemic potential of different fractions from *C. paliurus* leaves has never been compared side-by-side on the same model. The chemical principles responsible for the anti-hyperglycemic effect of *C. paliurus* leaves remain controversial. Herein, different fractions were prepared from *C. paliurus* leaves. Total flavonoids (CPF) were purified using polyamide resin and D101 macroporous adsorption resin, characterized by UPLC-Q-TOF-MS [19]. *C. paliurus* polysaccharides (CPP) were obtained by the water-extraction and alcohol-precipitation method [20]. Total triterpenoids (CPT) were extracted by ethanol and then purified using AB-8 macroporous resin and a gradient ethanol elution [21]. Their anti-hyperglycemic effects were evaluated on STZ-induced mice using glucose tolerance tests, insulin tolerance tests, and homeostasis model assessment of basal insulin resistance. Glucose-stimulated insulin secretion in MIN6 cells was performed to evaluate the protective effect of the fractions on pancreatic β-cell function. Diabetic nephropathy was evaluated by H&E staining and biochemical tests. Finally, molecular networking and Emperor analysis of the LC–MS/MS data provided chemical profiles of the active fraction. The purpose of the current study is to uncover the chemical constituents responsible for the anti-hyperglycemic effect of *C. paliurus* and further guide the clinical application of this herbal material.

2. Results

2.1. CPF and CPP Reverse Body Weight and Muscle Weight Losses in STZ-Induced Diabetic Mice

To evaluate the anti-hyperglycemic effect of the refined fractions from *C. paliurus* leaves, STZ-induced mice were recruited (Figure 1A). INN (15 mg/kg) was used as a positive control [22]. The body weight from the STZ group of mice was significantly decreased in comparison to that of the control group (Figure 1B); 25 days treatment of CPC or CPT slightly reversed body weight loss compared to the STZ group, and treatment of CPF, CPP, or CPF+CPP reversed STZ-induced body weight loss (Figure 1B). Next, the weights of kidney, liver, quadriceps, and gastrocnemius from each group were compared, and the corresponding tissue indexes were calculated. The weights of quadriceps and gastrocnemius from STZ group were significantly reduced, but not kidney or liver, when compared with those of the control group (Figure 1C–F). The weight losses of quadriceps and gastrocnemius from CPP, CPF, and CPP + CPF groups were obviously lower than those of the STZ group, which were comparable with those of the control group (Figure 1E,F). The weights of quadriceps and gastrocnemius from CPT and CPC groups were similar to those of the STZ group (Figure 1E,F). The same trend was observed for the tissue indexes of quadriceps and gastrocnemius (Figure 1I,J). On the other hand, the weights or indexes of liver and kidneys were unchanged in all treatments (Figure 1C,D,G,H). The above results suggest that CPF, CPP, or CPF + CPP treatment reversed STZ-induced body weight and muscle mass losses.

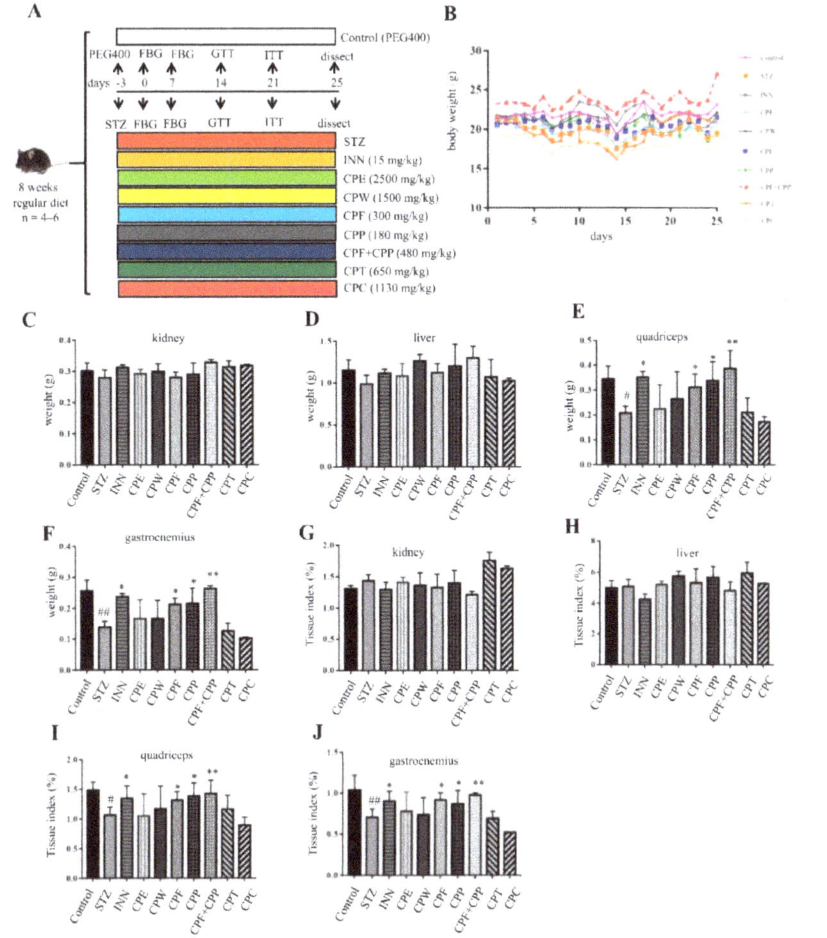

Figure 1. CPF (flavonoids fraction from *C. paliurus* leaves) and CPP (polysaccharide fraction from *C. paliurus* leaves) reverse body weight and muscle weight losses in STZ (streptozotocin)-induced diabetic mice. (**A**) The experiment procedure for STZ-induced diabetic mice. The male C57BL/6J mice were ip administrated with 150 mg/kg STZ, and then the mice were orally administrated with or without *C. paliurus* fractions once a day for 24 days. INN (glibenclamide, 15 mg/kg); CPE (ethanol extract from *C. paliurus* leaves, 2500 mg/kg); CPW (water extract from *C. paliurus* leaves, 1500 mg/kg); CPF (300 mg/kg); CPP (180 mg/kg); CPF+CPP (flavonoids and polysaccharide fractions from *C. paliurus* leaves, 480 mg/kg); CPT (triterpenoids fraction from *C. paliurus* leaves, 650 mg/kg); CPC (flavonoids, polysaccharide, and triterpenoids fractions from *C. paliurus* leaves, 1130 mg/kg). (**B**) Bodyweight of mice. (**C–F**) Kidneys, liver, quadriceps, and gastrocnemius weights of each group. (**G–J**) Tissue indexes of kidneys, liver, quadriceps, and gastrocnemius. $n = 6$. # $p < 0.05$, ## $p < 0.01$, STZ vs. control, * $p < 0.05$, ** $p < 0.01$, STZ vs. treatment.

2.2. CPF and CPP Improve Insulin Sensitivity in STZ-Induced Diabetic Mice

In STZ-treated mice, the fasting blood glucose (FBG) levels were significantly increased in comparison to those in the control group (Figure 2A), indicating impaired β-cell function. After one week's treatment, the FBG in CPF group was significantly reduced when compared with STZ group (Figure 2A). After three weeks' treatment, the FBGs in CPF, CPP, CPF+CPP, and INN groups were significantly reduced when compared with STZ group;

and the FBG in CPW group was slightly decreased, but not the CPE, CPT, or CPC group (Figure 2A). During GTT, the glucose clearance rates in the STZ, CPT, and CPC groups were greatly interrupted when compared with that of the control group, indicating an impaired pancreatic β-cell function in these groups of mice (Figure 2B). The CPF, CPP, and CPF+CPP treatment improved glucose disposal rate, comparable with that of INN-treated mice (Figure 2B). During ITT, the blood glucose levels of CPF, CPP, and CPF + CPP treated mice were significantly reduced under insulin stimulation compared to the STZ group (Figure 2C). After STZ treatment, the insulin levels of all groups were decreased, and the serum insulin level did not change in each treatment, indicating that these treatments did not rescue β-cell function (Figure 2D). HOMA-IR calculation results indicate that CPF, CPP, CPF + CPP, and INN treatment groups markedly attenuated insulin resistance compared to the STZ-induced diabetic mice group (Figure 2E). Taking the above results together, CPF, CPP, and CPF + CPP treatment showed anti-hyperglycemic effect in STZ-induced mice.

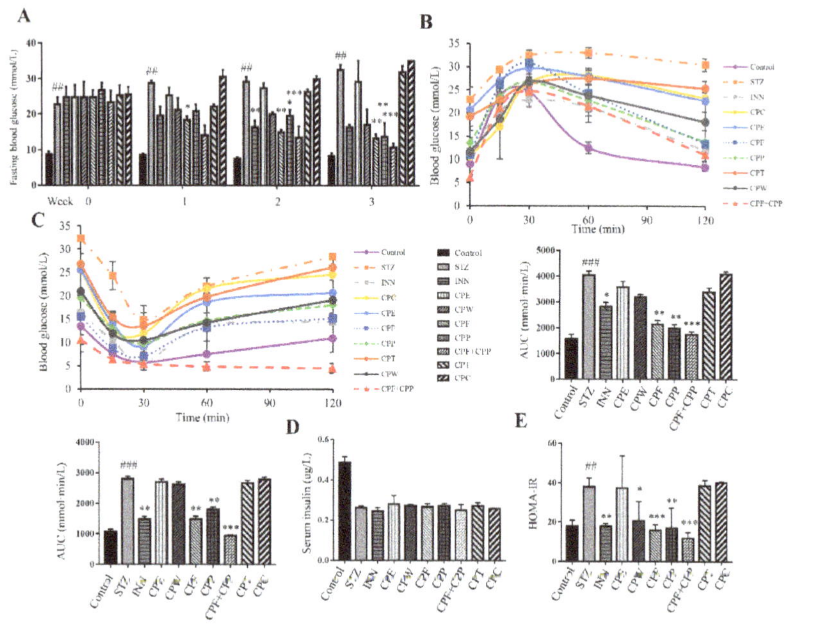

Figure 2. CPF (flavonoids fraction from *C. paliurus* leaves) and CPP (polysaccharide fraction from *C. paliurus* leaves) improve insulin sensitivity in STZ (streptozotocin)-induced diabetic mice. (**A**) Fasting blood glucose levels of all groups. (**B**) Glucose tolerance test was performed after 2-week CP fractions treatment. AUC of each group was calculated. (**C**) Insulin tolerance test was performed after 3-week CP fractions treatment. AUC (area under curve) of each group was calculated. (**D**) The serum insulin levels were determined after 18 h fasting. (**E**) The homeostasis model assessment of basal insulin resistance (HOMA-IR). $n = 6$. ## $p < 0.01$, ### $p < 0.001$, STZ vs. control, * $p < 0.05$, ** $p < 0.01$, *** $p < 0.001$, STZ vs. treatment.

2.3. *CPF Enhances Glucose-Stimulated Insulin Secretion in MIN6 Cells and Insulin-Stimulated Glucose Uptake in C2C12 Myotubes*

The flavonoids from *C. paliurus* leaves have been reported to improve insulin sensitivity [17,23]. Herein, we evaluated the effects of different CP fractions on glucose-induced insulin secretion in MIN6 cells. The MTT assay determined the maximum safe concentration of each fraction (Figure 3A). Only CPF (25 μg/mL) enhanced insulin secretion in both 5.5 mM- and 16.7 mM-stimulated MIN6 cells, which was comparable with INN-treated cells (Figure 3B). CPF enhances insulin secretion in normal pancreatic β cells.

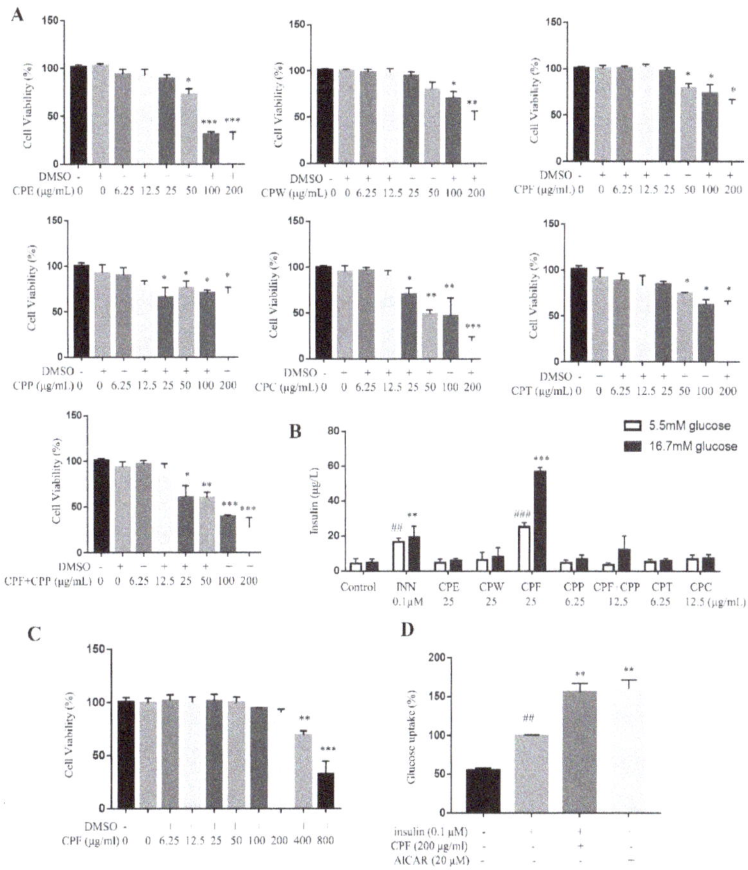

Figure 3. CPF (flavonoids fraction from *C. paliurus* leaves) enhanced glucose-stimulated insulin secretion in MIN6 cells and insulin-stimulated glucose uptake in C2C12 myotubes. (**A**) Cytotoxicity of different CP fractions on MIN6 cells, determined by MTT assay. $n = 6$, * $p < 0.05$, ** $p < 0.01$, *** $p < 0.001$, vs. DMSO. (**B**) CPF promoted glucose-stimulated insulin secretion in MIN6 cells. $n = 6$, ## $p < 0.01$, ### $p < 0.001$, vs. control cells in 5.5 mM glucose; ** $p < 0.01$, *** $p < 0.001$, vs. control cells in 16.7 mM glucose. (**C**) Cytotoxicity of CPF on C2C12 cells. (**D**) CPF promoted insulin-stimulated glucose uptake in C2C12 myotubes. $n = 6$, ## $p < 0.01$ insulin vs. control, ** $p < 0.01$, *** $p < 0.001$, CPF vs. insulin.

The anti-hyperglycemic effect of CPF might be due to enhanced insulin sensitivity of skeletal muscle. Firstly, the cytotoxicity of CPF on C2C12 myotubes was evaluated to determine the maximum safe dosage. The results show CPF did not show obvious cytotoxicity on C2C12 myotubes up to 200 µg/mL (Figure 3C). Next, CPF was found to enhance insulin-stimulated glucose uptake on C2C12 myotubes under the dosage of 200 µg/mL, comparable with the positive control AICAR (5-aminoimidazole-4-carboxamide1-β-D-ribofuranoside, Figure 3D). Thus, CPF enhances insulin sensitivity on C2C12 myotubes.

2.4. CPF and CPP Suppressed Inflammatory Cytokine Levels in STZ-Induced Diabetic Mice

Inflammation can be triggered by structural components of gut bacteria, resulting in a cascade of inflammatory pathways involving interleukins and other cytokines [24]. Previous studies showed that pro-inflammatory cytokines interleukin-6 (IL-6) and tumor necrosis factor-α (TNF-α) were associated with microbiota imbalance in diabetic mice [25,26]. Thus, the anti-hyperglycemic effect of CPF and CPP might be due to microbiota modulation. As shown in Figure 4A,B, the serum TNF-α and IL-6 levels were increased in STZ-treated mice, and the CPF, CPP, and CPF + CPP treatment significantly decreased the serum TNF-α and IL-6 levels compared to the STZ group. These results indicate that CPF and CPP possess anti-inflammation effect on STZ-induced mice.

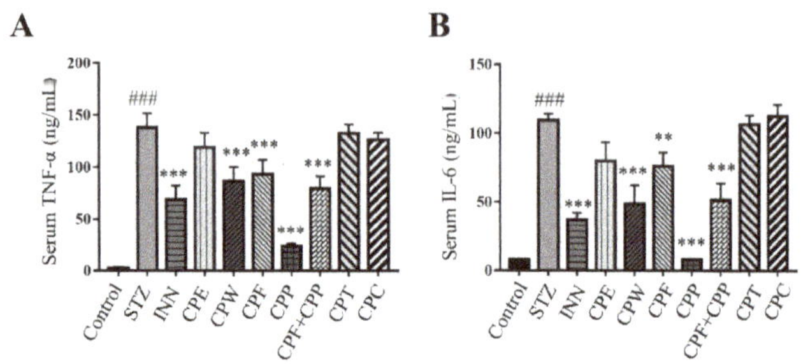

Figure 4. CPF (flavonoids fraction from *C. paliurus* leaves) and CPP (polysaccharide fraction from *C. paliurus* leaves) suppressed inflammatory cytokine levels in STZ (streptozotocin)-induced diabetic mice. (**A**) The levels of serum tumor necrosis factor-α (TNF-α) were determined by ELISA. (**B**) The levels of serum interleukin-6 (IL-6) were determined by ELISA. $n = 6$, ### $p < 0.001$, STZ vs. control, ** $p < 0.01$, *** $p < 0.001$, STZ vs. treatment.

2.5. CPF and CPP Improved Nephropathy in STZ-Induced Diabetic Mice

Diabetic nephropathy is a severe complication of diabetes, due to long-term high blood glucose levels [27,28]. Diabetic nephropathy further progresses to kidney failure, a life-threatening condition [29]. To study the influence of C. paliurus fractions on diabetic nephropathy, blood urea nitrogen and urine creatinine levels were tested. The blood urea nitrogen and urine creatinine levels were greatly elevated in STZ group and significantly reversed in CPW, CPF, CPP, CPF + CPP, and INN groups (Figure 5A,B). CPE, CPT, and CPC treatment did not change the blood urea nitrogen or urine creatinine levels (Figure 5A,B). In H&E staining of kidney, it appeared in the STZ group that the epithelial cells of renal tubules were exfoliated, the basement membrane of renal tubules was exposed, the renal tubules were compensatory dilated and incised, the glomeruli were hypertrophic, and the basement membrane became thicker; these symptoms were reversed in CPF, CPP, and CPF + CPP treatment groups (Figure 5C). The kidney damage was even worse in CPT and CPC treatment groups compared with the STZ group (Figure 5C). Taken together, CPF and CPP protect against STZ-induced diabetic nephropathy.

2.6. The Bioinformatics Predict Results Indicate CPF with Potential Anti-Diabetic Effects

Based on the results from liquid chromatography–mass spectrometry (LC–MS) and GNPS and Emperor analysis, the ion fragments of CPF from diverse solvents were pooled together, such as 100% acetonitrile, acetonitrile–water (7:3), acetonitrile–methanol (1:1), 100% ethanol, 90% ethanol, and 100% methanol; some ion fragments were not collected or specified based on the databases (Figure 6A). The predicted ion fragments with potential anti-diabetic activity are highlighted with red circles (Figure 6B). The results indicate that the fragments in CPF with potential anti-diabetic activity were mainly distributed in 90% ethanol (highlighted with a red circle).

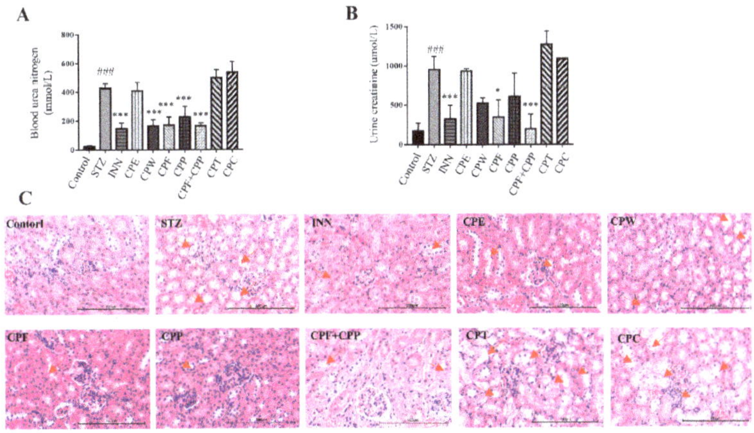

Figure 5. CPF (flavonoids fraction from *C. paliurus* leaves) and CPP (polysaccharide fraction from *C. paliurus* leaves) improved nephropathy in STZ (streptozotocin)-induced diabetic mice (**A**) The blood urea nitrogen levels of all CP treatment group. (**B**) The urine creatinine levels of all CP treatment groups. (**C**) Representative H&E staining of the kidney. Scale bar = 100 μm. The exfoliated renal tubes are indicated by the red arrows. $n = 6$, ### $p < 0.001$, STZ vs. control, * $p < 0.05$, *** $p < 0.001$, STZ vs. treatment.

2.7. Characterization of Chemical Constituents in CPE, CPF, CPP, and CPT

Before data processing, an in-house formula database involving compound name, molecular formula, chemical structure, accurate mass, and related product ions of the compounds in *Cyclocarya paliurus* was established by searching from databases such as Reaxys (https://www.reaxys.com/ accessed on 8 March 2020), PubMed (http://www.ncbi.nlm.nih.gov/pubmed accessed on 8 March 2020), and CNKI (http://www.cnki.net accessed on 8 March 2020).

The typical total ion chromatography (TIC) profiles of CPE in the negative ion mode were presented in Figure 7A. A total of 29 compounds were identified, including 3 phenolic acids, 10 flavonoids, and 16 triterpenoids (Table 1). These compounds were 3-*O*-caffeoylquinic acid, 4-*O*-caffeoylquinic acid, catechin, isoquercetin, quercetin-3-*O*-glucuronide, kaempferol-3-*O*-galactoside, kaempferol-3-*O*-glucuronide, kaempferol-3-*O*-glucoside, quercetin-3-*O*-rhamnoside, myricetin, kaempferol-3-*O*-rhamnoside, quercetin, kaempferol, 3β,19α,23-trihydroxy-1-oxo-olean-12-en-28-oic acid, arjunolic acid, cyclocaric acid B, asiatic acid, pterocaryoside B, cyclocarioside I, cyclocarioside K, pterocaryoside A, cyclocariol C, (20*S*,24*R*)-20,24-epoxy-25-hydroxy-12*β*-(α-L-arabinopyranosyloxy)-3,4-seco-dammara-4-(28)-en-3-oic acid, cyclocarioside B, hederagenin, cyclocarioside H, cyclocarioside X, cypaliuruside E, and ursolic acid [17,30–40].

The TIC profiles of CPF in the negative ion mode were presented in Figure 7B. A total of 12 compounds were identified, including 3 phenolic acids and 9 flavonoids (Table 2). They were 3-*O*-caffeoylquinic acid, 4-*O*-caffeoylquinic acid, catechin, isoquercetin, quercetin-3-*O*-glucuronide, kaempferol-3-*O*-galactoside, kaempferol-3-*O*-glucuronide, kaempferol-3-*O*-glucoside, quercetin-3-*O*-rhamnoside, kaempferol-3-*O*-rhamnoside, quercetin, and kaempferol.

The GPC analysis result for CPP is shown in Figure 7C,D. The molecular weights of CPP were in the range of 30,000–50,000 Da (Figure 7D). The *Mn*, *Mw*, and *Mz* of the peak were 1.4844×10^4, 3.6649×10^4, and 1.31810×10^5, respectively.

The TIC profiles of CPT in the negative ion mode are presented in Figure 7E. A total of 16 triterpenoids were identified (Table 3), including 3β,19α,23-trihydrox1-oxo-olean-12-en-28-oic acid, arjunolic acid, cyclocaric acid B, asiatic acid, pterocaryoside B, cyclocarioside I, cyclocarioside K, pterocaryoside A, cyclocariol C, (20S,24R)-20,24-epoxy-25-hydroxy-

12β-(α-L-arabinopyranosyloxy)-3,4-seco-dammara-4(28)-en-3-oic acid, cyclocarioside B, hederagenin, cyclocarioside H, cyclocarioside X, cypaliuruside E, and ursolic acid.

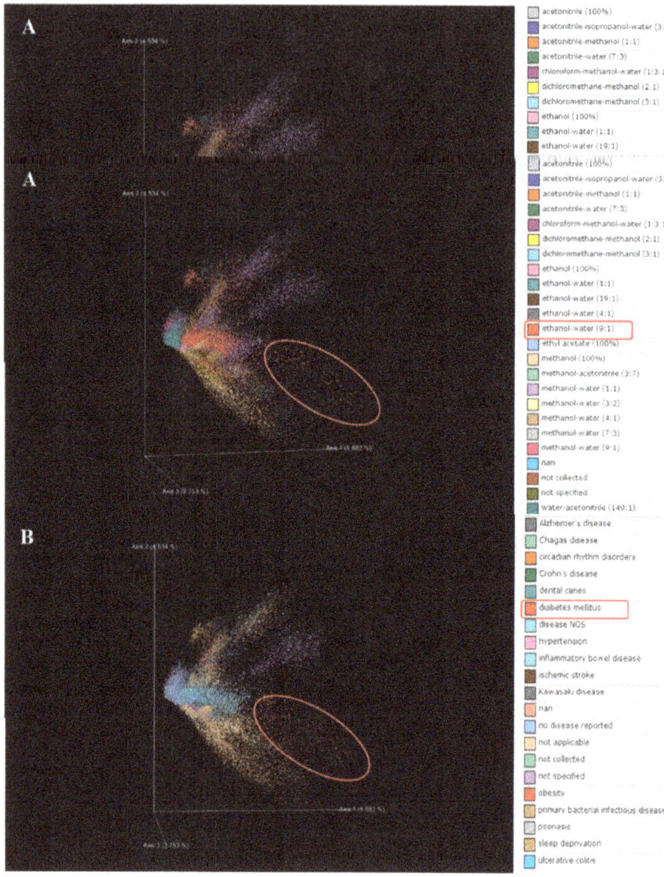

Figure 6. CPF (flavonoids fraction from *C. paliurus* leaves) ion fragments were collected from diverse solvent sources and showed potential anti-diabetic effects using bioinformatics prediction. (**A**) LC–MS^2 (liquid chromatography–mass spectrometry) to GNPS (Global Natural Product Social Molecular Networking) and Emperor analysis results show diverse solvent sources. (**B**) LC–MS^2 to GNPS and Emperor analysis results show potential anti-diabetic effect.

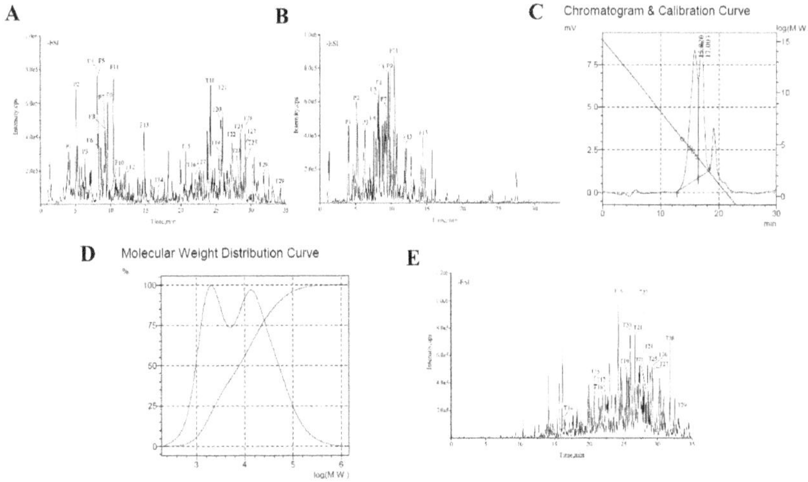

Figure 7. (**A**) Total ion chromatograms of CPE (ethanol extract from *C. paliurus* leaves, (**B**) CPF (flavonoids fraction from *C. paliurus* leaves), and (**C**) CPT (triterpenoids fraction from *C. paliurus* leaves), in negative ion mode. (**D**) Chromatogram and calibration curve of CPP. (**E**) Molecular weight distribution curve of CPP.

Table 1. Identification of the chemical constituents in CPE by HPLC-ESI-Q-TOF-MS/MS.

No.	t_R (min)	Identification	Formula	Selected Ion	Measured (*m/z*)	Calculated (*m/z*)	Error (ppm)	MS[2] (*m/z*)
P1	4.004	3-*O*-caffeoylquinic acid [41]	$C_{16}H_{18}O_9$	$[M-H]^-$	353.0884	353.0878	1.7	191.0556 (100) $[M-H-CA+H_2O]^-$ 179.0347 (40) $[CA-H]^-$ 135.0450 (48) $[CA-H-CO_2]^-$
P2	5.084	4-*O*-caffeoylquinic acid [41]	$C_{16}H_{18}O_9$	$[M-H]^-$	353.0885	353.0878	2.0	191.0557 (100) $[M-H-CA+H_2O]^-$
P3	6.180	catechin [30]	$C_{15}H_{14}O_6$	$[M-H]^-$	289.0724	289.0718	2.2	289.0725 (100) $[M-H]^-$ 271.0615 (8) $[M-H-H_2O]^-$
F4	8.094	isoquercetin [31,41]	$C_{21}H_{20}O_{12}$	$[M-H]^-$	463.0881	463.0882	−0.2	463.0910 (13) $[M-H]^-$ 301.0347 (31) $[M-H-glc]^-$
F5	8.145	quercetin-3-*O*-glucuronide [35,41]	$C_{21}H_{18}O_{13}$	$[M-H]^-$	477.0675	477.0675	0.1	477.0693 (13) $[M-H]^-$ 301.0353 (100) $[M-H-glcA]^-$ 283.0248 (10) $[M-H-glcA-H_2O]^-$
F6	8.872	kaempferol-3-*O*-galactoside [17,41]	$C_{21}H_{20}O_{11}$	$[M-H]^-$	447.0934	447.0933	0.3	447.0937 (56) $[M-H]^-$ 285.0393 (31) $[M-H-gal]^-$
F7	9.126	kaempferol-3-*O*-glucuronide [17,41]	$C_{21}H_{18}O_{12}$	$[M-H]^-$	461.0728	461.0725	0.5	461.0741 (5) $[M-H]^-$ 285.0397 (100) $[M-H-glcA]^-$
F8	9.198	kaempferol-3-*O*-glucoside [17,41]	$C_{21}H_{20}O_{11}$	$[M-H]^-$	447.0932	447.0933	−0.2	447.0926 (44) $[M-H]^-$ 285.0390 (71) $[M-H-glc]^-$
F9	9.297	quercetin-3-*O*-rhamnoside [17,41]	$C_{21}H_{20}O_{11}$	$[M-H]^-$	447.0928	447.0933	−1.1	447.0953 (19) $[M-H]^-$ 301.0347 (56) $[M-H-rha]^-$
F10	10.024	myricetin [34]	$C_{15}H_{10}O_8$	$[M-H]^-$	317.0303	317.0303	0.0	317.0299 (100) $[M-H]^-$ 178.9975 (35) $[M-H-C_7H_6O_3]^-$
F11	10.396	kaempferol-3-*O*-rhamnoside [35,41]	$C_{21}H_{20}O_{10}$	$[M-H]^-$	431.0981	431.0984	−0.6	431.1008 (24) $[M-H]^-$ 285.0401 (99) $[M-H-rha]^-$
F12	12.071	quercetin [35]	$C_{15}H_{10}O_7$	$[M-H]^-$	301.0354	301.0354	0.1	301.0341 (100) $[M-H]^-$ 273.0392 (18) $[M-H-CO]^-$
F13	14.152	kaempferol [36]	$C_{15}H_{10}O_6$	$[M-H]^-$	285.0410	285.0405	1.9	285.0401 (100) $[M-H]^-$
T14	16.778	3β,19α,23-trihydroxy-1-oxo-olean-12-en-28-oic acid	$C_{30}H_{46}O_6$	$[M-H]^-$	501.3226	501.3222	0.9	501.3275 (100) $[M-H]^-$ 455.3210 (71) $[M-H-FA]^-$
T15	20.616	arjunolic acid	$C_{30}H_{48}O_5$	$[M-H]^-$	487.3451	487.3429	4.5	487.3462 (100) $[M-H]^-$
T16	21.990	cyclocaric acid B	$C_{30}H_{46}O_5$	$[M-H]^-$	485.3276	485.3272	0.7	485.3300 (100) $[M-H]^-$
T17	22.128	asiatic acid	$C_{30}H_{48}O_5$	$[M-H]^-$	487.3424	487.3429	−1.0	487.3458 (100) $[M-H]^-$ 469.3347 (10) $[M-H-H_2O]^-$
T18	24.206	pterocaryoside B [37]	$C_{35}H_{58}O_9$	$[M-H]^-$	621.4007	621.4008	−0.2	621.4078 (100) $[M-H]^-$ 489.3624 (15) $[M-H-ara]^-$

Table 1. *Cont.*

No.	t_R (min)	Identification	Formula	Selected Ion	Measured (*m*/*z*)	Calculated (*m*/*z*)	Error (ppm)	MS[2] (*m*/*z*)
T19	25.341	cyclocarioside I	$C_{41}H_{70}O_{12}$	$[M-HCOO]^-$	799.4865	799.4849	2.0	799.4962 (19) $[M+HCOO]^-$ 753.4893 (100) $[M-H]^-$ 621.4482 (5) $[M-H-ara]^-$
T20	25.613	cyclocarioside K	$C_{41}H_{70}O_{12}$	$[M-HCOO]^-$	799.4864	799.4849	1.8	799.4962 (19) $[M+HCOO]^-$ 753.4893 (100) $[M-H]^-$
T21	25.892	pterocaryoside A [37]	$C_{36}H_{60}O_9$	$[M-H]^-$	635.4166	635.4165	0.2	635.4249 (100) $[M-H]^-$ 489.3639 (13) $[M-H-qui]^-$
T22	28.175	cyclocariol C [38]	$C_{30}H_{50}O_5$	$[M-HCOO]^-$	535.3636	535.3640	−0.8	489.3615 (100) $[M-H]^-$
T23	28.499	(20S,24R)-20,24-epoxy-25-hydroxy-12β-(α-L-arabinopyranosyloxy)-3,4-*seco*-dammara-4(28)-en-3-oic acid	$C_{35}H_{58}O_9$	$[M-H]^-$	621.4010	621.4008	0.3	621.4072 (100) $[M-H]^-$ 489.3617 (20) $[M-H-ara]^-$
T24	28.763	cyclocarioside B	$C_{42}H_{72}O_{12}$	$[M-HCOO]^-$	813.5020	813.5006	1.7	813.5129 (44) $[M+HCOO]^-$ 767.5066 (100) $[M-H]^-$
T25	29.008	hederagenin	$C_{30}H_{48}O_4$	$[M-H]^-$	471.3475	471.3480	−1.0	471.3493 (100) $[M-H]^-$ 453.3401 (10) $[M-H-H_2O]^-$
T26	29.171	cyclocarioside H	$C_{43}H_{72}O_{13}$	$[M-HCOO]^-$	841.4969	841.4955	1.7	841.5088 (76) $[M+HCOO]^-$ 795.5014 (100) $[M-H]^-$ 753.4900 (57) $[M-H-C_2H_2O]^-$
T27	29.245	cyclocarioside X [39]	$C_{37}H_{62}O_9$	$[M-H]^-$	649.4324	649.4321	0.4	649.4394 (100) $[M-H]^-$ 517.3948 (12) $[M-H-ara]^-$
T28	31.812	cypaliuruside E [40]	$C_{36}H_{58}O_8$	$[M-H]^-$	617.4058	617.4059	−0.2	617.4131 (100) $[M-H]^-$ 471.3508 (12) $[M-H-qui]^-$
T29	33.431	ursolic acid	$C_{30}H_{48}O_3$	$[M-H]^-$	455.3527	455.3531	−0.8	455.3580 (100) $[M-H]^-$

glc: glucosyl; gal: galactosyl; rha: rhamnosyl; ara: arabinosyl; qui: quinovosyl; glcA: glucuronic acid; CA: caffeic acid; FA: formic acid; AcOH: acetic acid.

Table 2. Identification of the chemical constituents in CPF by HPLC-ESI-Q-TOF-MS/MS.

No.	t_R (min)	Identification	Formula	Selected Ion	Measured (*m/z*)	Calculated (*m/z*)	Error (ppm)	MS[2] (*m/z*)
P1	3.938	3-*O*-caffeoylquinic acid [41]	$C_{16}H_{18}O_9$	$[M-H]^-$	353.0884	353.0878	1.7	191.0556 (100) $[M-H-CA+H_2O]^-$ 179.0347 (38) $[CA-H]^-$ 135.0454 (41) $[CA-H-CO_2]^-$
P2	5.083	4-*O*-caffeoylquinic acid [41]	$C_{16}H_{18}O_9$	$[M-H]^-$	353.0885	353.0878	2.0	191.0554 (100) $[M-H-CA+H_2O]^-$
P3	6.146	catechin [30]	$C_{15}H_{14}O_6$	$[M-H]^-$	289.0725	289.0718	2.6	289.0711 (88) $[M-H]^-$ 271.0600 (7) $[M-H-H_2O]^-$
F4	8.009	isoquercetin [31,41]	$C_{21}H_{20}O_{12}$	$[M-H]^-$	463.0888	463.0882	1.3	463.0919 (30) $[M-H]^-$ 301.0358 (44) $[M-H-glc]^-$
F5	8.125	quercetin-3-*O*-glucuronide [35,41]	$C_{21}H_{18}O_{13}$	$[M-H]^-$	477.0684	477.0675	2.0	477.0709 (11) $[M-H]^-$ 301.0362 (100) $[M-H-glcA]^-$ 283.0258 (7) $[M-H-glcA-H_2O]^-$
F6	8.841	kaempferol-3-*O*-galactoside [35,41]	$C_{21}H_{20}O_{11}$	$[M-H]^-$	447.0942	447.0933	2.0	447.0949 (55) $[M-H]^-$ 285.0391 (32) $[M-H-gal]^-$
F7	9.128	kaempferol-3-*O*-glucuronide [35,41]	$C_{21}H_{18}O_{12}$	$[M-H]^-$	461.0733	461.0725	1.6	461.0741 (5) $[M-H]^-$ 285.0397 (100) $[M-H-glcA]^-$
F8	9.198	kaempferol-3-*O*-glucoside [35,41]	$C_{21}H_{20}O_{11}$	$[M-H]^-$	447.0940	447.0933	1.6	447.0961 (29) $[M-H]^-$ 285.0404 (55) $[M-H-glc]^-$
F9	9.332	quercetin-3-*O*-rhamnoside [35,41]	$C_{21}H_{20}O_{11}$	$[M-H]^-$	447.0938	447.0933	1.2	447.0950 (18) $[M-H]^-$ 301.0351 (60) $[M-H-rha]^-$
F11	10.358	kaempferol-3-*O*-rhamnoside [35,41]	$C_{21}H_{20}O_{10}$	$[M-H]^-$	431.0987	431.0984	0.8	431.1008 (26) $[M-H]^-$ 285.0405 (88) $[M-H-rha]^-$ 257.0458 (25) $[M-H-rha-CO]^-$
F12	12.108	quercetin [35]	$C_{15}H_{10}O_7$	$[M-H]^-$	301.0366	301.0354	4.1	301.0360 (49) $[M-H]^-$ 273.0406 (16) $[M-H-CO]^-$ 245.0448 (8) $[M-H-2CO]^-$
F13	14.125	kaempferol [36]	$C_{15}H_{10}O_6$	$[M-H]^-$	285.0418	285.0405	4.7	285.0408 (100) $[M-H]^-$

glc: glucosyl; gal: galactosyl; rha: rhamnosyl; ara: arabinosyl; qui: quinovosyl; glcA: glucuronic acid; CA: caffeic acid; FA: formic acid; AcOH: acetic acid.

Table 3. Identification of the chemical constituents in CPT by HPLC-ESI-Q-TOF-MS/MS.

No.	t_R (min)	Identification	Formula	Selected Ion	Measured (*m*/*z*)	Calculated (*m*/*z*)	Error (ppm)	MS^2 (*m*/*z*)
T14	16.787	3β,19α,23-trihydroxy-1-oxo-olean -12-en-28-oic acid	$C_{30}H_{46}O_6$	$[M-H]^-$	501.3235	501.3222	2.6	501.3294 (100) $[M-H]^-$ 455.3225 (73) $[M-H-FA]^-$
T15	20.616	arjunolic acid	$C_{30}H_{48}O_5$	$[M-H]^-$	487.3451	487.3429	4.5	487.3482 (100) $[M-H]^-$
T16	22.005	cyclocaric acid B	$C_{30}H_{46}O_5$	$[M-H]^-$	485.3291	485.3272	3.9	485.3341 (100) $[M-H]^-$
T17	22.160	asiatic acid	$C_{30}H_{48}O_5$	$[M-H]^-$	487.3454	487.3429	5.1	487.3483 (100) $[M-H]^-$
T18	24.134	pterocaryoside B [37]	$C_{35}H_{58}O_9$	$[M-H]^-$	621.4029	621.4008	3.4	621.4104 (100) $[M-H]^-$ 489.3639 (26) $[M-H-ara]^-$
T19	25.328	cyclocarioside I	$C_{41}H_{70}O_{12}$	$[M-HCOO]^-$	799.4879	799.4849	3.8	799.5019 (25) $[M+HCOO]^-$ 753.4958 (100) $[M-H]^-$ 621.4500 (7) $[M-H-ara]^-$
T20	25.683	cyclocarioside K	$C_{41}H_{70}O_{12}$	$[M-HCOO]^-$	799.4866	799.4849	2.1	799.5024 (20) $[M+HCOO]^-$ 753.4952 (100) $[M-H]^-$ 607.4322 (5) $[M-H-qui]^-$
T21	25.895	pterocaryoside A [37]	$C_{36}H_{60}O_9$	$[M-H]^-$	635.4182	635.4165	2.7	635.4263 (100) $[M-H]^-$ 489.3628 (20) $[M-H-qui]^-$
T22	27.669	cyclocariol C [38]	$C_{30}H_{50}O_5$	$[M-HCOO]^-$	535.3671	535.3640	5.7	489.3636 (100) $[M-H]^-$
T23	28.486	(20S,24R)-20,24-epoxy-25-hydroxy-12β -(α-L-arabinopyranosyloxy)-3,4-*seco*-dammara-4(28)-en-3-oic acid	$C_{35}H_{58}O_9$	$[M-H]^-$	621.4034	621.4008	4.2	621.4118 (100) $[M-H]^-$ 489.3669 (16) $[M-H-ara]^-$
T24	28.800	cyclocarioside B	$C_{42}H_{72}O_{12}$	$[M-HCOO]^-$	813.5038	813.5006	3.9	813.5178 (35) $[M+HCOO]^-$ 767.5112 (100) $[M-H]^-$
T25	29.028	hederagenin	$C_{30}H_{48}O_4$	$[M-H]^-$	471.3500	471.3480	4.3	471.3535 (100) $[M-H]^-$ 453.3406 (11) $[M-H-H_2O]^-$ 425.3450 (9) $[M-H-H_2O-CO]^-$
T26	29.224	cyclocarioside H	$C_{43}H_{72}O_{13}$	$[M-HCOO]^-$	841.4989	841.4955	4.0	841.5109 (63) $[M+HCOO]^-$ 795.5039 (100) $[M-H]^-$ 735.4818 (47) $[M-H-AcOH]^-$
T27	29.252	cyclocarioside X [39]	$C_{37}H_{62}O_9$	$[M-H]^-$	649.4344	649.4321	3.5	649.4446 (100) $[M-H]^-$ 517.4000 (11) $[M-H-ara]^-$
T28	31.790	cypaliuruside E [40]	$C_{36}H_{58}O_8$	$[M-H]^-$	617.4076	617.4059	2.8	617.4163 (100) $[M-H]^-$ 471.3536 (13) $[M-H-qui]^-$
T29	33.510	ursolic acid	$C_{30}H_{48}O_3$	$[M-H]^-$	455.3546	455.3531	3.4	455.3571 (100) $[M-H]^-$

glc: glucosyl; gal: galactosyl; rha: rhamnosyl; ara: arabinosyl; qui: quinovosyl; glcA: glucuronic acid; CA: caffeic acid; FA: formic acid; AcOH: acetic acid.

2.8. The Molecular Networking for CPF Using GNPS Analysis

Public spectral libraries facilitate the dereplication of known molecules, and molecular networks allow annotation propagation of unknown related molecules. The aggregation of reference compounds in different clusters was based on different structural features. Compounds that were annotated as analogs were represented with a circular-colored node. Many flavonoids were identified from CPF. Some of them were reported to have anti-hyperglycemic effect, including epicatechin, myricetin-3-*O*-β-D-galactopyranoside, isoquercitrin, kaempferol 3-glucuronide, catechin, afzelin, quercetin, cynarine, kaempferol 3-α-L-arabinopyranoside, guajavarin, and quercetin 3-*O*-glucuronide, while other flavonoids' pharmacological activities remained unknown, such as 3,4-di-*O*-caffeoylquinic acid and 5,7-dihydroxy-2-(4-hydroxyphenyl)-3-[(2*S*,3*R*,4*R*,5*R*,6*S*)-3,4,5-trihydroxy-6-methyloxan-2-yl]oxychromen-4-one,(1*R*,3*R*,4*S*,5*S*)-4-{[(2*E*)-3-(3,4-dihydroxyphenyl)-2-propenoyl]oxy}-1,3,5-trihydroxycyclohexanecarboxylic acid (Figure 8). Some previous undescribed compounds were predicted according to the analysis results. These results indicate that diverse flavonoids exist in CPF, which might contribute to the anti-diabetic property of CPF.

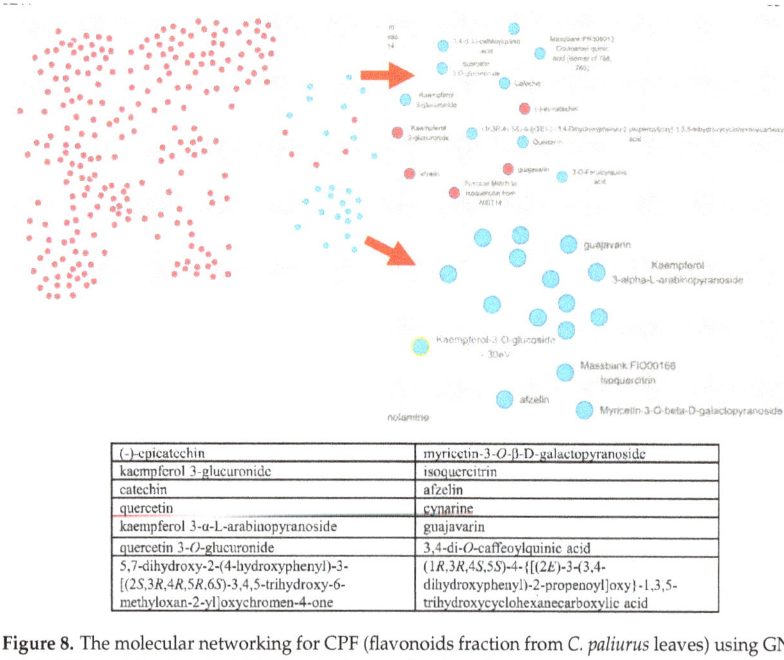

(-)-epicatechin	myricetin-3-*O*-β-D-galactopyranoside
kaempferol 3-glucuronide	isoquercitrin
catechin	afzelin
quercetin	cynarine
kaempferol 3-α-L-arabinopyranoside	guajavarin
quercetin 3-*O*-glucuronide	3,4-di-*O*-caffeoylquinic acid
5,7-dihydroxy-2-(4-hydroxyphenyl)-3-[(2*S*,3*R*,4*R*,5*R*,6*S*)-3,4,5-trihydroxy-6-methyloxan-2-yl]oxychromen-4-one	(1*R*,3*R*,4*S*,5*S*)-4-{[(2*E*)-3-(3,4-dihydroxyphenyl)-2-propenoyl]oxy}-1,3,5-trihydroxycyclohexanecarboxylic acid

Figure 8. The molecular networking for CPF (flavonoids fraction from *C. paliurus* leaves) using GNPS (Global Natural Product Social Molecular Networking) analysis.

3. Discussion

C. paliurus is named as "sweet tea tree" in China because of the flavor of its leaves, which have been used as an herbal tea to treat hyperglycemia and obesity [42]. *C. paliurus* is enriched with flavonoids, triterpenoids, and polysaccharides. So far, over 200 compounds have been isolated and identified from *C. paliurus* [43], including 27 polysaccharides [44], 137 flavonoids [45], and 43 triterpenoids [18,46]. These isolates showed various pharmacological activities including anti-inflammation, anti-oxidation, anti-microbial, anti-cancer, and anti-diabetes activities [47–49]. Different extracts from *C. paliurus* were reported with anti-diabetic potential on different animal models [17,23,24,48,50]. Until now, no random-controlled clinical trial of *C. paliurus* has been reported. More studies are needed to identify potential therapeutic targets and chemical principles of *C. paliurus*, and clinical trials are necessary to verify the anti-hyperglycemic effect of *C. paliurus*.

The polysaccharides from *C. paliurus* leaves have attracted much research attention in search of an anti-hyperglycemic principle. Around 20 polysaccharides were isolated from *C. paliurus*, and their monosaccharide compositions were identified [47]. The molecular weights and compositions of polysaccharides isolated by different methods varied greatly. Based on the GPC results, the molecular weights of CPP were in the range of 30,000–50,000 Da. Considering polysaccharides could hardly pass through the gastrointestinal tract intact and enter into circulation, gut microbiota and their fermentative products, short chain fatty acids, are thought to medicate the physiological function of polysaccharides. Polysaccharides from *C. paliurus* leaves were found to reduce blood glucose levels and improve glucose tolerance and serum lipid parameters in high-fat-diet-induced diabetic mice through modulating gut microbiota balance and short-chain fatty acids content [21]. In high-fat-diet-induced diabetic rats, polysaccharides from *C. paliurus* leaves markedly attenuated the symptoms of diabetes, inhibited the protein expression of Bax, increased the expression of Bcl-2 in the pancreas, normalized hormones secretion, and alleviated the levels of inflammatory mediators, which contributed to the regeneration of pancreatic β-cell and attenuation of insulin resistance [24]. Herein, we found CPP reverses body weight and muscle weight losses, ameliorates inflammatory responses, and lowers blood glucose in STZ-induced mice. This evidence suggests the polysaccharides from *C. paliurus* could partially contribute to the anti-hyperglycemic effect, through regulating gut microbiota balance, protecting gut epithelial barrier, and ameliorating inflammatory responses in the hosts. Whereas the CPP was not deproteinized in the isolation procedure, and a few flavonoids were identified in the CPP fraction, including quercetin 3-*O*-glucuronide, afzelin, and kaempferol 3-glucuronide, the bioactivity of CPC might partially be contributed to by proteins and/or flavonoids. A refined total polysaccharides fraction should be prepared and re-evaluated in the future.

Flavonoids are the major constituents of *C. paliurus* leaves and contribute to various pharmacological activities of *C. paliurus*, including anti-diabetic effect. The major flavonoids, including quercetin-3-*O*-glucuronide, kaempferol-3-*O*-glucuronide, kaempferol-7-*O*-α-l-rhamnoside, kaempferol, and quercetin, were identified from *C. paliurus* leaves with potent anti-oxidative activity [36]. Quercetin was reported to activate insulin receptor and glucose transporter 4, which in turn elevate glucose uptake in Caco-2E, Caco-2, C2C12 murine skeletal myoblast, and H4IIE murine hepatocytes [51,52]. The anti-hyperglycemic effect of kaempferol and quercetin glycosides from *C. paliurus* was determined on high-fat-diet-fed male C57BL/6J mice [17,53]. CPF treatment obviously reduced blood glucose levels in STZ-induced diabetic mice, which might be due to its effect of enhancing glucose uptake in skeletal muscle. CPF could enhance glucose-stimulated insulin secretion on normal MIN6 cells but did not change serum insulin level in STZ-induced mice, which suggested CPF might promote pancreatic β-cell function in normal objects. Additionally, CPF administration greatly decreased the levels of TNF-α and IL-6, indicating its anti-inflammatory activity. Pro-inflammatory cytokines can cause insulin resistance in adipose tissue, skeletal muscle, and liver by inhibiting insulin signal transduction [54,55]. The combination of CPP and CPF showed more pronounced anti-hyperglycemic effect in STZ-treated mice, which suggests CPP and CPF might function through different mechanisms and possess additive or synergic effects.

Although our previous study indicated several triterpenoids from *C. paliurus* leaves enhance insulin-stimulated glucose uptake on myotubes and adipocytes, the treatment of CPT and CPC did not show anti-diabetic effect in STZ-induced diabetic mice. This might be because triterpenoids cannot pass through the gastrointestinal tract and reach the target organs. Some improvements could be achieved, such as using drug-wrapping materials to improve intestinal absorption and the bioavailability of triterpenes, and the mode of administration can be changed from intragastric injection to intraperitoneal injection.

The clinically used anti-diabetic medicines mainly include insulin sensitizers, α-glucosidase inhibitors, and insulin-secreting agents. In recent years, glucagon-like peptide 1 receptor agonists, dipeptidyl peptidase-4 inhibitors, and sodium-glucose cotransporter-2 inhibitors have been successfully developed as anti-diabetic agents. Gut microbiota imbalance and, subsequently, systematic inflammation and metabolic disorders are positively correlated with insulin resistance, which has attracted more and more interest. Although many studies have indicated different extracts from *C. paliurus* possess anti-diabetic activity, refined fractions have never been compared on the same model. The dosage of each fraction was determined by the extraction rate, which was equal to the same amount of herbal material. Our study revealed for the first time that the polysaccharides and flavonoids from *C. paliurus* are responsible for its anti-hyperglycemic effect. This will benefit the clinical application of *C. paliurus* in the future. Bioinformatic tools including GNPS and Emperor were applied in the current study, which can help us better understand and predict bioactive principles from a complex extract and shorten the isolation and purification procedure.

We investigated the anti-hyperglycemic activity of refined fractions from *C. paliurus* on a high-dosage injection of STZ-induced diabetic mice model, which is more likely a type 1 diabetic model. Alternatively, high-fat diet feeding plus low-dosage STZ injection could induce type 2 diabetic mellitus phenotypes, which is more suitable for evaluating the anti-hyperglycemic effect of *C. paliurus*. Second, the glucose-stimulated insulin secretion in isolated islets of CP-fraction-treated mice should be performed, it can help us better understand the role of CP fraction in rescuing β-cell function. Additionally, some of the predicted compounds with potential anti-diabetic function have not been verified, and the systemic evaluation of these compounds should be carried out in the future.

4. Materials and Methods

4.1. Plant Material

The leaves of *C. paliurus* (Batal.) Iljinsk were collected in Sangzhi County, Zhangjiajie City, Hunan Province, China, in March 2018, which were identified by Jian-Xia Mo from Zhejiang University. A voucher specimen (accession number CP-2018-I) was deposited in the Institute of Modern Chinese Medicine, Zhejiang University (Hangzhou, China).

4.2. Extraction and Preparation

The extraction and purification procedures are shown in Figure 9. Air-dried leaves of *C. paliurus* (3 kg) were extracted with 70% ethanol (3 × 30 L) under reflux and evaporated to produce an ethanol extract (CPE, 534.0 g). The residue leaves were then extracted again with boiling distilled water (3 × 30 L), and a water extract (CPW, 220.9 g) was obtained under reduced pressure. A part of CPW was then dissolved with water and added with 5 parts of 95% ethanol. After sitting at room temperature overnight, the precipitate was collected by filtration and dried under vacuum to obtain the crude polysaccharides (CPP, 27.6 g). A part of CPE was suspended in water (1 L) and successively partitioned with petroleum ether (PE) and ethyl acetate (EtOAc) to obtain the EtOAc fraction. The EtOAc fraction (175.4 g) was first subjected to column chromatography over polyamide resin eluted with PE-acetone (1:1) to obtain the crude triterpenoids (87.6 g), followed by aqueous EtOH (70%) to obtain the crude flavonoids (45.8 g). Next, the crude triterpenoids were purified on a D-101 macroporous resin column eluted with aqueous EtOH (40%) to remove impurities. Then, 70% ethanol eluent was collected and subsequently evaporated under reduced pressure to yield the total triterpenoids (CPT, 40.2 g). Similarly, the crude flavonoids were dispersed in water and purified on a D-101 macroporous resin column. The water-soluble impurities were first removed with water, and 40% ethanol eluent was collected. The total flavonoids (CPF, 10.2 g) were obtained after removal of the solvent under reduced pressure.

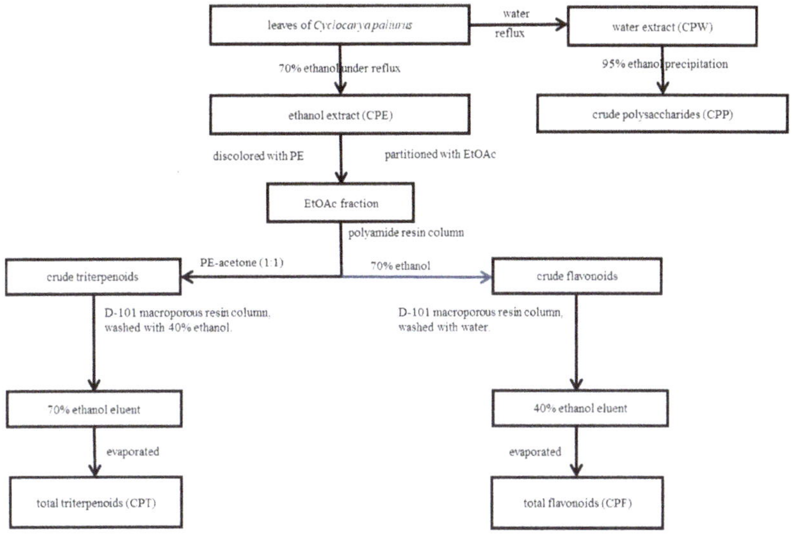

Figure 9. Extraction and purification procedures of *C. paliurus* leaves.

4.3. Animals

All animal care and experimental procedures followed the guidelines and regulations approved by the Animal Ethical and Welfare Committee of University of Macau (No. UMARE-033-2017). Male C57BL/6J mice (8–10 weeks old) were obtained from the animal facility of the Faculty of Health Science, University of Macau (Macau, China). The mice were housed at 22 ± 1 °C with 12-h light–dark cycles and fed with a regular chow diet (Guangdong Medical Lab Animal Center, Guangzhou, Guangdong, China) and water ad libitum under standard conditions (specific-pathogen-free) with air filtration.

4.4. STZ-Induced Diabetic Mice

The diabetes mice model was induced by intraperitoneal injection (i.p) with STZ (150 mg/kg) once. After 3 days, the mice with fasting blood glucose level between 11–28 mmol/L were considered as diabetic mice. The diabetic mice were then randomly divided into nine groups, including STZ (model), CPE (ethanol extract from *C. paliurus* leaves), CPW (water extract from *C. paliurus* leaves), CPF (flavonoids fraction from *C. paliurus* leaves), CPP (polysaccharide fraction from *C. paliurus* leaves), CPT (triterpenoids fraction from *C. paliurus* leaves), CPF + CPP (flavonoids and polysaccharide fractions from *C. paliurus* leaves), CPC (flavonoids, polysaccharide, and triterpenoids fractions from *C. paliurus* leaves), and Glibenclamide (INN) used as a positive control [43,56–58]. The mice were orally administrated the same volume of 60% PEG400 solution (control and STZ), 15 mg/kg INN (INN), 2500 mg/kg CPE (CPE), 1500 mg/kg CPW (CPW), 300 mg/kg CPF (CPF), 180 mg/kg CPP (CPP), 650 mg/kg CPT (CPT), 300 mg/kg CPF + 180 mg/kg CPP (CPF + CPP), and 300 mg/kg CPF + 180 mg/kg CPP + 650 mg/kg CPT (CPC), respectively, once a day (Figure 2A). At the end of the experiment (day 25), the mice were dissected after carbon dioxide inhalation, and the kidneys, liver, and skeletal muscles (quadriceps and gastrocnemius) were collected.

4.5. Glucose Tolerance Tests (GTT) and Insulin Tolerance Tests (ITT)

GTT and ITT were performed at 2 and 3 weeks post-treatment, respectively, as described previously [59]. After 18 h fasting, the tail blood glucose was measured using the OneTouch Ultra blood glucose meter and LifeScan test strips. Then, the mice received an intraperitoneal injection of glucose solution (Sigma-Aldrich, St. Louis, MO, USA) at a dose of 1.5 g/kg body weight. The tail blood glucose was measured at 15, 30, 60, and 120 min after injections. For the ITT, the tail blood glucose concentration was measured after 6 h fasting. Mice then received an intraperitoneal injection of human insulin (Eli Lilly, Indianapolis, IN, USA) at a dose of 1.0 U/kg body weight. Tail blood glucose concentration was measured at 15, 30, 60, and 120 min after injections.

4.6. Blood Sample Collection and Preparation

About 0.5 mL blood was collected in heparin tube. Plasma was obtained after a 3000 rpm centrifuge for 30 min at room temperature. Plasma insulin was determined using the Insulin assay kit (Sigma-Aldrich, St. Louis, MO, USA) according to the manufacturer's instructions.

4.7. Homeostasis Model Assessment of Basal Insulin Resistance

The homeostasis model assessment of basal insulin resistance (HOMA-IR) index was calculated as follows to assess insulin resistance: fasting serum glucose × fasting serum insulin/22.5. Lower HOMA-IR values indicated greater insulin sensitivity, whereas higher HOMA-IR values indicated insulin resistance.

4.8. Cell Viability

Cell viability was determined by 3-(4,5-dimethylthiazol-2-yl)-2,5-diphenyltetrazolium bromide (MTT, Sigma-Aldrich, St. Louis, MO, USA) assay. C2C12 cells were seeded in 96-well plates at a density of 1×10^4 cells per well. The fully differentiated myotubes were treated with the indicated concentrations of CPF for 24 h. Then, cell viability was determined by incubation with DMEM containing MTT (1 mg/mL) for 4 h, followed by dissolving the formazan crystals with 100 μL dimethyl sulfoxide (DMSO). The absorbance at 570 nm was measured by a SpectraMax M5 microplate reader (Molecular Devices, CA, USA). The calculation equation for relative cell viability was as follows: cell viability (%) = (As − A0)/(Ac − A0) × 100%, where As, A0, and Ac were the absorptions of the test sample, blank control, and negative control (DMSO), respectively.

4.9. Glucose-Stimulated Insulin Secretion

MIN6 cells were cultured in Dulbecco's modified eagle medium (DMEM, Gibco, Carlsbad, CA, USA) supplemented with 10% fetal bovine serum (FBS, Gibco, Carlsbad, CA, USA) and 1% penicillin–streptomycin (P/S, Gibco, Carlsbad, CA, USA), in a humidified incubator with 5% CO_2 at 37 °C. The experiments were performed between passages 16 and 24 [60]. The cells were cultured in 24-well plates with high-glucose DMEM (25 mM) and then treated with CP fractions for 24 h. INN (glibenclamide, 0.1 μM, Sigma–Aldrich, St. Louis, MO, USA) was used as a positive control [61]. Subsequently, the cells were washed twice with Krebs–Ringer bicarbonate buffer (KRBB: $CaCl_2$ 2.5 mM; KCl 4.7 mM; KH_2PO_4 1.2 mM; $MgCl_2$ 1.2 mM; NaCl 120 mM; HEPES 10 mM; $NaHCO_3$ 25 mM; pH = 7.4) and incubated with KRBB with 3 mM glucose for 30 min. The cells were washed twice with KRBB and then incubated with KRBB with 5.5 or 16.7 mM glucose for 1 h. The supernatants were collected, and insulin was measured by mouse insulin ELISA kit (Mercodia, Winston Salem, NC, USA).

4.10. C2C12 Cell Culture and Differentiation

Mouse C2C12 myoblasts were obtained from American Type Culture Collection (Manassas, VA, USA) and maintained in DMEM supplemented with 10% FBS and 1% P/S. C2C12 cells were differentiated as described previously [62]. In brief, C2C12 cells were grown to 90% confluence and incubated with DMEM containing 2% heat-inactivated horse serum (Gibco, Carlsbad, CA, USA) and 1% P/S for 6 days. Media were refreshed every other day. The fully differentiated myotubes were used for the following experiments.

4.11. *Insulin-Stimulated Glucose Uptake*

Insulin-stimulated glucose uptake was performed as described previously [63]. C2C12 myotubes were treated with different concentrations of the total flavonoids (CPF) for 24 h. Then, cells were washed with Krebs–Ringer phosphate (KRP) buffer (20 mmol/L HEPES, 137 mmol/L NaCl, 4.7 mmol/L KCl, 1.2 mmol/L $MgSO_4$, 1.2 mmol/L KH_2PO_4, 2.5 mmol/L $CaCl_2$, and 2 mmol/L pyruvate; pH 7.4) and incubated in KRP buffer with 0.2% bovine serum albumin (BSA, Sigma-Aldrich, St. Louis, MO, USA) for 3 h. To stimulate glucose uptake, cells were incubated with KRP buffer containing 0.1 μmol/L insulin (Sigma-Aldrich, St. Louis, MO, USA) for another 30 min. After being washed with KRP buffer once, cells were incubated in KRP containing 100 μmol/L 2-(*N*-(7-nitrobenzene-2-oxa-1,3-diazol-4-yl)amino)-2-deoxyglucose (2-NBDG, Sigma-Aldrich, St. Louis, MO, USA) for 30 min. The intracellular amount of 2-NBDG was measured at an excitation wavelength of 475 nm and an emission wavelength of 550 nm. Glucose uptake was further normalized by protein content.

4.12. *Determination of Cytokines*

The cytokines in mice serum were determined by using commercial ELISA kits (Neobioscience Technology Co., Ltd., Shenzhen, China), following the manufacturer's instructions.

4.13. *H&E Staining of Kidney*

Whole-mount histochemical analysis of the kidney was performed as previously described [64]. After fixation in 4% paraformaldehyde, the kidney samples were embedded in paraffin. A total of 5 μm sections were deparaffinized and rehydrated, followed by hematoxylin and eosin (H&E) staining.

4.14. *Sample Preparation*

The CPE powder (0.25 g) was dissolved by adding methanol in a 25 mL volumetric flask. The sample solution was centrifuged at 13,000 rpm for 10 min. The supernatant was stored at −4 °C before use. The CPF and CPT solution was prepared in the same way.

4.15. *Emperor Analysis*

Emperor is a tool for analyzing, visualizing, and understanding high-throughput microbial ecology data sets [65]. Due to its customized graphical user interface, it has never been easier to drill down into new data sets to illuminate patterns hidden in the data. Emperor can be integrated into any microbial ecological quantitative analysis (QIIME) or scikit bio compatible data set because its high-quality of modifications and customizations or sciKit biocompatible datasets. Throughout the lightweight data files and hardware-accelerated graphics, emperor is more and more often used in N-dimensional analyzing. [66].

4.16. *Chromatography and MS Conditions*

The samples were analyzed using a Waters UPLC system (Waters, Milford, MA, USA) equipped with a Waters XSelect® HSS T3 column (4.6 mm × 250 mm, 5 μm). The sample injection volume was set at 10 μL. The optimal mobile phase consisted of acetonitrile solution (A) and 0.1% formic acid aqueous solution (B) at a flow rate of 1.0 mL/min at 30 °C. The solvent gradient was used as follows: 5–30% A at 0–10 min, 30–70% A at 10–30 min, 70–95% A at 30–40 min. The MS analysis was carried out by a Triple TOF 5600 plus mass spectrometer (AB SCIEX, Framingham, MA, USA) equipped with an ESI source (AB SCIEX, Framingham, MA, USA). The mass spectrometer was operated in both positive and negative ion mode. The following operating parameters were used: scan range, m/z 100–2000 for TOF-MS scan and m/z 50–2000 for TOF-MS/MS scan; ion spray voltage, +5.5 kV for positive ion mode and −4.5 kV for negative ion mode; ion source heater, 600 °C for positive ion mode and 550 °C for negative ion mode; nebulizing gas (Gas 1, Air), 50 psi; Tis gas (Gas 2, Air), 50 psi; curtain gas (CUR, N_2), 30 psi; collision energy,

40 V with a collision energy spread of ± 20 V; maximum allowed error, ±5 ppm. The experiments were run with 250 ms accumulation time for TOF-MS and 100 ms accumulation time for TOF-MS/MS. The exact mass calibration was performed automatically before each analysis by employing the Automated Calibration Delivery System. The accurate mass and composition for the precursor ions and fragment ions were analyzed using the Peakview software (AB SCIEX, version 1.2.0.3, Framingham, MA, USA) integrated with the instrument.

4.17. Determination of Homogeneity and Molecular Weight for CPP

The molecular weight distribution of CPP was determined using gel permeation chromatography (GPC) method on a Shimadzu LC20 instrument equipped with a TSK GMPWXL gel filtration column (7.6 × 300 mm) and a differential refractive index (DRI) detector (Shimadzu RID-20A). CPP was dissolved in ultrapure water (5 mg/mL) and filtered through a 0.22 μm membrane prior to sample injection. Then, 20 μL of sample solution was injected in each run with water containing 0.1 N sodium nitrate and 0.06% sodium azide as the mobile phase at a flow rate of 0.6 mL/min in 35 ∘C. The molecular weight of polysaccharide sample was estimated by reference to the calibration curve made using dextran standards. The number-average molecular weight (*Mn*), weight-average molecular weight (*Mw*), Z-average molecular weight (*Mz*), and polydispersity were calculated using molecular weight calculation software connected to the GPC integration system, as indicated before [67].

4.18. Molecular Networking Workflow Description

A molecular network was created using an online workflow application (https://ccms-ucsd.github.io/GNPSDocumentation/ accessed on 25 March 2021) on the Global Natural Product Social Molecular Networking (GNPS) website (http://gnps.ucsd.edu accessed on 25 March 2021) [68]. The data were filtered by removing all MS/MS fragment ions within +/− 17 Da of the precursor *m/z*. MS/MS spectra were window-filtered by choosing only the top 6 fragment ions in the +/− 50 Da window throughout the spectrum. The precursor ion mass tolerance was set to 0.01 Da and an MS/MS fragment ion tolerance of 0.0075 Da. A network was then created in which edges were filtered to have a cosine score above 0.7 and more than 6 matched peaks. Further, edges between two nodes were kept in the network if and only if each of the nodes appeared in each other's respective top 10 most similar nodes. Finally, the maximum size of a molecular family was set to 100, and the lowest-scoring edges were removed from molecular families until the molecular family size was below this threshold. The spectra in the network were then searched against GNPS's spectral libraries. The library spectra were filtered in the same manner as the input data. All matches kept between network spectra and library spectra were required to have a score above 0.7 and at least 6 matched peaks.

4.19. Statistical Analysis

Data were analyzed using GraphPad Prism 7.0 (GraphPad Software, San Diego, CA, USA). All experimental data were expressed as mean ± S.D., and the sample size for each experiment corresponds to three biological replicates. Significant differences between groups were determined using a one-way analysis of variance (ANOVA) with Dunnet's multiple comparisons test, considering $p < 0.05$ as significant differences. Where statistical significance is evaluated, the variance between groups is confirmed to be similar between comparison groups (control vs. experimental), and the statistical analysis is considered appropriate.

5. Conclusions

In summary, polysaccharides and flavonoids are responsible for the anti-diabetic effect of *C. paliurus* leaves, but not triterpenoids. For clinical application, the polysaccharides and flavonoids in *C. paliurus* leaves need to be enriched to enhance anti-hyperglycemic activity.

Author Contributions: Z.F. (Zheling Feng): Investigation, Methodology, Writing—original draft, Data curation. Z.F. (Zhujun Fang): Investigation. C.C.: Review. C.T.V.: Investigation, Review. J.C.: Review. R.L.: Review. M.P.M.H.: Review. L.G.: Supervision. L.L.: Project administration, Conceptualization, Resources, Writing—Review and editing. All authors have read and agreed to the published version of the manuscript.

Funding: This research was funded by the National Natural Science Foundation of China (82073715, 22077111, 81872756), The Science and Technology Development Fund, Macau SAR (File no. FDCT 0031/2019/A1), and the Research Fund of University of Macau (MYRG2018-00037-ICMS and MYRG2020-00091-ICMS).

Institutional Review Board Statement: The study was conducted according to the guidelines of the Declaration of Helsinki, and approved by the Animal Ethical and Welfare Committee of University of Macau (No. UMARE-033-2017, 11 July 2018).

Informed Consent Statement: Not applicable.

Data Availability Statement: Not applicable.

Conflicts of Interest: The authors declare no conflict of interest.

Sample Availability: Samples of the fractions are available from the authors.

References

1. Chen, L.; Magliano, D.J.; Zimmet, P.Z. The worldwide epidemiology of type 2 diabetes mellitus-present and future perspectives. *Nat. Rev. Endocrinol.* **2012**, *8*, 228–236. [CrossRef]
2. WHO. World Health Organization Diabetes. Available online: https://www.who.int/news-room/fact-sheets/detail/diabetes (accessed on 9 October 2021).
3. Harding, J.L.; Pavkov, M.E.; Magliano, D.J.; Shaw, J.E.; Gregg, E.W. Global trends in diabetes complications: A review of current evidence. *Diabetologia* **2019**, *62*, 3–16. [CrossRef]
4. Eizirik, D.L.; Pasquali, L.; Cnop, M. Pancreatic beta-cells in type 1 and type 2 diabetes mellitus: Different pathways to failure. *Nat. Rev. Endocrinol.* **2020**, *16*, 349–362. [CrossRef] [PubMed]
5. Balaji, R.; Duraisamy, R.; Kumar, M. Complications of diabetes mellitus: A review. *Drug Invent. Today* **2019**, *12*, 98–103.
6. Testa, R.; Bonfigli, A.R.; Prattichizzo, F.; La Sala, L.; De Nigris, V.; Ceriello, A. The "Metabolic Memory" theory and the early treatment of hyperglycemia in prevention of diabetic complications. *Nutrients* **2017**, *9*, 437. [CrossRef]
7. Umegaki, H. Sarcopenia and diabetes: Hyperglycemia is a risk factor for age-associated muscle mass and functional reduction. *J. Diabetes Invest.* **2015**, *6*, 623. [CrossRef] [PubMed]
8. Hirata, Y.; Nomura, K.; Senga, Y.; Okada, Y.; Kobayashi, K.; Okamoto, S.; Minokoshi, Y.; Imamura, M.; Takeda, S.I.; Hosooka, T. Hyperglycemia induces skeletal muscle atrophy via a WWP1/KLF15 axis. *JCI Insight* **2019**, *4*, e124952. [CrossRef] [PubMed]
9. Sudamrao Garud, M.; Anant Kulkarni, Y. Hyperglycemia to nephropathy via transforming growth factor beta. *Curr. Diabetes Rev.* **2014**, *10*, 182–189. [CrossRef] [PubMed]
10. Yusuff, K.B.; Obe, O.; Joseph, B.Y. Adherence to anti-diabetic drug therapy and self management practices among type-2 diabetics in Nigeria. *Pharm. World Sci.* **2008**, *30*, 876–883. [CrossRef]
11. Ruscica, M.; Baldessin, L.; Boccia, D.; Racagni, G.; Mitro, N. Non-insulin anti-diabetic drugs: An update on pharmacological interactions. *Pharmacol. Res.* **2017**, *115*, 14–24. [CrossRef] [PubMed]
12. Kuang, R.; Li, Q. *Cyclocarya paliurus* (Batal.) Iljinsk. In *Flora Reipublicae Popularis Sinicae*; Editorial Committee of Flora of China; Chinese Academy of Sciences Press: Beijing, China, 1979; p. 19.
13. Jiang, C.; Yao, N.; Wang, Q.; Zhang, J.; Sun, Y.; Xiao, N.; Liu, K.; Huang, F.; Fang, S.; Shang, X. *Cyclocarya paliurus* extract modulates adipokine expression and improves insulin sensitivity by inhibition of inflammation in mice. *J. Ethnopharmacol.* **2014**, *153*, 344–351. [CrossRef] [PubMed]
14. Yoshitomi, H.; Tsuru, R.; Li, L.; Zhou, J.; Kudo, M.; Liu, T.; Gao, M. *Cyclocarya paliurus* extract activates insulin signaling via Sirtuin1 in C2C12 myotubes and decreases blood glucose level in mice with impaired insulin secretion. *PLoS ONE* **2017**, *12*, e0183988. [CrossRef] [PubMed]
15. Zhao, L.; Wang, X.; Li, J.; Tan, X.; Fan, L.; Zhang, Z.; Leng, J. Effect of *Cyclocarya paliurus* on hypoglycemic effect in type 2 diabetic mice. *Med. Sci. Monitor.* **2019**, *25*, 2976. [CrossRef]
16. Shi, L.; ShangGuan, X.; Wang, W.; Shen, Y.; Jiang, Y.; Yin, Z. Effects of polysaccharide of *Cyclocarya paliurus* on alloxan-induced diabetic mice. *Acta Nutr. Sin.* **2009**, *31*, 263–266.
17. Liu, Y.; Cao, Y.; Fang, S.; Wang, T.; Yin, Z.; Shang, X.; Yang, W.; Fu, X. Antidiabetic effect of *Cyclocarya paliurus* leaves depends on the contents of antihyperglycemic flavonoids and antihyperlipidemic triterpenoids. *Molecules* **2018**, *23*, 1042. [CrossRef]
18. Fang, Z.-J.; Shen, S.-N.; Wang, J.-M.; Wu, Y.-J.; Zhou, C.-X.; Mo, J.-X.; Lin, L.-G.; Gan, L.-S. Triterpenoids from *Cyclocarya paliurus* that enhance glucose uptake in 3T3-L1 adipocytes. *Molecules* **2019**, *24*, 187. [CrossRef] [PubMed]

19. Jiang, W.; Si, L.; Li, P.; Bai, B.; Qu, J.; Hou, B.; Zou, H.; Fan, X.; Liu, Z.; Liu, Z.; et al. Serum metabonomics study on antidiabetic effects of fenugreek flavonoids in streptozotocin-induced rats. *J. Chromatogrt B* **2018**, *1092*, 466–472. [CrossRef]
20. Cui, C.; Yang, Y.; Zhao, T.; Zou, K.; Peng, C.; Cai, H.; Wan, X.; Hou, R. Insecticidal activity and insecticidal mechanism of total saponins from *Camellia oleifera*. *Molecules* **2019**, *24*, 4518. [CrossRef]
21. Yao, Y.; Yan, L.; Chen, H.; Wu, N.; Wang, D. *Cyclocarya paliurus* polysaccharides alleviate type 2 diabetic symptoms by modulating gut microbiota and short-chain fatty acids. *Phytomedicine* **2020**, *77*, 153268. [CrossRef]
22. Marre, M.; Howlett, H.; Lehert, P.; Allavoine, T. Improved glycaemic control with metformin–glibenclamide combined tablet therapy (Glucovance®) in Type 2 diabetic patients inadequately controlled on metformin. *Diabet. Med.* **2002**, *19*, 673–680. [CrossRef]
23. Wang, Q.; Jiang, C.; Fang, S.; Wang, J.; Ji, Y.; Shang, X.; Ni, Y.; Yin, Z.; Zhang, J. Antihyperglycemic, antihyperlipidemic and antioxidant effects of ethanol and aqueous extracts of *Cyclocarya paliurus* leaves in type 2 diabetic rats. *J. Ethnopharmacol.* **2013**, *150*, 1119–1127. [CrossRef] [PubMed]
24. Li, Q.; Hu, J.; Nie, Q.; Chang, X.; Fang, Q.; Xie, J.; Li, H.; Nie, S. Hypoglycemic mechanism of polysaccharide from *Cyclocarya paliurus* leaves in type 2 diabetic rats by gut microbiota and host metabolism alteration. *Sci. China Life Sci.* **2021**, *64*, 117–132. [CrossRef] [PubMed]
25. Li, K.; Zhang, L.; Xue, J.; Yang, X.; Dong, X.; Sha, L.; Lei, H.; Zhang, X.; Zhu, L.; Wang, Z. Dietary inulin alleviates diverse stages of type 2 diabetes mellitus via anti-inflammation and modulating gut microbiota in db/db mice. *Food Funct.* **2019**, *10*, 1915–1927. [CrossRef] [PubMed]
26. Liu, G.; Bei, J.; Liang, L.; Yu, G.; Li, L.; Li, Q. Stachyose improves inflammation through modulating gut microbiota of high-fat diet/streptozotocin-induced type 2 diabetes in rats. *Mol. Nutr. Food Res.* **2018**, *62*, 1700954. [CrossRef]
27. Nosadini, R.; Tonolo, G. Relationship between blood glucose control, pathogenesis and progression of diabetic nephropathy. *J. Am. Soc. Nephrol.* **2004**, *15*, S1–S5. [CrossRef] [PubMed]
28. Warren, A.M.; Knudsen, S.T.; Cooper, M.E. Diabetic nephropathy: An insight into molecular mechanisms and emerging therapies. *Expert. Opin. Ther. Tar.* **2019**, *23*, 579–591. [CrossRef]
29. Umanath, K.; Lewis, J.B. Update on diabetic nephropathy: Core curriculum 2018. *Am. J. Kidney Dis.* **2018**, *71*, 884–895. [CrossRef] [PubMed]
30. Jiang, X.; Chen, S.; Guo, Z.; Gu, D.; Liang, X. Components characterisation of *Berchemia lineata* (L.) DC. by UPLC-QTOF-MS/MS and its metabolism with human liver microsomes. *Nat. Prod. Res.* **2021**, *35*, 521–524. [CrossRef] [PubMed]
31. El-Hawary, S.S.; Fathy, F.I.; Sleem, A.A.; Morsy, F.A.; Khadar, M.S.; Mansour, M.K. Anticholinesterase activity and metabolite profiling of *Syagrus romanzoffiana* (Cham.) Glassman leaves and fruits via UPLC–QTOF–PDA–MS. *Nat. Prod. Res.* **2021**, *35*, 1671–1675. [CrossRef]
32. Ning, Z.-N.; Zhai, L.-X.; Huang, T.; Peng, J.; Hu, D.; Xiao, H.-T.; Wen, B.; Lin, C.-Y.; Zhao, L.; Biao, Z.-X. Identification of α-glucosidase inhibitors from *Cyclocarya paliurus* tea leaves using UF-UPLC-Q/TOF-MS/MS and molecular docking. *Food Funct.* **2019**, *10*, 1893–1902. [CrossRef] [PubMed]
33. Sadowska-Bartosz, I.; Galiniak, S.; Bartosz, G. Kinetics of glycoxidation of bovine serum albumin by glucose, fructose and ribose and its prevention by food components. *Molecules* **2014**, *19*, 18828–18849. [CrossRef] [PubMed]
34. de Peredo, A.V.G.; Vázquez-Espinosa, M.; Piñeiro, Z.; Espada-Bellido, E.; Ferreiro-González, M.; Barbero, G.F.; Palma, M. Development of a rapid and accurate UHPLC-PDA-FL method for the quantification of phenolic compounds in grapes. *Food Chem.* **2021**, *334*, 127569. [CrossRef] [PubMed]
35. Ning, Z.-W.; Zhai, L.-X.; Peng, J.; Zhao, L.; Huang, T.; Lin, C.-Y.; Chen, W.-H.; Luo, Z.; Xiao, H.-T.; Bian, Z.-X. Simultaneous UPLC–TQ-MS/MS determination of six active components in rat plasma: Application in the pharmacokinetic study of *Cyclocarya paliurus* leaves. *Chin. Med.* **2019**, *14*, 1–11. [CrossRef] [PubMed]
36. Xie, J.-H.; Dong, C.-J.; Nie, S.-P.; Li, F.; Wang, Z.-J.; Shen, M.-Y.; Xie, M.-Y. Extraction, chemical composition and antioxidant activity of flavonoids from *Cyclocarya paliurus* (Batal.) Iljinskaja leaves. *Food Chem.* **2015**, *186*, 97–105. [CrossRef] [PubMed]
37. Lin, Z.; Wu, Z.-F.; Jiang, C.-H.; Zhang, Q.-W.; Ouyang, S.; Che, C.-T.; Zhang, J.; Yin, Z.-Q. The chloroform extract of *Cyclocarya paliurus* attenuates high-fat diet induced non-alcoholic hepatic steatosis in Sprague Dawley rats. *Phytomedicine* **2016**, *23*, 1475–1483. [CrossRef] [PubMed]
38. Chen, Y.-j.; Na, L.; Fan, J.; Zhao, J.; Hussain, N.; Jian, Y.-q.; Yuan, H.; Li, B.; Liu, B.; Choudhary, M.I. Phytochemistry Seco-dammarane triterpenoids from the leaves of *Cyclocarya paliurus*. *Phytochemistry* **2018**, *145*, 85–92. [CrossRef] [PubMed]
39. Liu, W.; Deng, S.; Zhou, D.; Huang, Y.; Li, C.; Hao, L.; Zhang, G.; Su, S.; Xu, X.; Yang, R. 3, 4-Seco-dammarane triterpenoid saponins with anti-inflammatory activity isolated from the leaves of *Cyclocarya paliurus*. *J. Agric. Food Chem.* **2020**, *68*, 2041–2053. [CrossRef]
40. Zhou, X.-L.; Li, S.-B.; Yan, M.-Q.; Luo, Q.; Wang, L.-S.; Shen, L.-L.; Liao, M.-L.; Lu, C.-H.; Liu, X.-Y.; Liang, C.-Q. Bioactive dammarane triterpenoid saponins from the leaves of *Cyclocarya paliurus*. *Phytochemistry* **2021**, *183*, 112618. [CrossRef]
41. Cao, Y.; Fang, S.; Yin, Z.; Fu, X.; Shang, X.; Yang, W.; Yang, H. Chemical fingerprint and multicomponent quantitative analysis for the quality evaluation of *Cyclocarya paliurus* leaves by HPLC–Q–TOF–MS. *Molecules* **2017**, *22*, 1927. [CrossRef]
42. Xie, J.-H.; Xie, M.-Y.; Nie, S.-P.; Shen, M.-Y.; Wang, Y.-X.; Li, C. Isolation, chemical composition and antioxidant activities of a water-soluble polysaccharide from *Cyclocarya paliurus* (Batal.) Iljinskaja. *Food Chem.* **2010**, *119*, 1626–1632. [CrossRef]

43. McCoy, S.; Kabadi, M.; Kabadi, U.; Birkenholz, M. More uniform diurnal blood glucose control and a reduction in daily insulin dosage on addition of glibenclamide to insulin in type 1 diabetes mellitus: Role of enhanced insulin sensitivity. *Diabet. Med.* **1995**, *12*, 880–884. [CrossRef] [PubMed]
44. Wang, R.; Qin, Y.; Zhou, J. Research progress in *Cyclocarya paliurus* (Batal.) Iljinsk. polysaccharide in China in the latest decade. *Chin. J. Inf. Trad Chin. Med.* **2017**, *24*, 133–136.
45. Sheng, X.; Chen, H.; Wang, J.; Zheng, Y.; Li, Y.; Jin, Z.; Li, J. Joint transcriptomic and metabolic analysis of flavonoids in *Cyclocarya paliurus* leaves. *ACS Omega.* **2021**, *6*, 9028–9038. [CrossRef] [PubMed]
46. Richard, A.J.; White, U.; Elks, C.M.; Stephens, J.M. Adipose tissue: Physiology to metabolic dysfunction. *Endotext* **2020**.
47. Kakar, M.U.; Naveed, M.; Saeed, M.; Zhao, S.; Rasheed, M.; Firdoos, S.; Manzoor, R.; Deng, Y.; Dai, R. A review on structure, extraction, and biological activities of polysaccharides isolated from *Cyclocarya paliurus* (Batalin) Iljinskaja. *Int. J. Biol. Macromol.* **2020**, *156*, 420–429. [CrossRef]
48. Kurihara, H.; Fukami, H.; Kusumoto, A.; Toyoda, Y.; Shibata, H.; Matsui, Y.; Asami, S.; Tanaka, T. Hypoglycemic action of *Cyclocarya paliurus* (Batal.) Iljinskaja in normal and diabetic mice. *Biosci Biotech Bioch.* **2003**, *67*, 877–880. [CrossRef]
49. Liu, W.; Wu, Y.; Hu, Y.; Qin, S.; Guo, X.; Wang, M.; Wu, L.; Liu, T. Effect of *Cyclocarya paliurus* aqueous and ethanol extract on glycolipid metabolism and the underlying mechanisms: A meta-analysis and systematic review. *Front Nutr.* **2020**, *7*, 289. [CrossRef]
50. Li, J.; Luo, M.; Luo, Z.; Guo, A.-Y.; Yang, X.; Hu, M.; Zhang, Q.; Zhu, Y. Transcriptome profiling reveals the anti-diabetic molecular mechanism of *Cyclocarya paliurus* polysaccharides. *J. Funct. Foods.* **2019**, *55*, 1–8. [CrossRef]
51. Aguirre, L.; Arias, N.; Teresa Macarulla, M.; Gracia, A.; P Portillo, M. Beneficial effects of quercetin on obesity and diabetes. *Open Nutr. J.* **2011**, *4*, 189–198.
52. Shi, G.-J.; Li, Y.; Cao, Q.-H.; Wu, H.-X.; Tang, X.-Y.; Gao, X.-H.; Yu, J.-Q.; Chen, Z.; Yang, Y. In vitro and in vivo evidence that quercetin protects against diabetes and its complications: A systematic review of the literature. *Biomed. Pharmacother.* **2019**, *109*, 1085–1099. [CrossRef]
53. Zang, Y.; Zhang, L.; Igarashi, K.; Yu, C. The anti-obesity and anti-diabetic effects of kaempferol glycosides from *Unripe soybean* leaves in high-fat-diet mice. *Food Funct.* **2015**, *6*, 834–841. [CrossRef]
54. De Luca, C.; Olefsky, J.M. Inflammation and insulin resistance. *FEBS Lett.* **2008**, *582*, 97–105. [CrossRef] [PubMed]
55. Shoelson, S.E.; Lee, J.; Goldfine, A.B. Inflammation and insulin resistance. *J. Clin. Invest.* **2006**, *116*, 1793–1801. [CrossRef] [PubMed]
56. Cai, J.; Lu, S.; Yao, Z.; Deng, Y.-P.; Zhang, L.-D.; Yu, J.-W.; Ren, G.-F.; Shen, F.-M.; Jiang, G.-J. Glibenclamide attenuates myocardial injury by lipopolysaccharides in streptozotocin-induced diabetic mice. *Cardiovasc. Diabetol.* **2014**, *13*, 1–11. [CrossRef] [PubMed]
57. Lamprianou, S.; Gysemans, C.; Bou Saab, J.; Pontes, H.; Mathieu, C.; Meda, P. Glibenclamide prevents diabetes in NOD mice. *PLoS ONE* **2016**, *11*, e0168839. [CrossRef] [PubMed]
58. Zhai, L.; Ning, Z.-W.; Huang, T.; Wen, B.; Liao, C.-H.; Lin, C.-Y.; Zhao, L.; Xiao, H.-T.; Bian, Z.-X. *Cyclocarya paliurus* leaves tea improves dyslipidemia in diabetic mice: A lipidomics-based network pharmacology study. *Front. Pharmacol.* **2018**, *9*, 973. [CrossRef] [PubMed]
59. Shen, S.; Liao, Q.; Zhang, T.; Pan, R.; Lin, L. Myricanol modulates skeletal muscle-adipose tissue crosstalk to alleviate high-fat diet-induced obesity and insulin resistance. *Br. J. Pharmacol.* **2019**, *176*, 3983–4001. [CrossRef]
60. Vong, C.T.; Tseng, H.H.L.; Kwan, Y.W.; Lee, S.M.-Y.; Hoi, M.P.M. Antrodia camphorata increases insulin secretion and protects from apoptosis in MIN6 Cells. *Front. Pharmacol.* **2016**, *7*, 67. [CrossRef] [PubMed]
61. Nishida, A.; Takizawa, T.; Matsumoto, A.; Miki, T.; Seino, S.; Nakaya, H. Inhibition of ATP-sensitive K+ channels and L-type Ca^{2+} channels by amiodarone elicits contradictory effect on insulin secretion in MIN6 cells. *Trends Pharmacol. Sci.* **2011**, *116*, 73–80. [CrossRef]
62. Shen, S.; Liao, Q.; Liu, J.; Pan, R.; Lee, S.M.Y.; Lin, L. Myricanol rescues dexamethasone-induced muscle dysfunction via a sirtuin 1-dependent mechanism. *J. Cachexia Sarcopenia Muscle* **2019**, *10*, 429–444. [CrossRef]
63. Shen, S.; Liao, Q.; Feng, Y.; Liu, J.; Pan, R.; Lee, S.M.-Y.; Lin, L. Myricanol mitigates lipid accumulation in 3T3-L1 adipocytes and high fat diet-fed zebrafish via activating AMP-activated protein kinase. *Food Chem.* **2019**, *270*, 305–314. [CrossRef] [PubMed]
64. Li, D.; Liu, Q.; Lu, X.; Li, Z.; Wang, C.; Leung, C.-H.; Wang, Y.; Peng, C.; Lin, L. α-Mangostin remodels visceral adipose tissue inflammation to ameliorate age-related metabolic disorders in mice. *Aging* **2019**, *11*, 11084. [CrossRef] [PubMed]
65. Vázquez-Baeza, Y.; Pirrung, M.; Gonzalez, A.; Knight, R. EMPeror: A tool for visualizing high-throughput microbial community data. *Gigascience* **2013**, *2*, 2047–2217X. [CrossRef] [PubMed]
66. Vázquez-Baeza, Y.; Gonzalez, A.; Smarr, L.; McDonald, D.; Morton, J.T.; Navas-Molina, J.A.; Knight, R. Bringing the dynamic microbiome to life with animations. *Cell Host Microbe* **2017**, *21*, 7–10. [CrossRef] [PubMed]
67. Xu, W.; Guan, R.; Shi, F.; Du, A.; Hu, S. Structural analysis and immunomodulatory effect of polysaccharide from *Atractylodis macrocephalae* Koidz. on bovine lymphocytes. *Carbohydr. Polym.* **2017**, *174*, 1213–1223. [CrossRef]
68. Wang, M.; Carver, J.J.; Phelan, V.V.; Sanchez, L.M.; Garg, N.; Peng, Y.; Nguyen, D.D.; Watrous, J.; Kapono, C.A.; Luzzatto-Knaan, T. Sharing and community curation of mass spectrometry data with Global Natural Products Social Molecular Networking. *Nat. Biotechnol.* **2016**, *34*, 828–837. [CrossRef]

molecules

MDPI

Article

Caralluma tuberculata N.E.Br Manifests Extraction Medium Reliant Disparity in Phytochemical and Pharmacological Analysis

Muhammad Waleed Baig [1], Madiha Ahmed [1,2], Nosheen Akhtar [3], Mohammad K. Okla [4], Bakht Nasir [1], Ihsan-Ul Haq [1,*], Jihan Al-Ghamdi [4], Wahidah H. Al-Qahtani [5] and Hamada AbdElgawad [6]

1 Department of Pharmacy, Faculty of Biological Sciences, Quaid-i-Azam University, Islamabad 45320, Pakistan; mwbg7@yahoo.com (M.W.B.); pharmacist_madiha@hotmail.com (M.A.); bakhtnasir61@yahoo.com (B.N.)
2 Shifa College of Pharmaceutical Sciences, Shifa Tameer-e-Millat University, Islamabad 45320, Pakistan
3 Department of Biological Sciences, National University of Medical Sciences, Rawalpindi 46000, Pakistan; nosheenakhtar@numspak.edu.pk
4 Botany and Microbiology Department, College of Science, King Saud University, P.O. Box 2455, Riyadh 11451, Saudi Arabia; Okla103@yahoo.com (M.K.O.); jalghamdi@ksu.edu.sa (J.A.-G.)
5 Department of Food and Nutrition, College of Food and Agriculture Sciences, King Saud University (KSU), Riyadh 11451, Saudi Arabia; wahida@ksu.edu.sa
6 Integrated Molecular Plant Physiology Research, Department of Biology, University of Antwerp, 2020 Antwerpen, Belgium; hamada.abdelgawad@uantwerpen.be
* Correspondence: ihsn99@yahoo.com or ihaq@qau.edu.pk; Tel.: +92-51-90644143

Citation: Baig, M.W.; Ahmed, M.; Akhtar, N.; Okla, M.K.; Nasir, B.; Haq, I.-U.; Al-Ghamdi, J.; Al-Qahtani, W.H.; AbdElgawad, H. *Caralluma tuberculata* N.E.Br Manifests Extraction Medium Reliant Disparity in Phytochemical and Pharmacological Analysis. *Molecules* **2021**, *26*, 7530. https://doi.org/10.3390/molecules26247530

Academic Editors: Ana Paula Duarte, Eugenia Gallardo and Ângelo Luís

Received: 7 November 2021
Accepted: 9 December 2021
Published: 13 December 2021

Abstract: Solubility of phytoconstituents depends on the polarity of the extraction medium used, which might result in the different pharmacological responses of extracts. In line with this, ethnomedicinally important food plant (i.e., *Caralluma tuberculata* extracts) have been made in fourteen distinct solvent systems that were then analyzed phytochemically via total phenolic amount estimation, total flavonoid amount estimation, and HPLC detection and quantification of the selected polyphenols. Test extracts were then subjected to a battery of in vitro assays i.e., antioxidants (DDPH scavenging, antioxidant capacity, and reducing power estimation), antimicrobial (antibacterial, antifungal, and antileishmanial), cytotoxic (brine shrimps, THP-1 human leukemia cell lines and normal lymphocytes), and protein kinase inhibition assays. Maximum phenolic and flavonoid contents were computed in distilled water–acetone and acetone extracts (i.e., 16 ± 1 μg/mg extract and 8 ± 0.4/mg extract, respectively). HPLC-DAD quantified rutin (0.58 μg/mg extract) and gallic acid (0.4 μg/mg extract) in methanol–ethyl acetate and methanol extracts, respectively. Water–acetone extract exhibited the highest DPPH scavenging of 36 ± 1%. Total reducing potential of 76.0 ± 1 μg/mg extract was shown by ethanol chloroform while maximum total antioxidant capacity was depicted by the acetone extract (92.21 ± 0.70 μg/mg extract). Maximal antifungal effect against *Mucor sp.*, antileishmanial, brine shrimp cytotoxicity, THP-1 cell line cytotoxicity, and protein kinase inhibitory activities were shown by ethyl acetate-methanol (MIC: 50 μg/disc), n-hexane (IC_{50}: 120.8 ± 3.7 μg/mL), ethyl acetate (LD_{50}: 29.94 ± 1.6 μg/mL), distilled water–acetone (IC_{50}: 118 ± 3.4 μg/mL) and methanol–chloroform (ZOI: 19 ± 1 mm) extracts, respectively. Our findings show the dependency of phytochemicals and bioactivities on the polarity of the extraction solvent and our preliminary screening suggests the *C. tuberculata* extract formulations to be tested and used in different ailments, however, detailed studies remain necessary for corroboration with our results.

Keywords: *Caralluma*; phytochemical; antimicrobial; cytotoxicity; antioxidant; protein kinase

1. Introduction

The "one disease one drug" concept in the modern drug discovery system is at the verge of losing its importance because of the development of complex, interrelated, and multitargeted disease outcomes. In this scenario, plant extracts have emerged as a source to target multiple diseases at once. A plethora of secondary metabolites in extracts work synergistically, and rarely is a unit phytochemical entity responsible for their pharmacological response [1]. Broad interdisciplinary approaches play a pivotal role in determining the full medicinal potential of any plant. These approaches include phytochemistry, bioassays, and mechanism identification strategies, in addition to ethnopharmacology [2].

Plants including vegetables can be exercised as therapeutic agents. Remedial properties of vegetables and fruits have been protected in the form of ethno-botanical tradition and as ancient heritage [3]. According to the WHO, people are eating less than 20–50% of vegetables, which has raised many concerns [4]. One third of the population have lifelong and serious health issues such as diabetes, allergy, fatigue, arthritis, heart problems, and cancer. A long list of phytonutrients in vegetables is beneficial in disorders such as anthocyanidins for myocardial infarction, catechins for chemoprevention, glucosinolates to boost immunity, phenols for inflammation, thiols for infections, and tochopherols to lowering cholesterol [3]. From head pain to cardiovascular diseases, plant based (vegetables and fruits) consumption is considered to be healthy in curing or preventing ailments [5] and scientific evidence supports this claim.

Genus *Caralluma* R. Br. of the Apocynaceae family includes 120 species, all of which are xerophytes and branched herbs [6]. *Caralluma tuberculata* N. E. Br. is an edible, juicy, leafless stiff plant that is propagated in dry, undomesticated regions of Pakistan and its neighboring countries Saudi Arabia, Nigeria, and Iran [7]. Ethnomedicinal applications include its usage in dysentery, jaundice, constipation, stomach pain, freckles and pimples, hepatitis B and C, diabetes, blood purification, liver ailments, rheumatism, febrifuge, hypertension, gastric problems, paralysis, inflammation, and cancer [8]. A literature survey showed antioxidant, antibacterial [7], hypolipidemic [9], anti-hyperglycemic [10], and in vitro anticancer potentials [11].

A comprehensive and retrospective review of the literature showed that there is still a gap for further research. No systematized studies have been performed, which encompass a chain of bioassays to evaluate the maximum therapeutic potential of *C. tuberculata*. Solvents that are routinely used for extraction (i.e., methanol and ethanol) do not extract all bioactive constituents; therefore, therapeutic profiling is not complete using a unipolarity extraction system approach. The current study is an attempt to provide a systemized and comparative report of fourteen wide polarity range extracts of *C. tuberculata* in varied solvents, either alone or in 1:1 combination, on a battery of antioxidant and biological assays (i.e., phosphomolybdenum based antioxidant capacity, reducing potential, HPLC fingerprinting, antifungal, antileishmanial, THP-1 cytotoxicity, and protein kinase inhibition assays).

2. Material and Methods

The methodology followed in the study are standard protocols that have been cited properly.

2.1. Reagents and Solvents

Reagent and solvent utilized were of analytical grade and bought from Sigma-Aldrich, USA and Germany. Solvents employed in the extraction process were acetone, methanol, ethyl acetate, ethanol, n-hexane, and chloroform while dimethylsulfoxide (DMSO) was used for sample preparation. Cefixime, clotrimazole ciprofloxacin, amphotericin B, and vincristine were standard drugs used. Remaining chemicals and reagents used in the research were: FC = Folin–Ciocalteu reagent; $AlCl_3$ = aluminum chloride; $FeCl_3$ = ferric chloride; TCA = trichloroacetic acid; DPPH = 2,2-diphenyl-1-picryl-hydrazyl-hydrate; sea salt; PBS = phosphate buffer saline; NaH_2PO_4 = monosodium dihydrogen phosphate; kaempferol; catechin; myricitin; gallic acid; quercetin; and rutin were purchased from Merck (Darmstadt, Germany) unless otherwise stated.

2.2. Plant Collection and Identification

The plant material (aerial portion) was collected from Kalabagh region, district Miyanwali, Punjab, Pakistan in June 2014. Prof. Dr. Rizwana Aleem Qureshi recognized and verified the field gathered plant. Dried plant was reserved in the Herbarium of Medicinal plants, Quaid-i-Azam University Islamabad under herbarium number 498.

2.3. Extraction

The putrefied plant part was removed, washed with tap water, dried in a shaded area, and finally pulverized. An amount of 50 g powder was weighed in each Erlenmeyer flask and macerated in the respective solvents with episodic shaking and ultrasonication. After three days, marc was separated and filtered (Whatman No. 1 filter paper). Finally, the extract was concentrated using a rotary evaporator (Buchi, 150 Switzerland) and dried to obtain fourteen different crude extracts.

Extract recovery was calculated as:

$$\text{Extract recovery} = \text{weight of dried extract/weight of powdered plant used} \times 100, \quad (1)$$

2.4. Phytochemical Analysis

2.4.1. Total Phenolics Content Estimation (TPC)

The Folin–Ciocalteu method was adopted. DMSO and gallic acid were used as positive and negative controls. The resultant phenolics (TPC) after triplicate analysis were expressed as µg gallic acid equivalent (GAE) per mg extract [12].

2.4.2. Total Flavonoids Content Estimation (TFC)

DMSO and quercetin (0, 2.5, 5, 10, 20 µg/mL) were used as negative and positive controls, respectively. After triplicate analysis, a calibration curve was drawn and the final TFC content was expressed as µg QE (quercetin equivalent) per mg extract of plant [13].

2.4.3. HPLC-DAD Quantitative Analysis

HPLC-DAD analysis was performed with slight adaptations [14]. The HPLC (Agilent Chem Station 1200 series, USA) system had a diode array DAD detector (Agilent technologies, Germany) along with a C8 analytical column. Standard solutions (50 µg/mL methanol) and sample solutions (10 mg/mL methanol) were freshly made and filtered using a 0.2 µm sartolon polyamide membrane filter. Available reference standards included kaempferol, caffeic acid, quercetin, rutin, myricitin, gallic acid, catechin, and apigenin. Two mobile phases, A and B, were used where mobile phase A had methanol:water:acetic acid:acetonitrile (10:85:1:5) and mobile phase B had acetic acid:acetonitrile:methanol (1:40:60). The flow rate was 1 mL/min. Twenty µL of the sample was injected into the column. Every sample analysis was followed by a 5 min column reconditioning phase. The gradient volume of mobile phase B was adjusted as shown below (Table 1).

Table 1. Gradient volume of mobile phase B (%) added in the system.

Time (min)	%B	Flow Rate (mL)	Max Pressure
0	0	1	350
20	50	1	350
25	100	1	350
30	100	1	350
35	0	1	350
40	0	1	350

Absorbance was measured at four different wavelengths for the respective compounds (i.e., 279 nm for catechin, 257 nm for gallic acid and rutin, 368 nm for quercetin kaempferol, and myricitin, and 325 nm for caffeic acid).

2.5. Biological Evaluation

2.5.1. Antioxidant Assays

Free radical scavenging assay (DPPH assay)

An established protocol was followed. Sample (4 mg/mL), ascorbic acid, and DMSO were used as positive and negative controls, respectively [15].

2.5.2. Total Antioxidant Capacity Determination

A well-established protocol was followed. Test samples, positive (ascorbic acid) and negative (DMSO) controls were used. Results are expressed as the number of μg equivalents of ascorbic acid per mg of dry extract (i.e., μg AAE/ mg extract) [15].

2.5.3. Total Reducing Power Estimation

A previously reported methodology was followed. DMSO and ascorbic acid as the negative and positive controls, respectively, were used. Reducing power was expressed as μg ascorbic acid equivalent (AAE) per mg extract after triplicate analysis [12].

2.5.4. Antimicrobial Assays

Test extracts were tested against five bacterial strains (i.e., *B. subtilis* (ATCC-6633), *S. aureus* (ATCC-6538), *K. pneumoniae* (ATCC-1705), *E. coli* (ATCC-25922), and *P. aeruginosa* (ATCC-15442) while fungal strains include *F. solani* (FCBP-0291), *Mucor sp.* (FCBP-0300), *A. flavus* (FCBP-0064), *A. fumigatus* (FCBP 66), and *A. niger* (FCBP-0198). The agar disc diffusion method was adopted. To prepare lawn on agar plates, 50 μL volume of refreshed culture was used. Positive and negative controls included cefixime, ciprofloxacin, clotrimazole, and DMSO. To a sterile filter paper disc, 5 μL (200 μg/disc) of test extract was applied, which was then settled on seeded agar plates and were incubated for 24 h. After the given time, emergence of the zone of inhibition around the disc was checked to the nearest mm with vernier calipers. Test extracts that gave ≥12 mm meaningful inhibition zones were further evaluated for MICs using the same method as in the antifungal assay [15]. Bacterial inoculum was made in pre-autoclaved nutrient broth under sterile conditions, density being adjusted to 5×10^2 CFU/mL, approximately. Final concentrations of test samples (200 66.66, 22.22, and 7.4 μg/mL) in corresponding wells were made. A total of 190 μL bacterial culture was added to each well, incubated for 30 min at 37 °C, zero time reading taken at 600 nm, again incubated for 24 h at 37 °C, absorbance checked, and results calculated as:

$$\% \text{ Growth inhibition} = 1 - \ \mathrm{T_s} \ / \ \mathrm{T_c} \times 100, \quad (2)$$

T_s and T_c = turbidity of the sample and negative control, respectively.

2.5.5. In Vitro Antileishmanial Potential Determination

Serial dilutions of test samples (40 mg/mL DMSO) were prepared in 96-well plates so that final concentrations were 200, 66.66, and 22.22 μg/mL, respectively. To each well, 1×10^6 *L. tropica* kwh23 promastigotes were added and incubated at 25 °C for 72 h. Surviving promastigotes were counted using a light microscope. The same procedure was performed on Amphoterecin-B (positive control) and 1% DMSO in PBS (negative control). Finally, IC_{50} was calculated [15].

2.5.6. Cytotoxicity Assays

Lethality Testing in Brine Shrimps

Freshly hatched *Artemia salina* larvae in sea water were picked and transferred to each well of a 96-well plate. Calculated volume of four concentrations (200, 100, 50, and 25 μg/mL) of test samples was added to the corresponding wells. Positive and negative control wells include doxorubicin (4 mg/mL) and DMSO (<1%). Volume was made up to 300 μL by adding more sea water. Dead larvae that settled at the base of the wells were counted with a microscope after 24 h. This procedure was repeated twice. Finally, the LC_{50} of test samples was calculated [15].

2.5.7. Cytotoxicity Assay Using THP-1 Human Leukemia Cell Line

Depending upon the system suitability, a slightly different approach was followed as described by [15]. The THP-1 cell line was purchased from ATCC 10801 University Boulevard Manassas, VA 20110-2209 USA. Culturing involved THP-1 (ATCC-TIB202) human leukemia cell lines in RPMI 1640 buffered medium (pH 7.4) supplemented with fetal bovine serum (10%). Incubation was conducted with conditions maintained at: 5% CO_2 and 37 °C. Seeding density was adjusted to 5×10^5 cells/mL. A total of 190 μL of culture was added thereafter, and 10 μL (test samples (1% DMSO) in PBS) was added to the respective wells. The plate was then allowed to incubate (5% CO_2) for 72 h at 37 °C. The assay was also performed on 4 mg/mL DMSO solutions of 5-florouracil and vincristine and 1% DMSO in PBS (negative control). Surviving cells were counted by putting culture on an improved Neubauer chamber and observed under microscope, after which IC_{50} was calculated.

2.5.8. Cytotoxicity Assay Using Isolated Human Lymphocytes

Blood (3 mL) was collected via venipuncture from a healthy donor and was diluted with equal proportion of PBS. It was carefully layered over histopaque (2 mL) and centrifuged for 20 min at 800 g. A buffy coat layer was separated, added into 5 mL PBS, and centrifuged (4 min; 350 rpm). The formed pellet was suspended in RPMI-1640. For analysis, 20 μL (sample, positive, and negative control) was exposed to 180 μL of lymphocyte (1×10^5 cells/mL) suspension. Incubation was carried out for one day in an incubator at 37 °C. Phytohaemagglutinin was added to stimulate lymphocyte growth [13]. The study was approved by the Institutional Review board of Quaid-i-Azam University (letter number #BEC-FBS-QAU2019-135A). Written informed consent was obtained from the participant.

2.5.9. Protein Kinase Inhibition Assay

Refreshed *Streptomyces* culture obtained after 24 h incubation in tryptic soy broth was used for the growth in ISP4 media plates. Soaked paper (negative (DMSO), positive (surfactin), and sample (extracts; 100 μg/disc) discs were put on sterile and freshly inoculated plates. Display of bald zones surrounded discs after 72 h incubation time was interpreted as protein kinase inhibitory activity. Zones were noted by vernier calipers up to the nearest mm [15].

2.5.10. Statistical Analysis

Statistical analysis software included Statistx 8.1 (ANOVA; analysis of variance), table cure 2D (IC50 and LD50) v 5.01, Graph Pad Prism (significance level at $p < 0.05$), and Origin 8.5 (graphical representation after triplicate analysis of each experiment).

3. Results

The undertaken study was designed in a manner to evaluate phytochemical and biological analysis of *C. tuberculata* using 14 different solvent systems. Extract yield ranged between 25.16% w/w (for distilled water) and 0.86% (for n-hexane) (Table 2).

3.1. Phytochemical Analysis

Maximum TPC value was quantified in DA (16 ± 1) (i.e., in μg GAE/mg extract) followed by DM and A with corresponding values (i.e., 15.43 ± 0.60 and 14.04 ± 0.72 μg GAE/mg extract, respectively) (Figure 1). The highest TFC value was quantified in A (i.e., 8 ± 0.39 μg QE/mg extract), followed by EC and E test extracts with corresponding values 9 (i.e., 7.25 ± 0.33 and 6.80 ± 0.40 µg QE/mg extract, respectively) (Figure 1). HPLC chromatograms of standards and detected compounds are shown in Figure 2. Calculated results signify the presence of rutin in EthM extract with a value of 0.58 µg/mg extract followed by M extract with a value of 0.51 µg/mg extract (Table 3). A significant amount of gallic acid was found to be present in the M extract with a value of 0.4 µg/mg extract.

Table 2. Extract recovery.

S. No.	Solvent Extract Code	% Extract Yield
1	Nh	0.86
2	C	6.76
3	A	10.02
4	EthA	7.12
5	Eth	4.44
6	EC	9.04
7	MC	10.34
8	EthE	10.2
9	EthM	8.70
10	E	12.10
11	DA	16.08
12	M	13.78
13	DM	9.70
14	D	25.16

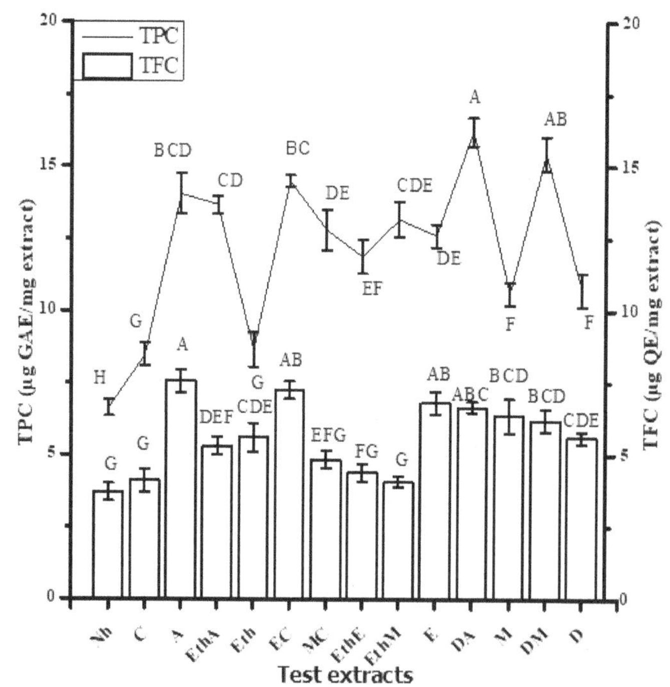

Figure 1. TPC (total phenolic content; mean ± standard deviation) and TFC (total flavonoid contents; mean ± standard deviation) are displayed after triplicate investigation. A–H represent the significance of results. Samples that display different alphabets were significantly different at $p < 0.05$.

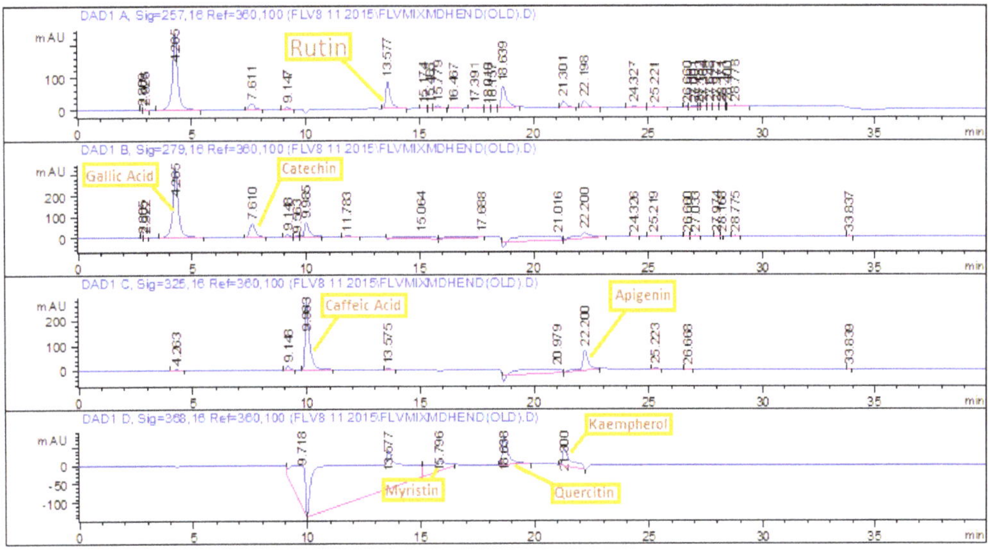

(a)

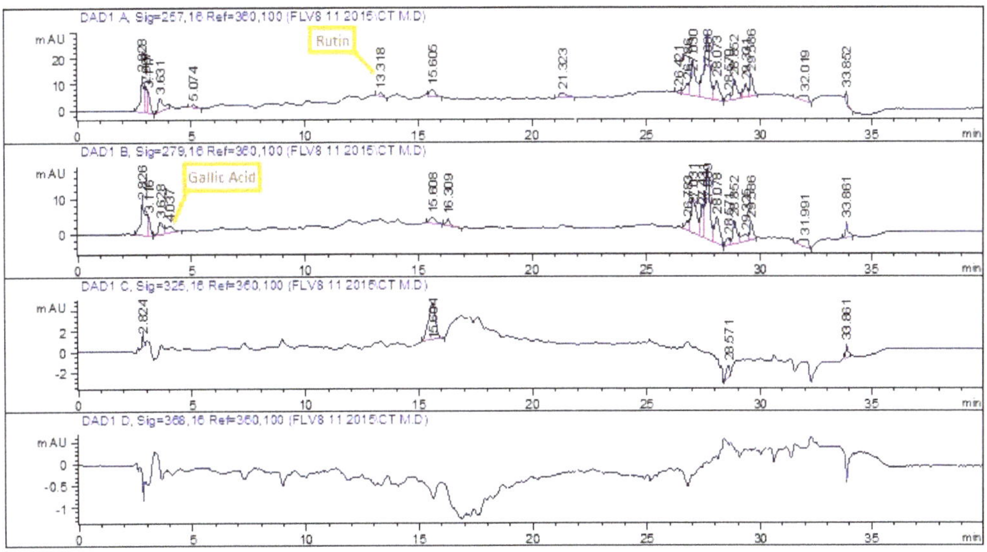

(b)

Figure 2. *Cont.*

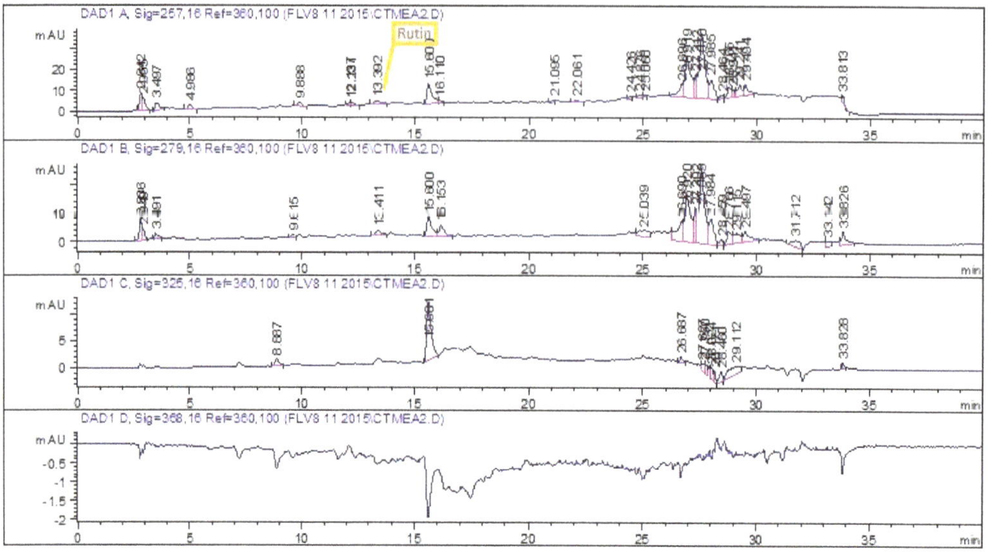

(c)

Figure 2. RP HPLC chromatograms of (**a**) standard polyphenols, (**b**) methanol, and (**c**) ethyl acetate–methanol crude extracts of *C. tuberculata*.

Table 3. RP HPLC analysis of various solvent extracts of *C. tuberculata*.

Extract Name	Polyphenols (μg/mg Extract)	
	Rutin	Gallic Acid
M	0.51	0.35
E	----	----
Eth	----	----
EthM	0.58	----

----: not detected.

3.2. Biological Evaluation

3.2.1. Antioxidant Assays

Among the tested extracts, the DA (200 μg/mL) extract exhibited the maximum free radical scavenging potentiality (i.e., 36.13% ± 0.84), followed by A and EthA with corresponding values of 33.712% ± 1 and 33.712% ± 1.32, respectively (Figure 3). Assay findings suggest marked antioxidant capacity in A (92.21 ± 0.70 μg AAE/mg extract), followed by MC and EaA with corresponding values of 87.53 ± 0.9 and 86.5 ± 0.80 μg AAE/mg extract, respectively (Figure 4). Computed results in the total reducing power assay symbolize meaningful activity in EC (76.01 ± 0.90 μg AAE/mg extract), followed by DA (73.94 ± 0.70 μg AAE/mg extract) and EaA (71.92 ± 0.93 μg AAE/mg extract) (Figure 4).

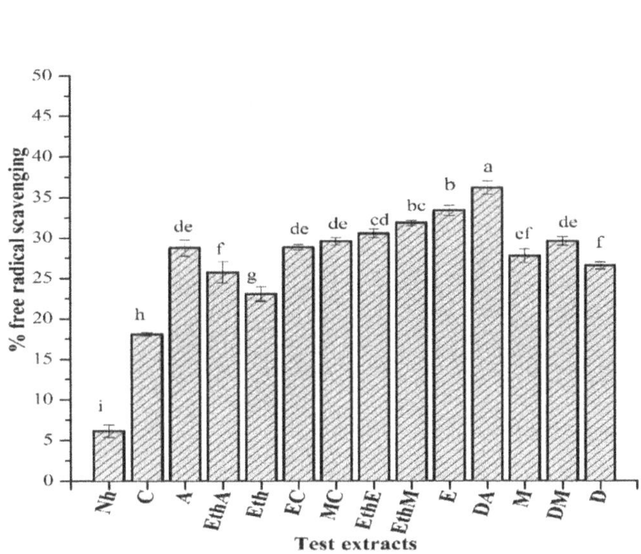

Figure 3. Percent free radical scavenging activity (mean ± standard deviation) after triplicate investigation is shown. a–i represent the significance of results. Samples that display different alphabets were significantly different at $p < 0.05$.

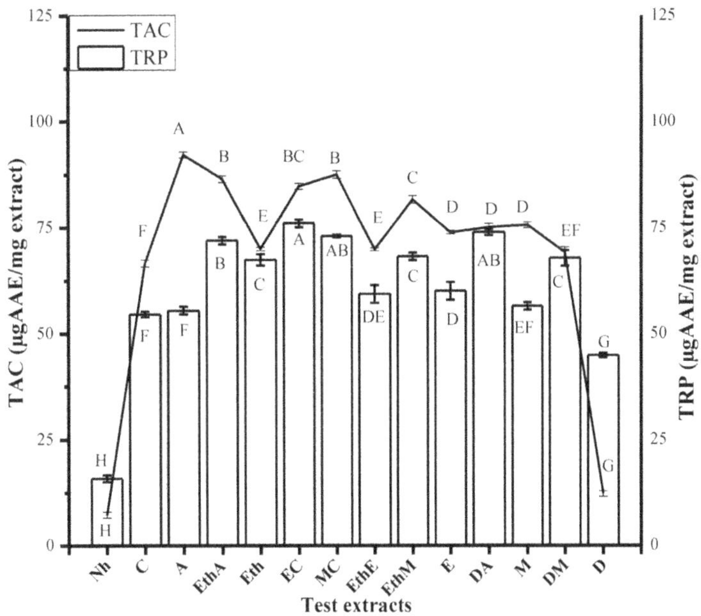

Figure 4. TAC (total antioxidant capacity; mean ± standard deviation) and TRP (total reducing power; mean ± standard deviation) results are displayed after experimenting thrice. A–F represent the significance of results. Samples that display different alphabets were significantly different at $p < 0.05$.

3.2.2. Antimicrobial Assays

Test extracts were tested against five bacterial species, but they have only shown activity against *P. aeruginosa*. Maximum zones of inhibition (ZOI) were displayed by EthA, Eth, and EthE (Figure 5) with corresponding values of 15 ± 2 mm, 15 ± 1 mm, and 15 ± 1 mm with MIC values of 66.66 µg/mL in all three extracts (Table 4). The DMSO impregnated disc did not form any ZOI.

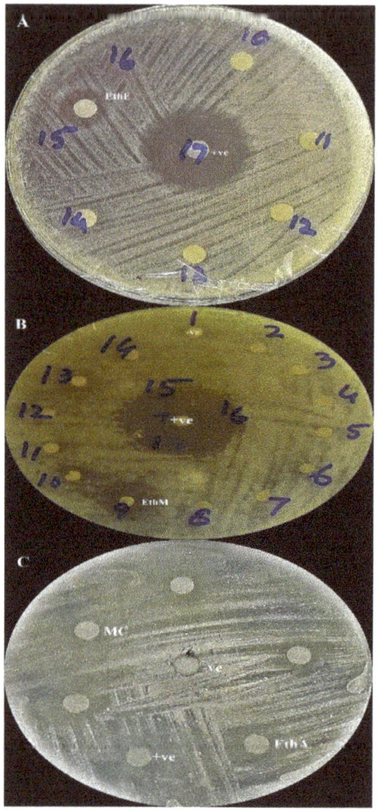

Figure 5. Antibacterial (**A**), antifungal (**B**), and protein kinase inhibition (**C**) representation of selected test samples and controls. EthE = ethyl acetate−ethanol, EthM = ethyl acetate−methanol, EthA = ethyl acetate–acetone, and MC = methanol–chloroform extracts, respectively.

Results of the antifungal assay are shown in Table 4. Largest ZOI was given by EthM (25 ± 2 mm and MIC of 50 µg/disc) (Figure 5) against *Mucor sp.* The M extract also showed good activity against *A. flavus* (12 ± 1 mm) and *A. fumigatus* (15 ± 2 mm) with a MIC of 100 µg/disc each. When tested against *F. solani*, maximum ZOI were given by MC and EthE with values 16 ± 1 mm and 16 ± 1 mm and MIC values of 100 µg/disc, respectively.

Screening revealed remarkable antileishmanial potential in the Nh extract with % inhibition of 97.5% and 50% mortality (IC_{50}) of 120.8 ± 3.7 µg/mL. Thirteen out of the total fourteen extracts showed promising results, which are given in the decreasing order as Nh > EthM > E > EthA > EC > C > DA > EthE > A > Eth > M > MC > DM (Table 4).

Table 4. Antibacterial, antifungal activity, and MIC values of test extracts.

Antibacterial Assay			**Antifungal Assay**							
Extract Names	**Diameter of ZOI mm (Mean ± SD) (MIC: μg/mL)**		**Diameter of ZOI in mm (Mean ± SD) (MIC: μg/disc)**							
	P. aeruginosa	MIC	*Mucor sp.*	MIC	*F. solani*	MIC	*A. Flavus*	MIC	*A. fumigatus*	MIC
Nh	11 ± 1^{cd}	--	--	--	--	--	--	--	--	--
C	9 ± 1^{d}	--	--	--	--	--	--	--	--	--
A	12 ± 1^{c}	200	--	--	--	--	--	--	--	--
EthA	15 ± 2^{b}	66.66	--	--	10 ± 1^{d}	--	--	--	--	--
Eth	15 ± 1^{b}	66.66	--	--	12 ± 1^{c}	100	11 ± 1^{bc}	--	12 ± 1^{c}	100
EC	9 ± 1^{d}	--	--	--	8 ± 1^{e}	--	9 ± 1^{cd}	--	8 ± 1^{d}	--
MC	12 ± 1^{c}	200	--	--	16 ± 1^{b}	100	8 ± 1^{d}	--	--	--
EthE	15 ± 1^{b}	66.66	--	--	16 ± 1^{b}	100	--	--	--	--
EthM	11 ± 1^{cd}	--	25 ± 2^{b}	50	11 ± 1^{cd}	--	--	--	--	--
E	8 ± 1^{d}	--	12 ± 2^{c}	100	--	--	10 ± 1^{c}	--	14 ± 2^{b}	100
DA	--	--	--	--	--	--	10 ± 1^{c}	--	12 ± 1^{c}	100
M	--	--	--	--	13 ± 1^{c}	100	12 ± 1^{b}	100	15 ± 2^{b}	100
DM	--	--	15 ± 1^{c}	100	10 ± 1^{d}	--	12 ± 1^{b}	100	12 ± 1^{c}	100
D	--	--	--	--	--	--	--	--	--	--
Cefixime	19 ± 1^{a}	1.11	--							
Cipro **	19.5 ± 1^{a}	3.33	--	--	--	--	--	--	--	--
Clotri *			35 ± 2^{a}							
DMSO	--	--	--	--	--	--	--	--	--	--

The sample concentration was 200 μg per disc. Values (mean ± SD) were the average of the triplicate of each plant extract ($n = 1 \times 3$). -- = No activity in disc diffusion assay or not applicable (zone < 12 mm) for MIC determination. Cefixime and ** Ciprofloxacin: antibacterial assay positive controls (20 μg/disc); * Clotrimazole: antifungal assay positive control, negative control: DMSO. $^{a–e}$ represent the significance of results. Samples that display different alphabets were significantly different at $p < 0.05$.

3.2.3. Cytotoxicity Assays

Cytotoxicity determination via brine shrimp lethality assay showed the Eth extract to be most potent with a LD_{50} of 29.94 ± 1.6 μg/mL. It was then followed by DA and EthA with a LD_{50} of 48.01 ± 3 μg/mL and 48.75 ± 2.9 μg/mL, respectively (Table 5).

An excellent cytotoxic activity paved the way to also test the plant extracts against the human leukemia (THP-1 ATCC# TIB-202) cell line. A noticeable inhibition in cell line proliferation was discovered in DA and D extracts at 200 μg/mL (Table 5). DA extract exhibited 80.3 ± 2.4% inhibition with IC_{50} of 118 ± 3.4 μg/mL, whereas the D extract showed 64.5 ± 1.4% with an IC_{50} of 140 ± 3.2 μg/mL.

Cytotoxicity in normal isolated human lymphocytes was also checked to assess if the test extracts were selectively toxic to cancer cells or not. None of the extract gave cytotoxicity against normal cells. All extracts provided IC_{50} values > 200 μg/mL (Table 5).

3.2.4. Protein Kinase Inhibition Assay

Inhibitors of protein kinases are represented as distinguished entities that might be facilitated to discover new cancer therapeutic agents. MC (Figure 5) extract exhibited maximum bald ZOI with 19 ± 1 (Table 5). Except for the pure n-hexane and water extracts, all other extracts acted as resourceful prospects to obtain propitious inhibitors of protein kinases.

Table 5. Leishmania, brine shrimp, THP-1, lymphocyte toxicity studies, and protein kinase inhibition assay.

Samples	Antileishmanial (µg/mL)		Brine Shrimp Cytotoxicity (µg/mL)		THP-1 Cytotoxicity(µg/mL)		Lymphocyte Cytotoxicity (µg/mL)		Protein Kinase Inhibition (µg/disc)	
	% Mortality	LC_{50}	% Mortality	LD_{50}	% Inhibition	IC_{50}	% Inhibition	IC_{50}	Diameter (mm) at 100 µg/disc	
	200		200		200		200		Clear Zone	Bald Zone
Nh	97.5 ± 1.1	120.8 ± 3.7 ^{b}	0	0	16.4 ± 1.1	>200	6.2 ± 1.1	>200	--	--
C	82.5 ± 1.3	139.1 ± 4.2 ^{cd}	100	50.09 ± 2.2 ^{c}	21.3 ± 1.7	>200	5.9 ± 1.3	> 200	--	14 ± 0.57 ^{c}
A	80.3 ± 2	142 ± 5.7 ^{d}	100	98.6 ± 3.4 ^{e}	30 ± 2.0	>200	6.6 ± 2.1	>200	--	17 ± 0.57 ^{b}
EthA	83.4 ± 2.1	137.6 ± 6.1 ^{cd}	100	48.75 ± 2.9 ^{c}	19.6 ± 2.1	>200	7.1 ± 1.5	>200	--	14 ± 0.57 ^{c}
Eth	80.4 ± 1.8	142 ± 5.9 ^{d}	100	29.94 ± 1.6 ^{b}	21.7 ± 0.7	>200	7.4 ± 1.1	>200	--	13 ± 1 ^{c}
EC	83 ± 1.2	137.6 ± 3.9 ^{cd}	100	50 ± 3.01 ^{c}	24.2 ± 1.5	>200	5.5 ± 0.9	>200	--	12 ± 1 ^{cd}
MC	79 ± 2.2	143.5 ± 6.2 ^{de}	100	74.22 ± 3.4 ^{d}	26.8 ± 0.8	>200	5.7 ± 1.1	>200	--	19 ± 1 ^{b}
EthE	81 ± 1.0	140.5 ± 3.5 ^{cd}	100	49.16 ± 2.8 ^{c}	22.1 ± 1.5	>200	6.7 ± 1.3	>200	--	10± 1 ^{d}
EthM	91 ± 2.2	127.4 ± 6.3 ^{bc}	100	49.15 ± 2.7 ^{c}	27.5 ± 1.3	>200	6.8 ± 1.6	>200	--	11 ± 0.57 ^{cd}
E	85 ± 2.1	134.9 ± 6.0 ^{c}	80 ± 5.77	102.24 ± 4.7 ^{e}	31 ± 2.8	>200	4.4 ± 0.83	>200	--	10 ± 1 ^{d}
DA	82 ± 1.2	139.1 ± 3.5 ^{cd}	100	48.01 ± 3.0 ^{c}	80.3 ± 2.4 ^{b}	118 ± 3.4 ^{b}	4.6 ± 1.3	>200	--	8 ± 1 ^{de}
M	80 ± 1.8	142 ± 4.3 ^{d}	90 ± 5.77	51.34 ± 3.5 ^{c}	27.8 ± 2.2	>200	5.8 ± 2.1	>200	--	8 ± 1 ^{de}
DM	78 ± 2.3	145.1 ± 6.3 ^{e}	90 ± 5.77	50.32 ± 3.3 ^{c}	31.7 ± 1.4	>200	4.9 ± 1.2	>200	--	7 ± 0.57 ^{e}
D	0	0	30 ± 10	≥200	64.5 ± 1.4 ^{c}	140 ± 3.2	4.1 ± 0.66	>200	--	--
Amphoterecin B	100	0.016 ^{a}								
Doxorubicin			100	6.24 ± 0.7 ^{a}						
5-Florouracil					100	5.3 ± 0.68				
Vincristine					100	8.5 ± 0.92	74.49 ± 3.35	6.6 ± 0.2		
Surfactin										29 ± 1 ^{a}
DMSO									-	-
1% DMSO in PBS/sea water	-		-		-		--			

Values are presented as mean ± standard deviation of triplicate analysis. --: no activity. $^{a–e}$ represent the significance of results. Samples that display different alphabets were significantly different at $p < 0.05$.

4. Discussion

We observed that the extract yield changed when changing the solvent (Table 1); one possible explanation is the solubility or insolubility of distinct phytochemical constituents in each characteristic solvent [16], which results in different pharmacological responses. First time execution of water–acetone extract regarding its phytochemical analysis was conducted and resulted in alignment with the reported findings where most polar methanol extracts of *C. tuberculata* displayed the maximal result [17]. The existence of various functional groups in the phenolics class (i.e., hydroxyl, methoxy, ketonic, or double bond conjugation) are considered answerable for antioxidant potential [18], in addition to other oxygenated derivatives that also have the same abilities [19]. Ethnomedicine also claims the use of the decoction and infusion of *C. tuberculata* for various ailments [20]. Equilibration of oxidative imbalance by phenols might be due to the presence of flavonoids. The main motive of HPLC analysis is to detect and quantify unreported polyphenols in the study plant. The presence of rutin and gallic acid are perhaps determinants of the known bio-assay results of our study (Table 3; Figure 2). Rutin is known for its compelling health benefits in obesity, diabetes, oxidant imbalance, cancer, and susceptibility to inhibit different bacteria while gallic acid has been proven to be a healing and defensive agent in ameliorating disturbance caused by disturbed redox reactions in cells causing neurodegeneration, cancer, aging, and cardiac problems. [21]. Its indigenous use in treatment of cancer, when taken orally as a dried and ground powder with water or as salad [22], is also strengthened by our findings. Rutin and gallic acid detected in *C. tuberculata* extracts might be considered responsible for antioxidant, antimicrobial, antileishmanial, cytotoxic, and protein kinase inhibitory potential. Polar extracts, due to their significantly higher detection of polyphenols, are recommended for further advance studies related to phytochemistry and biological activity determination, specifically acetone medium extracts alone or in combination with other solvents.

Defense and repair system in response to oxidative damage is innately present in almost every living creature, nevertheless, these defensive mechanisms are insufficient to provide complete protection by oxidative imbalance [23]. Antioxidant actions of an extract cannot be fully determined by a single technique, therefore, an array of different procedures have been used to uncover the antioxidant capacity based on different mechanisms (i.e., free radical scavenging, inhibition of chain initiation, and peroxide decomposition). The decolorization of the DPPH reagent is due to the presence of antioxidants in the test extracts, which can be quantified by computing changes in absorbance by a spectrophotometer [24]. The assay findings conform with previous studies in which the M (most polar solvent used) extract of *C. tuberculata* exhibited the maximum activity, C extract provided comparatively less results than the M extract, and Nh provided the least results [17]. In the phosphomolybdenum based antioxidant assay, the computed results strengthen previous reports that have documented the maximum antioxidant potential by other means in the M extract [25]. The reducing efficacy of plants has been documented to be correlated with the presence of phenolic constituents. The HPLC graphs denoted the presence of phenolics in the test samples. There are different types of compounds called reductones that are responsible for antioxidant potential. Their mechanism of action is either that their peroxide synthesis is hindered or free radical chain breakage by donation of a hydrogen atom [26].

In the antibacterial assay, ethyl acetate extracts, whether alone or in 1:1 proportion with other solvents, proved to be effective only against *P. aeruginosa*. The results might be due to inhibition of protein kinase activity noted in the ongoing investigation, which may act as new antimicrobials [27]. *P. aeruginosa* infections in subcutaneous nodules of immunocompromised patients portend higher risk of mortality and morbidity [28]. Ethnopharmacology claims the use of under discussed plants for skin diseases [29]. Therefore, ethyl acetate combinatorial extracts are suggested to scientifically augment ethnopharmacological data regarding derma. Results are also in accordance with previous studies in which food plants have shown antibacterial activity for Gram-negative bacteria in contrast to medicinal plants that are more potent against Gram-positive bacteria [30].

Fungal infections have drastically increased in immunocompromised patients, which has increased the urgency for newer and safer antifungal agents [31]. Test extracts exhibited well to moderate results when tested against fungal strains. Maximum result was shown against *F. solani*. *Fusarium* skin infection is associated with painful erythematous nodules and papules [32]. Polar extracts of the study plant orally or in the form of cream can be used for fungal infections. Previous studies not providing results against fungal strains is due to the utilization of a unipolarity solvent (i.e., ethanol). Ethanol extracts contain sugars, which promote fungus growth and interferes with the activity of antifungal constituents in extracts [33]. Detailed analysis is required to perceive an exact mode of action and the relationship between fractionation and its effect on biological activity.

Failure of first line therapy and propagation of leishmaniasis worldwide has persuaded researchers to discover new drugs. The findings conform with studies on the closely related specie *Caralluma sinaica*, whose methanolic extract is potent against *L. infantum* [34] and the methanolic extract of *Caralluma quadrangular*, which also showed meaningful activity against *L. infantum* [35]. The assay results also revealed and suggest under discussed extracts as having a possible preventive and supportive role in leishmaniasis. All extracts exhibited meaningful results against *L. tropica*. Oral administration of extracts and/or powder is suggested to be validated by in vivo studies to authenticate the preliminary claim.

It is commonly inferred that brine shrimps or *Artemia salina* larvae and carcinoma cells of mammals behave similarly in many regards, which is why the cytotoxic effects of the undertaken test extracts might become potential candidates for anti-tumor and anti-cancer activities [36]. It is an effective procedure to hypothesize possible biological activities of test extracts against malarial parasites, pests, tumors, and harmful microbes [37]. The pattern of results was also almost observed during the protein kinase inhibition assay. The findings correspond with previous reports in which significant in vitro anticancer results were exhibited by C, Eth (maximum activity), and M, while the Nh and D extracts did not exhibit any activity [11]. Acceptability of *C. tuberculata* as a food despite its high cytotoxic profile is due to the existence of flavonoids and saponins. Dietary flavonoids are known to be involved in reducing RNA and increasing cytotoxicity and mutagenicity. On the other hand, saponins possess strong antimutagenic activity (protection of DNA and protein biosynthesis) and might be considered responsible for the efficacy of herbal drugs. Previous studies have shown *C. tuberculata* as cytotoxic, but not clastogenic in mice [38]. Further research is needed to evaluate the exact mechanism behind the DNA protective effect of this plant specie.

Excellent cytotoxic activity has paved the way to also testing plant extracts against human leukemia (THP-1 ATCC# TIB-202) cell line. Cytotoxicity against THP-1 human leukemia cell line is in accordance with previous studies in which the M (same polarity as acetone) extract exhibited anticancer potential against Caco-2 cells, MCF-7, and MDA-MB-468 cells [11]. These findings can also be related to the undertaken extract testing against brine shrimp lethality and the protein kinase inhibition assay where DA and A extracts have shown prominent results. The findings also conform with the total reducing and total antioxidant capability assays in which the DA extract showed the maximum results, which leads to possible antioxidant cytotoxic activity. Rutin is a flavonoid known for its multispectral pharmacological responses (e.g., hypertension, diabetes, and cancer [39]) and its efficacy has been proven against THP-1 leukemia cells [40]. Therefore, the rutin detected in HPLC graphs of water and methanol extracts may be considered responsible for the noted cytotoxic activity. Moreover, the test extracts did not show cytotoxicity in normal lymphocytes, which ensures targeting the cancer cells only and limiting toxicity toward normal cells. Its indigenous use in the treatment of cancer, when taken orally as dried and ground powder with water or as salad [22], is also strengthened by our findings. Further evaluation needs to be conducted to examine the effects on gene expression/pathway in an animal/human model. However, the sequential extraction method is recommended to remove the non-polar constituents and use only the polar extracts for anticancer use, especially in leukemia.

Protein kinase inhibitory activity might be due to the presence of saponins previously detected in C, Eth, and M extracts [17], which have the potential to downregulate protein kinase C (PKC) [41]. Another possible reason is the presence of gallic acid, as confirmed in HPLC profiling (Figure 2; Table 3), which has anticancer properties against human osteosarcoma cells by interfering with signaling pathways involving protein kinases [42]. Over the years, discovery and development of protein kinase inhibitors have increased from plant sources due to their beneficial aspects [43]. The number of kinases coded in human genome has reached about 500. Test extracts that have the ability to adhere allosterically with either active or inactive site of any of these kinases can be regarded as a potential and crucial source to establish new therapy regimens against cancer [44]. This assay is advantageous in many regards as it quickly identifies a compound as cytotoxic and can also pinpoint potential inhibitors, which hinders signal transduction in infections and cancers [45].

5. Conclusions

Dependency of phytochemicals and bioactivities on the polarity of the extraction solvent was observed. Detection and quantification of phenolics, flavonoids, rutin, and gallic acid suggest that *C. tuberculata* is a useful prospect for antioxidants. Significant responses were given by test extracts in in vitro antifungal, antileishmanial, brine shrimp cytotoxicity, THP-1 leukemia cell line cytotoxicity, and protein kinase inhibition assays. Findings revealed pharmacological features of *C. tuberculate*, which authenticates its therapeutic potential in various ailments besides inferring its resourcefulness to isolate phytotherapeutical constituents through bioactivity guided isolation. Our results show that *C. tuberculate* is a worthy lead for advanced studies.

Author Contributions: M.W.B. performed the experiments, analyzed, and interpreted the data, wrote, and revised the manuscript. M.A., N.A. and B.N. contributed to the in vitro experiments, acquisition of data, and critical review of the manuscript. M.K.O. and J.A.-G. assisted in the data analysis, interpretation, and made critical revisions. W.H.A.-Q. and H.A. contributed to the phytochemical analysis, acquisition of the data, and critical review of the manuscript. I.-U.H. conceptualized and designed the study, supervised the execution of experiments, critically revised the manuscript, and approved the final version of this manuscript. All authors have read and approved the final manuscript.

Funding: The authors extend their appreciation to the researchers supporting project no. RSP-2021/374 at King Saud University, Riyadh, Saudi Arabia.

Institutional Review Board Statement: The study was conducted after ethical approval from the Institutional Animal Ethics Committee of Quaid-i-Azam University Islamabad, Pakistan (letter number # BEC-FBS-QAU2019-135-A).

Informed Consent Statement: Informed consent was obtained from the subject involved in the study. This study was also approved by the Institutional Review board of Quaid-i-Azam University, Islamabad (letter number # BEC-FBS-QAU2019-135-A.

Data Availability Statement: Data is contained within the article.

Acknowledgments: The authors are thankful to Rizwana Aleem Qureshi, Department of Plant Sciences, Faculty of Biological Sciences, Quaid-i-Azam University Islamabad, Pakistan for identifying the plant sample. HEC Pakistan is acknowledged for the funding through the Indigenous fellowship program for Muhammad Waleed Baig to execute the study. The University Research Fund from Quaid-i-Azam University Islamabad is acknowledged for the execution of the study. The authors extend their appreciation to the researchers supporting project no. RSP-2021/374 at King Saud University, Riyadh, Saudi Arabia.

Conflicts of Interest: The authors declare that they have no conflict of interest.

Sample Availability: Samples of the plant material are available from the authors.

References

1. Brusotti, G.; Cesari, I.; Dentamaro, A.; Caccialanza, G.; Massolini, G. Isolation and characterization of bioactive compounds from plant resources: The role of analysis in the ethnopharmacological approach. *J. Pharm. Biomed. Anal.* **2014**, *87*, 218–228. [CrossRef]
2. Atanasov, A.G.; Waltenberger, B.; Pferschy-Wenzig, E.-M.; Linder, T.; Wawrosch, C.; Uhrin, P.; Temml, V.; Wang, L.; Schwaiger, S.; Heiss, E.H. Discovery and resupply of pharmacologically active plant-derived natural products: A review. *Biotechnol. Adv.* **2015**, *33*, 1582–1614. [CrossRef] [PubMed]
3. Rahal, A.; Verma, A.K.; Kumar, A.; Tiwari, R.; Kapoor, S.; Chakraborty, S.; Dhama, K. Phytonutrients and nutraceuticals in vegetables and their multi-dimensional medicinal and health benefits for humans and their companion animals: A review. *J. Biol. Sci.* **2014**, *14*, 1. [CrossRef]
4. Rickman, J.C.; Barrett, D.M.; Bruhn, C.M. Nutritional comparison of fresh, frozen and canned fruits and vegetables. Part 1. Vitamins C and B and phenolic compounds. *J. Sci. Food Agric.* **2007**, *87*, 930–944. [CrossRef]
5. Slavin, J.L.; Lloyd, B. Health benefits of fruits and vegetables. *Adv. Nutr.* **2012**, *3*, 506–516. [CrossRef]
6. Rehman, R.; Chaudhary, M.; Khawar, K.; Lu, G.; Mannan, A.; Zia, M. In vitro propagation of Caralluma tuberculata and evaluation of antioxidant potential. *Biologia* **2014**, *69*, 341–349. [CrossRef]
7. Bibi, Y.; Tabassum, S.; Zahara, K.; Bashir, T.; Haider, S. Ethnomedicinal and pharmacological properties of Caralluma tuberculata NE Brown-A review. *Pure Appl. Biol.* **2015**, *4*, 503.
8. Adnan, M.; Jan, S.; Mussarat, S.; Tariq, A.; Begum, S.; Afroz, A.; Shinwari, Z.K. A review on ethnobotany, phytochemistry and pharmacology of plant genus Caralluma R. Br. *J. Pharm. Pharmacol.* **2014**, *66*, 1351–1368. [CrossRef] [PubMed]
9. Abdel-Sattar, E.; Harraz, F.M.; Ghareib, S.A.; Elberry, A.A.; Gabr, S.; Suliaman, M.I. Antihyperglycaemic and hypolipidaemic effects of the methanolic extract of Caralluma tuberculata in streptozotocin-induced diabetic rats. *Nat. Prod. Res.* **2011**, *25*, 1171–1179. [CrossRef] [PubMed]
10. Abdel-Sattar, E.A.; Abdallah, H.M.; Khedr, A.; Abdel-Naim, A.B.; Shehata, I.A. Antihyperglycemic activity of Caralluma tuberculata in streptozotocin-induced diabetic rats. *Food Chem. Toxicol.* **2013**, *59*, 111–117. [CrossRef] [PubMed]
11. Waheed, A.; Barker, J.; Barton, S.J.; Khan, G.-M.; Najm-us-Saqib, Q.; Hussain, M.; Ahmed, S.; Owen, C.; Carew, M.A. Novel acylated steroidal glycosides from Caralluma tuberculata induce caspase-dependent apoptosis in cancer cells. *J. Ethnopharmacol.* **2011**, *137*, 1189–1196. [CrossRef] [PubMed]

12. Nasir, B.; Baig, M.W.; Majid, M.; Ali, S.M.; Khan, M.Z.I.; Kazmi, S.T.B.; Haq, I.-u. Preclinical anticancer studies on the ethyl acetate leaf extracts of Datura stramonium and Datura inoxia. *BMC Complementary Med. Ther.* **2020**, *20*, 1–23. [CrossRef] [PubMed]
13. Ahmed, M.; Fatima, H.; Qasim, M.; Gul, B. Polarity directed optimization of phytochemical and in vitro biological potential of an indigenous folklore: *Quercus dilatata* Lindl. ex Royle. *BMC Complementary Altern. Med.* **2017**, *17*, 1–16. [CrossRef]
14. Fatima, H.; Khan, K.; Zia, M.; Ur-Rehman, T.; Mirza, B.; Haq, I.-u. Extraction optimization of medicinally important metabolites from *Datura innoxia* Mill.: An in vitro biological and phytochemical investigation. *BMC Complementary Altern. Med.* **2015**, *15*, 1–18. [CrossRef]
15. Zahra, S.S.; Ahmed, M.; Qasim, M.; Gul, B.; Zia, M.; Mirza, B.; Haq, I.-u. Polarity based characterization of biologically active extracts of *Ajuga bracteosa* Wall. ex Benth. and RP-HPLC analysis. *BMC Complementary Altern. Med.* **2017**, *17*, 1–16. [CrossRef]
16. Hsu, B.; Coupar, I.M.; Ng, K. Antioxidant activity of hot water extract from the fruit of the Doum palm, Hyphaene thebaica. *Food Chem.* **2006**, *98*, 317–328. [CrossRef]
17. Rauf, A.; Jan, M.; Rehman, W.; Muhammad, N. Phytochemical, phytotoxic and antioxidant profile of Caralluma tuberculata NE Brown. *Wudpecker J. Pharm. Pharmacol.* **2013**, *2*, 21–25.
18. Majid, M.; Khan, M.R.; Shah, N.A.; Haq, I.U.; Farooq, M.A.; Ullah, S.; Sharif, A.; Zahra, Z.; Younis, T.; Sajid, M. Studies on phytochemical, antioxidant, anti-inflammatory and analgesic activities of *Euphorbia dracunculoides*. *BMC Complementary Altern. Med.* **2015**, *15*, 1–15. [CrossRef]
19. Devasagayam, T.; Tilak, J.; Boloor, K.; Sane, K.S.; Ghaskadbi, S.S.; Lele, R. Free radicals and antioxidants in human health: Current status and future prospects. *Japi* **2004**, *52*, 4.
20. Bibi, T.; Ahmad, M.; Tareen, R.B.; Tareen, N.M.; Jabeen, R.; Rehman, S.-U.; Sultana, S.; Zafar, M.; Yaseen, G. Ethnobotany of medicinal plants in district Mastung of Balochistan province-Pakistan. *J. Ethnopharmacol.* **2014**, *157*, 79–89. [CrossRef]
21. Karamac, M.; Kosińska, A.; Pegg, R.B. Comparison of radical-scavenging activities for selected phenolic acids. *Pol. J. Food Nutr. Sci.* **2005**, *14*, 165–170.
22. Khan, I.; AbdElsalam, N.M.; Fouad, H.; Tariq, A.; Ullah, R.; Adnan, M. Application of ethnobotanical indices on the use of traditional medicines against common diseases. *Evid.-Based Complementary Altern. Med.* **2014**, *2014*. [CrossRef] [PubMed]
23. Yang, J.-H.; Lin, H.-C.; Mau, J.-L. Antioxidant properties of several commercial mushrooms. *Food Chem.* **2002**, *77*, 229–235. [CrossRef]
24. Prieto, P.; Pineda, M.; Aguilar, M. Spectrophotometric quantitation of antioxidant capacity through the formation of a phosphomolybdenum complex: Specific application to the determination of vitamin E. *Anal. Biochem.* **1999**, *269*, 337–341. [CrossRef]
25. Khan, M.; Khan, R.; Ahmed, M.; Muhammad, N.; Khan, M.; Khan, H.; Atlas, N.; Khan, F. Biological screening of methanolic crude extracts of Caralluma tuberculata. *Int. J. Indig. Med. Plants* **2013**, *46*, 2051–4263.
26. Wang, H.; Gao, X.D.; Zhou, G.C.; Cai, L.; Yao, W.B. In vitro and in vivo antioxidant activity of aqueous extract from *Choerospondias axillaris* fruit. *Food Chem.* **2008**, *106*, 888–895. [CrossRef]
27. Matsushita, M.; Janda, K.D. Histidine kinases as targets for new antimicrobial agents. *Bioorganic Med. Chem.* **2002**, *10*, 855–867. [CrossRef]
28. Spernovasilis, N.; Psichogiou, M.; Poulakou, G. Skin manifestations of Pseudomonas aeruginosa infections. *Curr. Opin. Infect. Dis.* **2021**, *34*, 72–79. [CrossRef] [PubMed]
29. Mahmood, A.; Mahmood, A.; Tabassum, A. Ethnomedicinal Survy of Plants From District Sialkot, Pakistan. *J. Appl. Pharm.* **2011**, *3*, 212–220. [CrossRef]
30. Ullah, M.O.; Haque, M.; Urmi, K.F.; Zulfiker, A.H.M.; Anita, E.S.; Begum, M.; Hamid, K. Anti–bacterial activity and brine shrimp lethality bioassay of methanolic extracts of fourteen different edible vegetables from Bangladesh. *Asian Pac. J. Trop. Biomed.* **2013**, *3*, 1–7. [CrossRef]
31. Anjum, K.; Abbas, S.Q.; Shah, S.A.A.; Akhter, N.; Batool, S.; ul Hassan, S.S. Marine sponges as a drug treasure. *Biomol. Ther.* **2016**, *24*, 347. [CrossRef]
32. Batista, B.G.; Chaves, M.A.d.; Reginatto, P.; Saraiva, O.J.; Fuentefria, A.M. Human fusariosis: An emerging infection that is difficult to treat. *Rev. Soc. Bras. Med. Trop.* **2020**, *53*. [CrossRef] [PubMed]
33. Arabi, Z.; Sardari, S. An investigation into the antifungal property of Fabaceae using bioinformatics tools. *Avicenna J. Med. Biotechnol.* **2010**, *2*, 93. [PubMed]
34. Al-Musayeib, N.M.; Mothana, R.A.; Al-Massarani, S.; Matheeussen, A.; Cos, P.; Maes, L. Study of the in vitro antiplasmodial, antileishmanial and antitrypanosomal activities of medicinal plants from Saudi Arabia. *Molecules* **2012**, *17*, 11379–11390. [CrossRef]
35. Al-Musayeib, N.M.; Mothana, R.A.; Matheeussen, A.; Cos, P.; Maes, L. In Vitro antiplasmodial, antileishmanial and antitrypanosomal activities of selected medicinal plants used in the traditional Arabian Peninsular region. *BMC Complementary Altern. Med.* **2012**, *12*, 1–7. [CrossRef] [PubMed]
36. Ullah, R.; Ibrar, M.; Shah, S.; Hameed, I. Phytotoxic, cytotoxic and insecticidal activities of *Calendula arvensis* L. L. *J. Biotechnol. Pharm. Res.* **2012**, *3*, 104–111.
37. Khan, I.; Yasinzai, M.M.; Mehmood, Z.; Ilahi, I.; Khan, J.; Khalil, A.; Saqib, M.S.; Rahman, W.U. Comparative study of green fruit extract of *Melia azedarach* Linn. with its ripe fruit extract for antileishmanial, larvicidal, antioxidant and cytotoxic activity. *Am. J. Phytomed. Clin. Ther.* **2014**, *2*, 442–452.
38. Al-Bekairi, A.; Qureshi, S.; Ahmed, M.; Qazi, N.; Khan, Z.; Shah, A. Effect of *Caralluma tuberculata* on the cytological and biochemical changes induced by cyclophosphamide in mice. *Food Chem. Toxicol.* **1992**, *30*, 719–722. [CrossRef]

39. Sharma, S.; Ali, A.; Ali, J.; Sahni, J.K.; Baboota, S. Rutin: Therapeutic potential and recent advances in drug delivery. *Expert Opin. Investig. Drugs* **2013**, *22*, 1063–1079. [CrossRef]
40. Park, M.H.; Kim, S.; Song, Y.-r.; Kim, S.; Kim, H.-J.; Na, H.S.; Chung, J. Rutin induces autophagy in cancer cells. *Int. J. Oral Biol.* **2016**, *41*, 45–51. [CrossRef]
41. Kim, H.-Y.; Yu, R.; Kim, J.-S.; Kim, Y.-K.; Sung, M.-K. Antiproliferative crude soy saponin extract modulates the expression of IκBα, protein kinase C, and cyclooxygenase-2 in human colon cancer cells. *Cancer Lett.* **2004**, *210*, 1–6. [CrossRef]
42. You, B.R.; Park, W.H. The effects of mitogen-activated protein kinase inhibitors or small interfering RNAs on gallic acid-induced HeLa cell death in relation to reactive oxygen species and glutathione. *J. Agric. Food Chem.* **2011**, *59*, 763–771. [CrossRef] [PubMed]
43. Cohen, P. Protein kinases—The major drug targets of the twenty-first century? *Nat. Rev. Drug Discov.* **2002**, *1*, 309–315. [CrossRef]
44. Smyth, L.A.; Collins, I. Measuring and interpreting the selectivity of protein kinase inhibitors. *J. Chem. Biol.* **2009**, *2*, 131–151. [CrossRef] [PubMed]
45. Waters, B.; Saxena, G.; Wanggui, Y.; Kau, D.; Wrigley, S.; Stokes, R.; Davies, J. Identifying protein kinase inhibitors using an assay based on inhibition of aerial hyphae formation in *Streptomyces*. *J. Antibiot.* **2002**, *55*, 407–416. [CrossRef] [PubMed]

molecules

MDPI

Article

HPLC-DAD Based Polyphenolic Profiling and Evaluation of Pharmacological Attributes of *Putranjiva roxburghii* Wall.

Adila Nazli [1,2], Muhammad Zafar Irshad Khan [3], Madiha Ahmed [4,*], Nosheen Akhtar [5], Mohammad K. Okla [6], Abdulrahman Al-Hashimi [6], Wahidah H. Al-Qahtani [7], Hamada Abdelgawad [8] and Ihsan-ul- Haq [1,*]

1 Department of Pharmacy, Faculty of Biological Sciences, Quaid-i-Azam University, Islamabad 45320, Pakistan; adilanazli44@gmail.com
2 Chongqing Key Laboratory of Natural Product Synthesis and Drug Research, School of Pharmaceutical Sciences, Chongqing University, Chongqing 401331, China
3 College of Pharmaceutical Sciences, Zhejiang University, Hangzhou 310058, China; m.zafarirshad@yahoo.com
4 Shifa College of Pharmaceutical Sciences, Shifa Tameer-e-Millat University, Islamabad 44000, Pakistan
5 Department of Biological Sciences, National University of Medical Sciences, Rawalpindi 43600, Pakistan; nosheenakhtar@numspak.edu.pk
6 Botany and Microbiology Department, College of Science, King Saud University, Riyadh 11451, Saudi Arabia; okla103@yahoo.com (M.K.O.); al-ghamd@gmail.com (A.A.-H.)
7 Department of Food Sciences & Nutrition, College of Food & Agriculture Sciences, King Saud University, Riyadh 11451, Saudi Arabia; wahida@ksu.edu.sa
8 Integrated Molecular Plant Physiology Research, Department of Biology, University of Antwerp, 2020 Antwerpen, Belgium; hamada.abdelgawad@uantwerpen.be
* Correspondence: madiha.scps@stmu.edu.pk (M.A.); ihaq@qau.edu.pk (I.-u.-H.)

Citation: Nazli, A.; Irshad Khan, M.Z.; Ahmed, M.; Akhtar, N.; Okla, M.K.; Al-Hashimi, A.; Al-Qahtani, W.H.; Abdelgawad, H.; Haq, I.-u.- HPLC-DAD Based Polyphenolic Profiling and Evaluation of Pharmacological Attributes of *Putranjiva roxburghii* Wall.. *Molecules* **2022**, *27*, 68. https://doi.org/10.3390/molecules27010068

Academic Editors: Ana Paula Duarte, Eugenia Gallardo and Ângelo Luís

Received: 9 November 2021
Accepted: 18 December 2021
Published: 23 December 2021

Publisher's Note: MDPI stays neutral with regard to jurisdictional claims in published maps and institutional affiliations.

Abstract: The current study was intended to explore the phytochemical profiling and therapeutic activities of *Putranjiva roxburghii* Wall. Crude extracts of different plant parts were subjected to the determination of antioxidant, antimicrobial, antidiabetic, cytotoxic, and protein kinase inhibitory potential by using solvents of varying polarity ranges. Maximum phenolic content was notified in distilled water extracts of the stem (DW-S) and leaf (DW-L) while the highest flavonoid content was obtained in ethyl acetate leaf (EA-L) extract. HPLC-DAD analysis confirmed the presence of various polyphenols, quantified in the range of 0.02 ± 0.36 to 2.05 ± 0.18 µg/mg extract. Maximum DPPH scavenging activity was expressed by methanolic extract of the stem (MeOH-S). The highest antioxidant capacity and reducing power was shown by MeOH-S and leaf methanolic extract (MeOH-L), respectively. Proficient antibacterial activity was shown by EA-L extract against *Bacillus subtilis* and *Escherichia coli*. Remarkable α-amylase and α-glucosidase inhibition potential was expressed by ethyl acetate fruit (EA-F) and n-Hexane leaf (nH-L) extracts, respectively. In case of brine shrimp lethality assay, 41.67% of the extracts ($LC_{50} < 50$ µg/mL) were considered as extremely cytotoxic. The test extracts also showed mild antifungal and protein kinase inhibition activities. The present study explores the therapeutic potential of *P. roxburghii* and calls for subsequent studies to isolate new bioactive leads through bioactivity-guided isolation.

Keywords: natural products; phenolic compounds; brine shrimps; protein kinase inhibition; antioxidants

1. Introduction

Plants are nature's gifts that have been employed for the management of various health-threatening diseases since the early ages [1–3]. It has been revealed that approximately 80% of the population across the globe depends upon plant-based therapeutics for healthcare needs [3]. Phytoconstituents are the compounds generated by plants as a defense system against pathogens and predators. They possess certain characteristics that are helpful for the treatment of diseases including antimicrobial, antioxidant, and stimulation as well as inhibition enzymes [4]. Plants contain a wide range of constituents such as alkaloids, tannins, polyphenolics, terpenoids, etc. which are attributable to therapeutic potential [5]. Phytochemicals either isolated compounds or crude extracts, providing opportunity for drug discovery. Human beings seem interested in natural therapeutics as synthetic drugs are associated with adverse effects [6]. Plant-based therapeutic agents are cost-effective and possess fewer side effects as compared to synthetic agents [2].

The current era has noticed the growing enthusiasm of pharmaceutical industries to explore plants for the discovery of new therapeutic moieties [7]. Various commercially available therapeutic agents have been derived from plants such as galegine isolated from *Galega officinalis* L. provided the base for synthesis of metformin which is an antidiabetic drug. Cinchona obtained from *Cinchona officinalis* lead to the development of chloroquine and mefloquine as antimalarial agents [8]. Moreover, vinblastine, vincristine isolated from *Vinca rosea* Linn., and paclitaxel derived from *Taxus brevifolia* are currently available anticancer agents [9,10].

Currently, HPLC is acquiring popularity for the identification of phytochemicals. Qualitative analysis generates a "fingerprint" chromatogram which is helpful for quality control of phytoconstituents. We can also use TLC; however, under certain circumstances it can give false results. HPLC is also helpful for chemosystematics and capable to characterize different species on the basis of their secondary metabolite contents. Reversed-phase HPLC is being widely used for analysis of flavonoids [11].

Putranjiva roxburghii Wall. is a moderate size, evergreen tree from the Euphorbiaceae family. It is mainly found in Thailand, Myanmar, Sri Lanka, Bangladesh, Nepal, and Indochina [12,13]. Folklore uses of *P. roxburghii* include treatment of rheumatism, muscle twisting, fever, arthralgia, pain, and inflammation [14,15]. Previous studies have revealed the presence of a wide range of triterpenoids including putranjivadione, putranjivanonol, putranjic acid, roxburghonic acid, and roxburgholone, etc. [16–19]. Despite the traditional use of *P. roxburghii* since ancient times for various ailments; scientific evidence regarding the therapeutic potential of this plant and phytochemical profiling is still deficient [16]. Reliability for the therapeutic potential of conventionally used plants requires scientifically valid data. Therefore, the therapeutic effectiveness of plants should be explored to provide a base for the isolation of bioactive leads thus facilitating the drug discovery process [20]. Lack of data on polyphenolic profiling and absence of scientific evidence for the therapeutic potential of *P. roxburghii* encouraged us to bridge this research gap. Hence, the present study was carried out to analyze the polyphenols present in different parts of *P. roxburghii* and to evaluate their potential implications for reducing oxidative stress. Moreover, the antibacterial, antifungal, α-amylase inhibition, α-glucosidase inhibition, protein-kinase inhibition, and cytotoxic potential of these extracts were also explored.

2. Materials and Methods

2.1. Acquisition and Identification of Selected Plants

Different plant parts (stems, leaves, and fruits) were acquired from the locality of Quaid-i-Azam University, Islamabad (33.747° N, 73.1356° E) in December 2018. After identification, an authorized sample of the plant was kept in the departmental Herbarium, under voucher number PHM 510 mL.

2.2. Reagents and Solvents for Biological Evaluation

PhosPhate buffer was acquired from Riedel-de Haen, Seelze, Germany while Sabouraud dextrose agar from Oxoid, England. Ferric chloride, Potassium ferricyanide, Trichloroacetic acid (TCA), Tryptone soy broth, Surfactin, Doxorubicin, Ascorbic acid, 2,2-diphenyl-1-picrylhydrazyl (DPPH), Nutrient agar, Sea salt, Standard antifungals (clotrimazole and amphotericin B) and Standard antibiotics (cefixime and roxithromycin) were purchased from Sigma Aldrich, Saint Louis, MO, USA. Dried instant yeast was acquired from Fermipan BDH, Poole, England and Medium ISP4 was formulated in laboratory. Brine shrimp "*Artemia salina*" eggs were acquired from Ocean star Int., Coral springs, FL, USA and Tween-20 from Merck-Schuchardt, USA. Microplate reader was purchased from Biotech, Minneapolis, MN, USA, microplate reader Elx 800 while Eppendorf tubes were acquired from Merck, Kenilworth, NJ, USA. Bacterial and fungal strains were acquired from (Microbiologics, Saint Cloud, MN, USA). Eppendorf tubes were taken from Merck, Kenilworth, NJ, USA.

2.3. Preparation of Crude Extracts

Plant parts were thoroughly washed and shade-dried for four weeks. After drying, plant parts were crumbled and subjected to the sonication aided maceration in 1000 mL Erlenmeyer flasks for three days. Solvents used for maceration include n-hexane (nH), ethyl acetate (EA), methanol (MeOH), and distilled water (DW), respectively. After the specified duration, filtration was done while marc was macerated in the same solvent for 1 day followed by filtration. All filtrates of the same solvent were merged and subjected to drying by a rotary evaporator (Buchi, Flawil, Switzerland). The crude extracts after complete drying were stored at −80 °C.

Following formula was used to calculate percent extract recovery:

$$\text{Percent extract recovery } (\%\ w/w) = (x/y) \times 100$$

x = Total weight of the dried extract, y = Total dried weight of the powdered plant material used in extraction, i.e., 500 g each part.

2.4. Phytochemical Analysis

Stock solutions of all test extracts were formulated as 4 mg/mL DMSO for phytochemical analysis.

2.4.1. Determination of Total Phenolic Content (TPC)

Initially, 20 µL test extract was shifted to the different wells of 96-well plate with subsequent inclusion of 90 µL Folin–Ciocalteu (FC reagent), 90 µL of sodium carbonate and absorbance was checked after incubation. Gallic acid was taken as positive control and calibration curve was plotted (y = 0.0738x + 0.086) while DMSO was taken as negative control. Results were depicted as the mean of µg gallic acid equivalent (GAE) /mg extract ± SD [21].

2.4.2. Determination of Total Flavonoid Content (TFC)

Initially, 20 µL test extract was shifted to different wells of the 96-well plate with subsequent inclusion of 10 µL (1M) potassium acetate solution, 10 µL (10%) aluminum chloride, and 160 µL distilled water and absorbance was checked after incubation. Quercetin was taken as a positive control and the calibration curve was plotted (y = 0.0535x − 0.0033) while DMSO was employed as negative control. Results were recorded as mean of µg quercetin equivalent (QE)/mg extract ± SD [21].

2.4.3. HPLC-DAD Analysis

High performance liquid chromatography (HPLC) was carried out by Agilent Chem station Rev. B.02-01-SR1 (260), Boulder, Colorado, USA [22]. Two mobile phases were employed including acetonitrile-methanol–water-acetic acid in a ratio of 5:10:85:1 (solvent A) and acetonitrile-methanol-acetic acid in a ratio of 40:60:1 (solvent B). Stock solutions of different polyphenols including cinnamic acid derivatives (caffeic acid, ferulic acid, coumaric acid), benzoic acid derivative (vanillic acid, gallic acid, syringic acid, gentisic acid), flavanol flavonoids (kaempferol, quercetin, myricitin), flavan-3-ol flavonoids (catechin), flavone flavonoid (luteolin, apigenin), naphthoquinone (plumbagin), anthraquinone (emodin), and benzoquinone (thymoquinone). Serial dilutions of stock solutions were prepared, i.e., 10, 20, 50, 100, 200 µg/mL, and a calibration curve was plotted for peak areas at different concentrations of standards. The absorption of samples was checked at 257 nm, 279 nm, 325 nm, and 368 nm and the results were depicted as µg/mg extract. Polyphenols were identified by comparing the retention time of test samples with standards and results were expressed as mean of polyphenolic concentration µg/mg ± S.D.

2.4.4. Reagents and Solvents for Phytochemical Analysis

Ethyl acetate, Methanol, Dimethyl sulfoxide (DMSO), and n-hexane were purchased from Sigma-Aldrich, Schnelldorf, Germany. Folin–Ciocalteu (FC) reagent was acquired from Riedel-de Haen, Germany. Quercetin, Potassium acetate, Gallic acid, and Aluminium chloride were purchased from Sigma Aldrich, Saint Louis, MO, USA. Rotary evaporator was acquired from Buchi, Flawil, Switzerl and while microplate reader was purchased from Biotech USA, microplate reader Elx 800. HPLC was acquired from Agilent Chem station Rev. B.02-01-SR1 (260), Boulder, Colorado, USA. Eppendorf tubes were taken from Merck, Kenilworth, NJ, USA.

2.5. Biological Evaluation

2.5.1. Antioxidant Assays

DPPH Free Radical Scavenging Assay

At first, each test extract (10 µL) was shifted to the corresponding wells of a 96 well-plate with subsequent inclusion of 190 µL DPPH solution and [23]. The absorbance was checked and % scavenging activity was determined as:

$$\% \text{ Scavenging activity} = (1 - \text{Abs}/\text{Abc}) \times 100$$

whereas Abs = Absorbance of sample, Abc = Absorbance of control. Ascorbic acid and DMSO were employed as positive and negative controls, respectively. Results were expressed as the mean of % DPPH scavenging activity ± SD and IC_{50} value was determined for the samples expressing more than 50% scavenging activity.

Determination of Total Antioxidant Capacity (TAC)

Initially, each test sample (100 µL) was shifted to Eppendorf tubes with subsequent addition of 900 µL TAC reagent and incubation at 95 °C for 90 min [24]. Ascorbic acid and DMSO were taken as positive control and blank, respectively. Absorbance of the reaction mixture was noted at 630 nm and a calibration curve was plotted (y = 0.0408x − 0.025) while results were illustrated as the mean of µg ascorbic acid equivalent (AAE)/mg extract ± SD.

Determination of Total Reducing Power (TRP)

TRP of test extracts was investigated by adopting a previously reported protocol [24]. At first, each test extract (100 µL) was shifted to Eppendorf tubes (Merck, Kenilworth, USA) with the subsequent addition of 200 µL phosphate buffer (0.2 mol/L, pH 6.6) and 250 µL of 1% potassium ferricyanide. Amalgam was incubated and 10% trichloroacetic acid (200 µL) was then added followed by centrifugation. Then, 150 µL aliquots from the supernatant was transferred into the wells of microtiter plate containing $FeCl_3$ (50 µL, 0.1%) and absorbance was recorded. Positive and negative controls were ascorbic acid and DMSO. A calibration curve was plotted (y = 0.0754x + 0.1034) while the results were demonstrated as a mean of µg AAE/mg extract ± S.D.

2.5.2. Antimicrobial Assays

Antibacterial Assay

The antibacterial activity of test samples was investigated by disc diffusion protocol [25]. Refreshed bacterial cultures [*Pseudomonas aeruginosa* (ATCC15442), *Staphylococcus aureus* (ATCC-6538), *Klebsiella pneumoniae* (ATCC-1705), *Escherichia coli* (ATCC-25922), and *Bacillus subtilis* (ATCC-6633)] (Microbiologics, Saint Cloud, MN, USA) were employed to prepare lawns on nutrient agar plates. Briefly, 5 µL aliquots of every sample were transferred from stock solution (20 mg/mL DMSO) to discs. Subsequently, cefixime and roxithromycin were chosen as positive control while DMSO as a negative control. The discs were implanted on agar plates followed by incubation and zones of inhibition (ZOI) were measured (mm) around each disc. Minimum inhibitory concentration (MIC) was investigated for all the test samples showing ≥12 mm zone of inhibition. For MIC determination, the bacterial inoculum was prepared with a pre-defined density (5×10^4 CFU/mL). Serial dilutions were prepared for each test sample in a 96-well plate by using nutrient broth. Subsequently, 195 µL of inoculum was included and the plate was then incubated. The absorbance of assay plate was recorded after 30 min (zero time reading) and 24 h of incubation and difference was determined. Percent inhibition of bacterial growth was calculated as:

$$\% \text{ inhibition} = (1 - T_s/T_c) \times 100$$

T_s is the turbidity of the sample well and T_c is the turbidity of the control well. The lowest concentration at which test samples showed ≥90% inhibition of bacterial growth was referred as MIC.

Antifungal Assay

Initially, refreshed inoculum was prepared by harvesting the fungal strains, i.e., *Mucor* species (FCBP 0300), *Aspergillus niger* (FCBP 0198), *Aspergillus flavus* (FCBP 0064), *Fusarium solani* (FCBP 0291), and *Aspergillus fumigatus* (FCBP 66) (Microbiologics, Saint Cloud, MN, USA) in Tween 20 solution. Subsequently, 20–25 mL sabouraud dextrose agar (SDA) was taken in petri plates and swabbed with refreshed inoculum (100 µL). The discs loaded with test extracts (5 µL, 20 mg/mL DMSO), clotrimazole (5 µL, 4 mg/mL DMSO) and DMSO (5 µL) were implanted on SDA plates. The petri plates were then incubated and ZOI (mm) around each disc was measured after incubation [26].

2.5.3. Enzyme Inhibition Assays

α-Amylase Inhibition Assay

This assay was conducted in accordance with a preceding protocol [24]. First of all, phosphate buffer (15 µL, pH 6.8), α-amylase enzyme (25 µL, 0.0175 U/mL), test sample (10 µL, 4 mg/mL DMSO) and starch solution (40 µL, 2 mg/mL potassium phosphate buffer) were transferred to the corresponding wells of 96 well plate. The plates were kept at 50 °C for 30 min after which HCl (20 µL, 1 M) and iodine reagent (90 µL) were added, and absorbance was checked at 540 nm. An identical protocol was adopted for the formation of negative and positive controls by substituting the test sample with a similar volume of DMSO and acarbose, respectively. Moreover, a blank was prepared by adding an equal quantity of buffer rather than test extract and α-amylase enzyme solution. %enzyme inhibition was calculated as:

$$\%\text{enzyme inhibition} = [(\text{Abs} - \text{Abn})/(\text{Abb} - \text{Abn}) \times 100]$$

where, Abs = absorbance of test sample, Abb = absorbance of blank, and Abn = absorbance of negative control.

α-Glucosidase Inhibition Assay

This assay was executed by adopting a protocol demonstrated by Nair et al. [27]. Initially, a substrate solution (25 µL, 20 mM), phosphate buffer (69 µL, 50 mM, pH 6.8), test sample (5 µL, 4 mg/mL DMSO), and enzyme solution (1 µL) were transferred to the corresponding wells of 96 well plate. First reading was taken at 405 nm immediately and the amalgam was placed at 37 °C for 30 min. The reaction was quenched by sodium bicarbonate solution (100 µL, 0.5 mM) and the absorbance was again checked. Negative and positive controls were prepared by following a similar protocol just by substituting the test extract with the same amount of DMSO and acarbose, respectively. Following formula was used for the calculation of %enzyme inhibition:

$$\%\text{enzyme inhibition} = \text{Ac} - \text{As}/\text{Ac} \times 100$$

Ac = Absorbance of control, As = Absorbance of sample

2.5.4. Preliminary Toxicity Test

Brine Shrimp Lethality Assay

This assay was executed by using brine shrimp (*Artemia salina* acquired from Ocean star Int., Coral Springs, FL, USA) larvae in a 96 well plate by following a preceding protocol [22]. *Artemia salina* eggs were allowed to hatch in a dual-chamber punctured tank containing simulated seawater. The chamber filled with eggs was enfolded with aluminum foil while another chamber was kept under a light source. The tank was kept at 30–32 °C for 24–48 h after which nauplii started moving towards the lightened chamber. Subsequently, 10 mature nauplii were taken in each well of a 96-well plate with subsequent inclusion of 150 µL seawater and corresponding volume of test samples to attain final concentrations of 200, 100, and 50 µg/mL. Seawater was added to make the final volume of each well up to 300 µL. Doxorubicin was chosen as positive control while 1% DMSO as negative control. Ultimately, 96 well plate was incubated after which dead nauplii were computed by using an inverted microscope. Following formula was used to calculate %mortality and LC_{50} was calculated by graph pad prism 5 software (GraphPad, San Diego, CA, USA).

$$\% \text{ mortality} = \text{no. of dead shrimps/total no. of shrimps} \times 100$$

Protein Kinase Inhibition Assay

Mycelia fragments of *Streptomyces* species were immersed into sterile tryptone soy broth and kept for 24 h. Petri plates containing ISP4 medium were swabbed with refreshed culture (100 µL). Subsequently, discs loaded with a test sample (5 µL, 20 mg/mL DMSO) were implanted on seeded plates. Surfactin and DMSO-loaded discs were employed as positive and negative controls, respectively. The plates were incubated and the diameter of the bald or clear zone (mm) of inhibition was measured as an indicator of protein kinase inhibition potential [26].

2.6. *Data Analysis*

All the assays were conducted in a triplicate manner and the data have been demonstrated as mean ± standard deviation of triplicate analysis. LC_{50} and IC_{50} values were determined through Graph pad prism 5 software (GraphPad, San Diego, CA, USA) while Post Hoc Tukey HSD (Honest significant difference) test was used to analyze the results using the statistical package SPSS (IBM SPSS, Armonk, NY, USA) and $p < 0.05$ (95% confidence interval) was considered as significant. Correlation coefficient (R^2) of phytochemical activities was calculated by employing the correlation and regression function of Microsoft Excel program (Microsoft, Redmond, WA, USA).

3. Results and Discussion

3.1. *Percent Extract Yield*

Percent recovery of test extracts is summarized in Figure 1. The results demonstrated that DW and MeOH extracts of leaf possess maximum extraction efficiency, i.e., 6.3 ± 0.4% and 5.3 ± 0.3%, respectively [28,29]. During the current study, polar solvents showed maximum extraction efficiency. Water-soluble phytoconstituents mainly present in plants include flavonoids, tannins, terpenoids, quinones, etc. [30,31]. The presence of some of these metabolites might be attributable to the maximum extraction efficiency of DW and MeOH leaf extracts. The results were in conformity with a previous study in which non-polar extracts of *Ajuga bracteosa* Wall. showed less extraction efficiency as compared to polar ones [25].

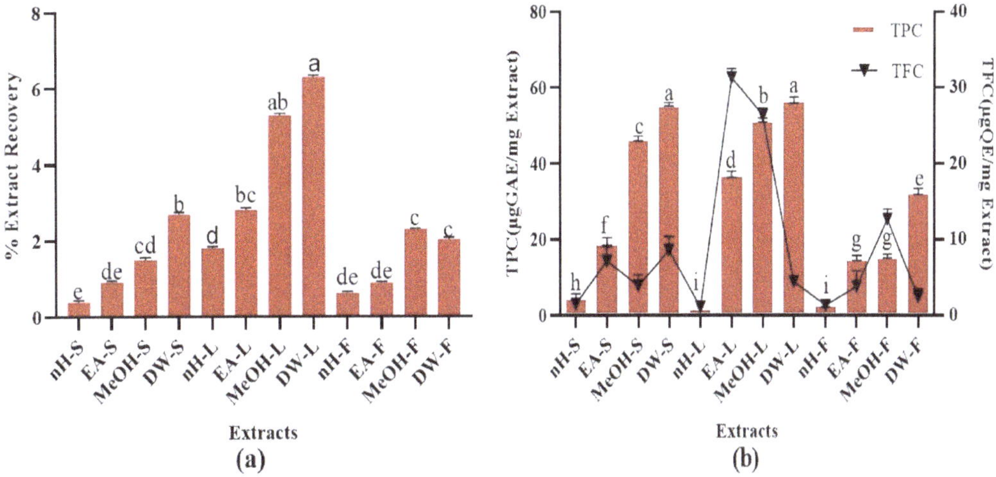

Figure 1. (**a**) Extraction efficiency, (**b**) TPC (Total phenolic content µg GAE/mg) and TFC (Total flavonoid content µg QE/mg) of *P. roxburghii* crude extracts. Assays were conducted in a triplicate manner and the data have been demonstrated as mean ± standard deviation. Significantly different means ($p < 0.05$) are represented by different superscripts ($^{a-i}$). nH-S; n-Hexane stem, EA-S; ethyl acetate stem, MeOH-S; methanol stem, DW-S; distilled water stem, nH-L; n-Hexane leaf, EA-L; ethyl acetate leaf, MeOH-L; methanol leaf, DW-L; distilled water leaf, nH-F; n-Hexane fruit, EA-F; ethyl acetate fruit, MeOH-F; methanol fruit, DW-F; distilled water fruit.

3.2. Phytochemical Analysis

3.2.1. Total Phenolic Content (TPC)

A remarkable TPC was noted in test extracts. Maximum TPC was notified in DW extracts of stem and leaf, i.e., 55.23 ± 0.10 µg GAE/mg extract while least TPC was found in nH extracts of leaf and fruit, i.e., 2.06 ± 0.03 µg GAE/mg extract, respectively (Figure 1). TPC notified in our research was somewhat higher than a recently published study which revealed that *P. roxburghii* leaf (hydroethanol, 30:70) extract contains 46.58 ± 2.52 mg/g GAE polyphenolic content [32]. Alterations in agro-climatic conditions accompanied with temperature and rainfall impart a significant impact on the amount of phytoconstituents within similar species of plants growing in different regions [33]. Hence, differences in solvent composition and plant collection sites might be responsible for the variations of estimated TPC in both studies. Current study investigated the phenolic content in fruit and stem parts of *P. roxburghii* as well and employed a wide range of solvents [32]. Phenolics are secondary metabolites of plants which help to counteract different pathogenic microbes and impart colors to the plants. They are widely distributed in all plant parts and hence, act as an essential component of the human diet [34]. Phenolics are highly effective to counteract oxidative stress-related ailments such as neurodegenerative diseases, cancer, diabetes, ageing, and cardiovascular diseases [35]. Phenolic compounds perform antioxidant activity by the virtue of hydrogen donating action and chelation of metal ions involved in the generation of free radicals [36,37]. The presence of hydrophobic benzenoid rings enables the phenolic constituents to interact with proteins and to inhibit enzymes involved in free radical production [36,38]. These compounds can also act as neuroprotective, gastroprotective, antipyreticm, antiatherosclerotic, cardioprotective, antiulcer [39], anti-inflammatory [35], antibacterial [40], antifungal [41], antiviral [42], and antitumor agents [35,43], antidiabetic, anxiolytic [39]. Hence, the presence of a remarkable quantity of phenolic content in *P. roxburghii* might be accountable to its promising therapeutic potential.

3.2.2. Total Flavonoid Content (TFC)

A significant TFC was observed in crude extracts. Highest TFC was found in EA and MeOH extracts of leaf, i.e., 30.62 ± 0.31, 27.25 ± 0.05 µg QE/mg extract, respectively (Figure 1). Minimum TFC was notified in nH extracts of stem, leaf, and fruit, i.e., 1.04 ± 0.02 µg QE/mg extract. The results were in close resemblance to a previous study which revealed that the hydroethanolic extract of *P. roxburghii* leaf part contains 29.0 ± 2 mg/g QE flavonoid content [32]. It was found that EA leaf extract contains a greater quantity of flavonoids than MeOH extract. Flavonoids are an important class of polyphenolics containing benzo-γ-pyrone moiety and possess a wide range of therapeutic activities [44]. Flavonoids contain a flavan core, consisting of 15 carbon atoms organized in three rings. Depending upon the oxidation state of the central core, flavonoids are categorized as: anthocyanins, flavanones, isoflavones, flavones, flavanols, and flavonols. The structural differences in each subclass are due to the extent and pattern of prenylation, methoxylation, glycosylation, and hydroxylation [34]. Flavonoids can act as antimicrobial, anti-inflammatory, immunomodulatory, enzyme inhibitor, antiviral, antiparasitic, cardioprotective, cytotoxic, anti-tumor, antiaging agents [45,46]. A wide range of flavonoids have been isolated from *roxburghii* species such as catechin and quercetin in *Rosa roxburghii* Tratt. [47], isorhamnetin, rhamnazin, roxburoside from *Anoectochilus roxburghii* Wall. [48], and putraflavone from *P. roxburghii* [49]. During the present study, the flavonoid presence in polar extracts of *P. roxburghii* referred it to be a potential reservoir of antioxidants.

3.2.3. HPLC-DAD Analysis of Polyphenols

Polyphenols were quantified in various extracts of *P. roxburghii* by HPLC-DAD analysis (using 16 references) (Figure 2). Caffeic Acid, ferulic acid, coumaric acid, vanillic acid, gallic acid, syringic acid, gentisic acid, catechin, and emodin were present in significant amounts (Table 1). Whereas kaempferol, quercetin, myricitin, luteolin, apigenin, plumbagin, and thymoquinone were not present in any crude extract. Maximum number of polyphenols was detected in DW-L extract including caffeic acid (0.35 ± 0.06 µg/mg extract), ferulic acid (1.02 ± 0.51 µg/mg extract), vanillic acid (0.26 ± 0.32 µg/mg extract), syringic acid (0.92 ± 0.52 µg/mg extract), gentisic acid (1.19 ± 0.09 µg/mg extract), and catechin (2.05 ± 0.18 µg/mg extract). It was followed by EA-F extract which contains coumaric acid (0.02 ± 0.36 µg/mg extract), syringic acid (0.76 ± 0.42 µg/mg extract), catechin (0.56 ± 0.56 µg/mg extract), and emodin (0.87 ± 23 µg/mg extract). Subsequently, MeOH-F extract contains ferulic acid (0.44 ± 0.26 µg/mg extract) and catechin (0.61 ± 0.67 µg/mg extract), DW-S extract contains caffeic acid (0.26 ± 0.03 µg/mg extract) and catechin (0.69 ± 0.18 µg/mg extract) while EA-S extract contains syringic acid (0.17 ± 0.04 µg/mg extract) and emodin (0.07 ± 0.16 µg/mg extract). Moreover, MeOH-S extract contains catechin (1.02 ± 0.52 µg/mg extract) while DW-F extract contains gentisic acid (0.49 ± 0.82 µg/mg extract). The detection of these polyphenols derives a direct correlation of plants potential with notified biological activities. All reported polyphenols possess a wide range of therapeutic activities. Caffeic acid has antioxidant [50], anticancer, anti-fibrosis, antihypertension [51,52], anti-hepatitis B virus [53] anti-inflammatory, anticoagulant activities [54]. Ferulic acid has anti-oxidant [55], cytotoxic [56], antiallergic, antiviral, anti-inflammatory, antimicrobial, anticoagulant, and hepatoprotective properties [57,58]. Reported activities of coumaric acid include antioxidant [59], anti-microbial [60], anti-leishmaniasis, cytotoxic [61]. Vanillic acid has hepato-protective [62] and anti-oxidant activities [63] while syringic acid has hepato-protective, neuro-protective, cardio-protective, anti-inflammatory, and hypoglycemic potential [64]. Gentisic acid possesses fibro growth factor inhibition [65], antimicrobial, antioxidant, anti-inflammatory, hepatoprotective, and neuroprotective activities [66]. Reported activities of catechin are anti-cancer [67], antimicrobial [68], hypolipidemic [69], vasodilator, antispasmodic, and bronchodilator activities [70]. Finally, emodin has anti-inflammatory, anti-cancer, antimicrobial, hepatoprotective potential [71,72]. Hence, detection of a wide range of polyphenols further manifests the therapeutic potential of *P. roxburghii*. Best to our knowledge, HPLC-DAD analysis of *P.*

roxburghii was reported for the first time in the present study. Previously, HPLC analysis of polyphenols was carried out for *Rosa roxburghii* Tratt. leaf extracts. Approximately, 30 polyphenols were detected, which mainly includes gallic acid, myricetin, (+)-catechin, quercetin-3-O-galactoside, arbutin, and 3-hydroxybenzoic acid [73]. HPLC analysis of *Pinus roxburghii* barks and needles was carried out and quercetin was identified as the most abundant flavonol [11].

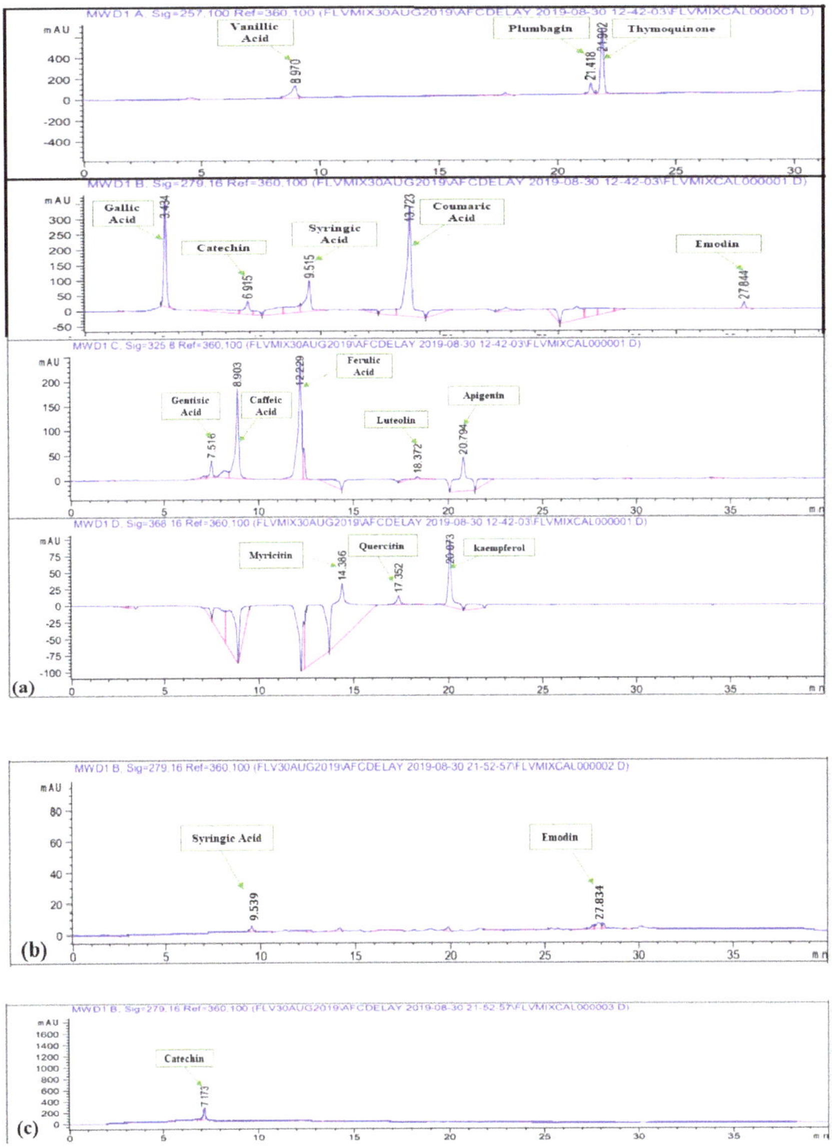

Figure 2. *Cont.*

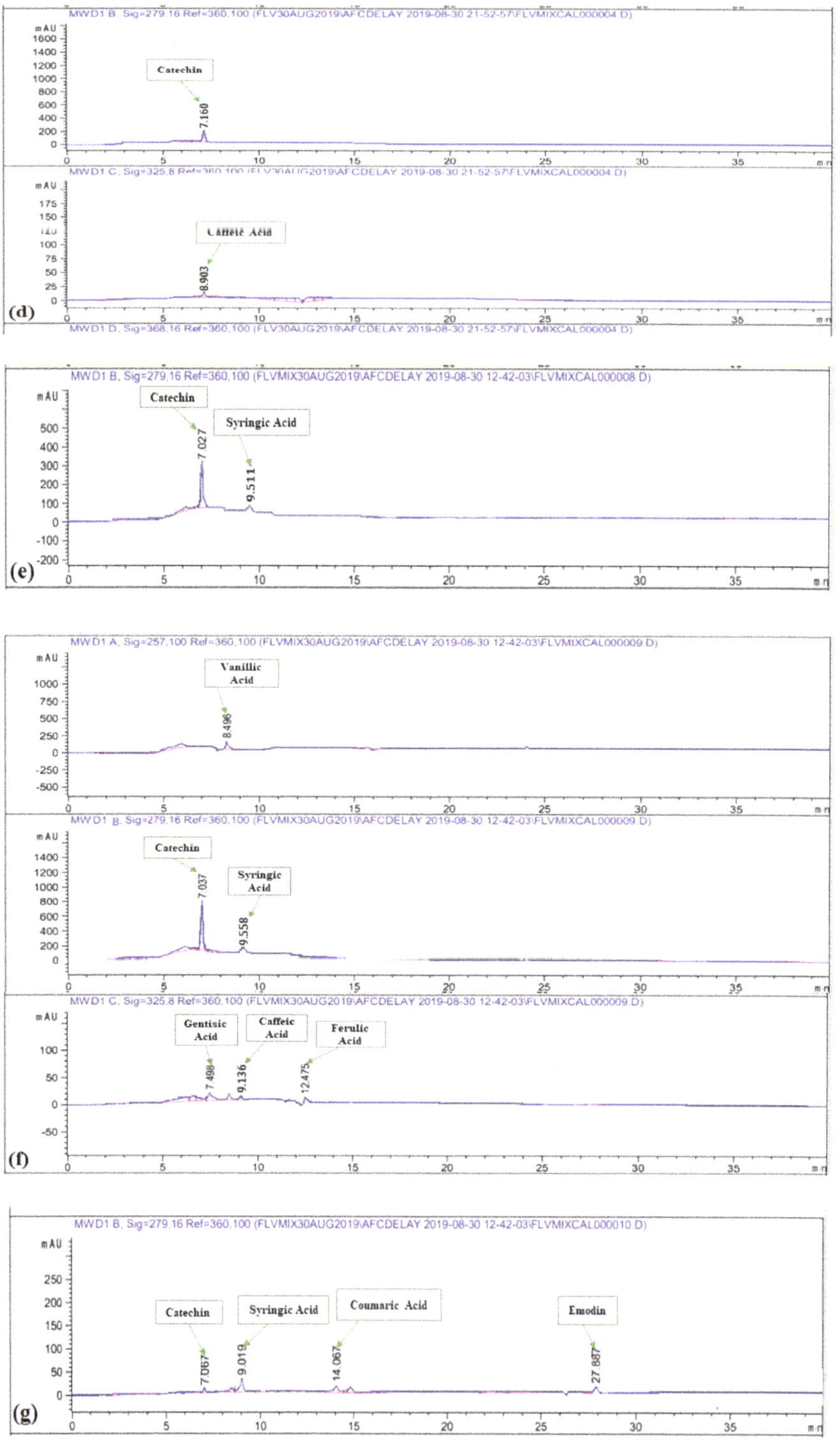

Figure 2. *Cont.*

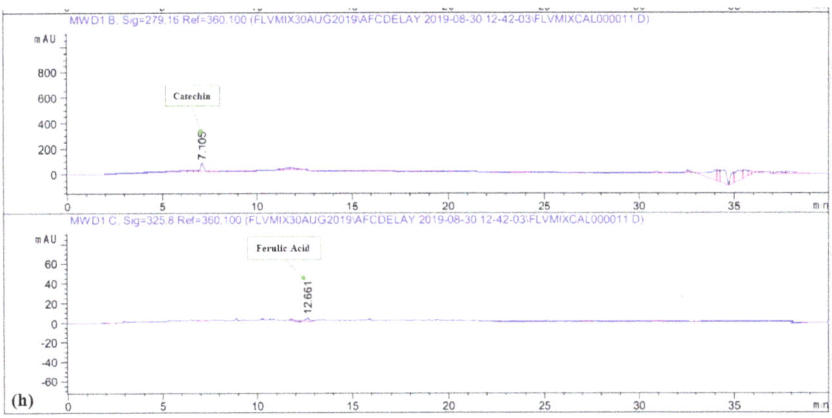

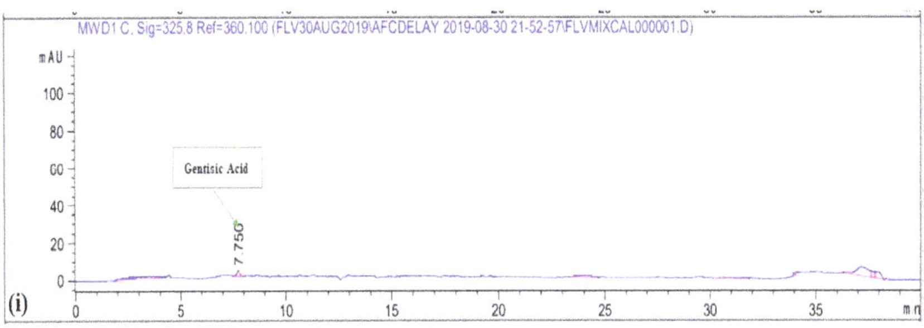

Figure 2. HPLC chromatograms of (**a**) standard polyphenols (**b**) EA-S (**c**) MeOH-S (**d**) DW-S (**e**) MeOH-L (**f**) DW-L (**g**) EA-F (**h**) MeOH-F (**i**) DW-F.

Table 1. Chemical profiling (μg/mg extract) of stem, leaf and fruit parts of *P. roxburghii* using HPLC-DAD.

Class	Polyphenols	EA-S	MeOH-S	DW-S	EA-L	MeOH-L	DW-L	EA-F	MeOH-F	DW-F
Cinnamic acid derivatives	Caffeic Acid	–	–	0.26 ± 0.03 ^{m}	–	–	0.35 ± 0.06 ^{l}	–	–	–
	Ferulic acid	–	–	–	–	–	1.02 ± 0.51 ^{c}	–	0.44 ± 0.26 ^{k}	–
	Coumaric Acid	–	–	–	–	–		0.02 ± 0.36 ^{p}	–	–
Benzoic acid derivative	Vanillic Acid	–	–	–	–	–	0.26 ± 0.32 ^{m}	–	–	–
	Gallic Acid	–	–	–	–	–	–	–	–	–
	Syringic Acid	0.17 ± 0.04 ^{n}	–	–	–	0.44 ± 0.21 ^{j}	0.92 ± 0.52 ^{d}	0.76 ± 0.42 ^{f}	–	–
	Gentisic Acid	–	–	–	–	–	1.19 ± 0.09 ^{b}	–	–	0.49 ± 0.82 ^{j}
Flavonol flavonoids	Kaempferol	–	–	–	–	–	–	–	–	–
	Quercetin	–	–	–	–	–	–	–	–	–
	Myricitin	–	–	–	–	–	–	–	–	–
Flavan-3-ol flavonoids	Catechin	–	1.02 ± 0.52 ^{c}	0.69 ± 0.18 ^{g}	–	0.74 ± 0.01 ^{fg}	2.05 ± 0.18 ^{a}	0.56 ± 0.56 ^{i}	0.61 ± 0.67 ^{h}	–
Flavone flavonoid	Luteolin	–	–	–	–	–	–	–	–	–
	Apigenin	–	–	–	–	–	–	–	–	–
Naphthoquinone	Plumbagin	–	–	–	–	–	–	–	–	–
Anthraquinone	Emodin	0.07 ± 0.16 ^{o}	–	–	–	–	–	0.87 ± 23 ^{e}	–	–
Benzoquinone	Thymoquinone	–	–	–	–	–	–	–	–	–

– = not detected. HPLC analysis was carried out in a triplicate manner and the data have been demonstrated as mean ± standard deviation. The values with different superscripts ($^{a–p}$) demonstrate significantly ($p < 0.05$) different mean values.

3.3. *Biological Evaluation*

3.3.1. Antioxidant Assays

DPPH Assay

Metabolic processes within the body and environmental factors produce free radicals. Free radicals mainly include reactive oxygen species (ROS) which cause various ailments including ageing, carcinogenesis, mutagenesis, and cardiovascular abnormalities. Antioxidants are the agents, which counteract the effects of free radicals and limit oxidative stress [74]. DPPH assay is a standard method to investigate the free radical scavenging ability of test samples [75]. The percent free radical scavenging activity (%FRSA) of test extracts was analyzed and results are given in Figure 3. Maximum %FRSA was shown by MeOH-S and DW-L extracts (86 ± 0.56%) with IC_{50} values of 68 ± 0.43, 149 ± 0.21 μg/mL, respectively. Whereas, nH extracts of stem, leaf, and fruit were unable to express any free radical scavenging activity. Moreover, ascorbic acid showed 78% FRSA with IC_{50} of 14.56 μg/mL. We calculated the correlation of %FRSA with TPC (R^2 = 0.8084) and TFC (R^2 = 0.0211) which demonstrates that phenolic compounds other than flavonoids are mainly attributable to free radical scavenging activity. Previously reported %FRSA of methanolic stem extracts of *P. roxburghii* was somewhat lower than calculated during our commenced research. Moreover, the current study revealed that DW-L extract of *P. roxburghii* also possesses a promising free radical scavenging potential. Highest extraction efficiency of DW-L extract provides another perspective to employ this extract as an antioxidant. Overall, our results were in accordance with a previous study which revealed that polar extracts possess better scavenging potential than non-polar extracts [25].

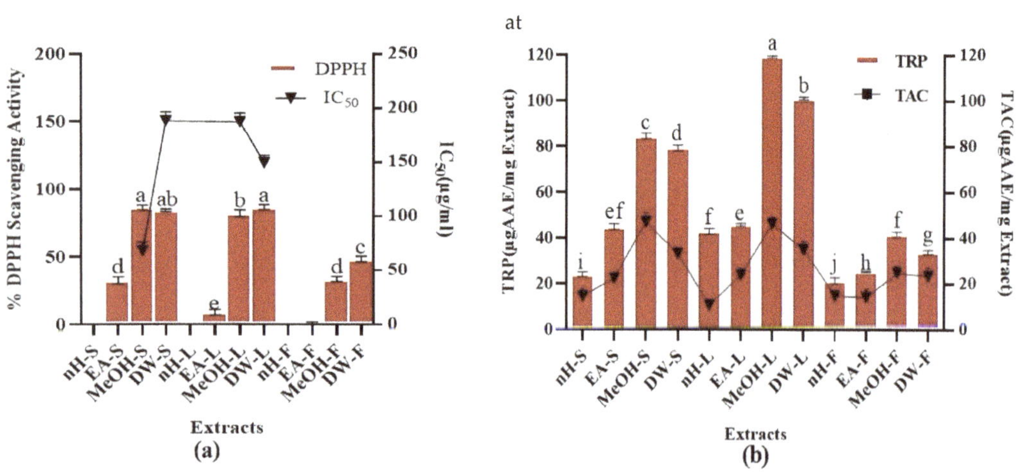

Figure 3. Demonstration of (**a**) %DPPH radical scavenging activity (at 40 μg/mL concentration of test extracts) (**b**) TAC and TRP (μg AAE/mg) of *P. roxburghii* crude extracts. Assays were conducted in a triplicate manner and the data have been demonstrated as mean ± standard deviation. Significantly different means ($p < 0.05$) are represented by different superscripts ($^{a–j}$).

Total Antioxidant Capacity (TAC)

TAC of test extracts was calculated as shown in Figure 3. Maximum TAC was shown by MeOH-S and MeOH-L extracts, i.e., 48.65 ± 0.52, 46.07 ± 0.07 μg AAE/mg extract, respectively. Minimum TAC (10.91 ± 0.05 μg AAE/mg extract) was expressed by nH-L extract. We noticed a remarkable relationship between TAC and TPC (R^2 = 0.744) which demonstrates that phytoconstituents producing antioxidant activity mainly belong to polyphenols. Moreover, the correlation between TAC and TFC (R^2 = 0.1419) shows that flavonoids are not the main contributor to the antioxidant activity of test extracts. During a previous study, the estimated TAC of *P. roxburghii* stem part was significantly lower than that of the current study. This might be due to the fact that varying agro-climatic conditions can impact the type and quantity of phytoconstituents [41]. Moreover, the antioxidant capacity of different parts of *P. roxburghii* except stems, has been investigated for the first time [9]. Plant-derived antioxidants are considered better than synthetic antioxidants as they possess better compatibility with the human body [41], hence provide a justification for the exploitation of herbal products for therapeutic appraisal.

Total Reducing Power (TRP)

TRP was investigated for crude extracts and results are depicted in Figure 3. Maximum TRP was expressed by MeOH-L and DW-L extracts, i.e., 118.59 ± 0.08, 101.28 ± 0,09 μg AAE/mg extract while nH-F extract showed the least TRP, i.e., 20.48 ± 0.52 μg AAE/mg extract, respectively. A positive relation was notified between TPC and TRP (R^2= 0.7251) in contrast to the correlation between TFC and TRP (R^2 = 0.1495) which demonstrates that phenolic constituents other than flavonoids are mainly accountable for reducing potential of plants. Reducing power of *P. roxburghii* stem part was previously investigated; however, the current study revealed that leaf part of this plant possesses better antioxidant potential [12]. Reducing power is mainly attributable to the presence of reductones which cleaves the free radical chain by hydrogen atom donation and participates in antioxidant action. Multiple studies have notified a close relationship between antioxidant activity and reducing power of crude extracts which is in support of our current research results [24,76].

3.3.2. Antimicrobial Assays

Antibacterial Assay

Antibacterial property of test extracts (100 μg per disc) was investigated against gram-negative (*K. pneumoniae, P. aeruginosa, E. coli*) and gram-positive (*B. subtilis, S. aureus*) bacteria as shown in Table 2. Test extracts exhibiting zone of inhibition ≥ 12 mm were further subjected for MIC evaluation by employing the broth microdilution method. EA-L extract was highly active against *B. subtilis* (MIC 3.7 μg/mL), *K. pneumoniae* (MIC 33.3 μg/mL), and *E. coli* (MIC 3.7 μg/mL) with 24 ± 0.5, 20 ± 0.50, 23 ± 0.76 mm ZOIs, respectively. The ZOIs generated by the test extracts were equivalent to those produced by standards, i.e., roxithromycin and cefixime. Moreover, DMSO taken as negative control was unable to show inhibition. Overall, polar extracts showed better antibacterial potential than non-polar extracts. In a previous study, antibacterial potential of *P. roxburghii* MeOH leaf part was investigated, and results were in somewhat resemblance to our study. However, EA leaf extract was investigated for the first time, and it was noted that this extract possessed stronger antibacterial activity than previously reported MeOH leaf extract. Phenolics impart toxic effects to microbes either through interaction with sulfhydryl groups or proteins resulting in enzyme inhibition. Moreover, polyphenols generate heavy complexes with proteins which interact with bacterial adherence and disrupt the receptors on cell surface [26,77]. During the current study, a proficient amount of TPC and TFC was notified which might contribute to the antibacterial activity.

Antifungal Assay

During present study, MeOH-F extract (100 μg/disc) showed slight activity against *A. flavus* and *A. fumigatus* with 7 ± 0.89 and 8 ± 0.98 mm ZOIs, respectively and MIC value of >100 μg/mL. MeOH-S extract was moderately active against *A. flavus* with 12 ± 0.98 ZOI and >100 μg/mL MIC while other extracts were unable to show antifungal activity. Clotrimazole (10 μg/disc) showed 20 ± 0.57 to 31 ± 1.1 mm ZOIs. A previous study reported the promising antifungal activity of seed and pericarp of *P. roxburghii*. However, during the present study; stem, leaf, and fruit parts of *P. roxburghii* showed slight to moderate antifungal active. Moreover, fungal strains used in current study were different from those used in previous study. Flavonoids and tannins can form complexes with extracellular proteins present in the cell wall of fungi and rupture the fungal membrane [78,79]. *P. roxburghii* contains tannins (ellagic acid, gallic acid, gallocatechin, ellagi- and gallo-tannins) and flavonoids (bioflavones) which might be a contributor to antifungal activity [49].

Table 2. Antibacterial activity and MIC values of *P. roxburghii* test extracts.

Extract Codes	Zone of Inhibition (mm) at 100 µg/disc and MIC (µg/mL)									
	S. A	MIC	B. S	MIC	P. A	MIC	K. P	MIC	E. C	MIC
nH-S	7 ± 0.29 [de]	–	11 ± 0.2 [de]	–	7 ± 0.15 [bc]	–	6 ± 0.87 [ef]	–	11 ± 0.51 [e]	–
EA-S	7 ± 0.01 [de]		9 ± 0.5 [e]	–	8 ± 0.31 [b]	–	7 ± 0.35 [e]	–	6 ± 0.50 [g]	–
MeOH-S	10 ± 0.36 [c]	–	12 ± 0.7 [d]	100 [a]	7 ± 0.10 [bc]	–	10 ± 0.76 [d]	–	13 ± 0.50 [d]	100 [a]
DW-S	7 ± 0.36 [de]	–	3 ± 0.5 [g]	–	6 ± 0.10 [c]	–	12 ± 0.17 [c]	100 [a]	20 ± 0.50 [b]	33.3 [b]
nH-L	8 ± 0.7 [d]	–	12 ± 0.5 [d]	100 [a]	6 ± 0.31 [c]	–	7 ± 0.55 [e]	–	7 ± 0.31 [fg]	–
EA-L	6 ± 0.3 [e]	–	24 ± 0.5 [a]	3.7 [c]	7 ± 0.15 [bc]	–	20 ± 0.50 [a]	33.3 [b]	23 ± 0.76 [a]	3.7 [c]
MeOH-L	8 ± 0.36 [d]	–	20 ± 0.45 [b]	33.3 [b]	6 ± 0.21 [c]	–	14 ± 0.76 [bc]	100 [a]	12 ± 0.31 [de]	100 [a]
DW-L	9 ± 0.31 [cd]	–	9 ± 0.5 [e]	–	7 ± 0.25 [bc]	–	12 ± 0.51 [c]	100 [a]	6 ± 0.76 [g]	–
nH-F	5 ± 0.15 [f]	–	7 ± 0.7 [f]	–	7 ± 0.15 [bc]	–	7 ± 0.50 [e]	–	10 ± 0.15 [ef]	–
EA-F	13 ± 0.12 [b]	100 [a]	8 ± 0.5 [ef]	–	6 ± 0.15 [c]	–	19 ± 0.58 [ab]	33.3 [b]	8 ± 0.58 [f]	–
MeOH-F	7 ± 0.31 [de]	–	12 ± 0.2 [d]	100 [a]	5 ± 0.25 [cd]	–	7 ± 0.31 [e]	–	18 ± 0.5 [c]	–
DW-F	10 ± 0.32 [c]	–	7 ± 0.3 [f]	–	5 ± 0.33 [cd]	–	18 ± 0.29 [b]	33.3 [b]	7 ± 0.50 [fg]	–
Rox	23 ± 0.54 [a]	1.11 [b]	17 ± 0.3 [c]	3.33 [c]	–	–	–	–	–	–
Cefix	–	–	–	–	22 ± 0.89 [a]	1.11	20 ± 1.2 [a]	1.11 [c]	20 ± 1.5 [b]	3.33 [c]
DMSO	–	–	–	–	–	–	–	–	–	–

– = no activity. S. A = *Staphylococcus aureus*, B. S = *Bacillus subtilis*, P. A = *Pseudomonas aeruginosa*, K. P = *Klebsiella pneumoniae*, E. C = *Escherichia coli*. Rox = Roxithromycin, Cefix = Cefixime. Assay was conducted in a triplicate manner and the data have been demonstrated as mean ± standard deviation. The values with different superscripts ($^{a–g}$) depict significantly ($p < 0.05$) different mean values.

3.3.3. Enzyme Inhibition Assays

α-Amylase Inhibition Assay

The suppression of carbohydrate cleaving enzymes, i.e., α-amylase and α-glucosidase is a promising approach to reduce the blood glucose concentration in case of diabetes [27]. Inhibition of both enzymes impedes carbohydrate digestion and enhance the time required for carbohydrate digestion, limits rate of glucose absorption and ultimately prevents post-prandial rise in plasma glucose level [80]. During the present study, α-amylase inhibition potential of crude extracts was investigated and results are depicted in Figure 4. Maximum α-amylase inhibition activity was expressed by EA-F and EA-S extracts, i.e., 67.37 ± 0.05%, 61.37 ± 0.06% with IC_{50} values, i.e., 14.28 ± 0.9, 33.26 ± 0.5µg/mL, respectively. Acarbose was employed as standard which showed 80.45 ± 0.76% α-amylase inhibition with IC_{50} value of 34.85 ± 0.21 µg/mL. Results were in accordance with previous studies which reported that less polar extracts exhibited improved enzyme inhibition activity than highly polar extracts. These studies also notified a direct relationship between the concentration of phenolics and flavonoids with α-amylase inhibition activity [81,82]. However, the current study did not notice any significant relation of TPC (R^2 = 0.343), and TFC (R^2 = 0.0008) with α-amylase inhibition potential.

α-Glucosidase Inhibition Assay

Antidiabetic activity of *P. roxburgii* was further confirmed by estimation of α-glucosidase inhibition potential of crude extracts and results are depicted in Figure 4. Maximum α-glucosidase inhibition (80 ± 0.78%) was shown by nH-L, nH-S, and DW-S extracts with IC_{50} values, i.e., 90.21 ± 0.02, 93 ± 0.078, 94.11 ± 0.99 µg/mL, respectively. Just like α-amylase inhibition, there was no positive relationship of TPC (R^2 = 0.1971) and TFC (R^2 = 0.1231) with α-glucosidase inhibition. Acarbose and miglitol are commercially available α-glucosidase inhibitors which hinder the carbohydrate absorption and reduce post-prandial escalation in glucose levels [83]. However, these agents possess severe gastrointestinal adverse effects such as diarrhea and flatulence [84]. α-amylase inhibitors have been reported to be present in plants which helps them to combat predators [85].

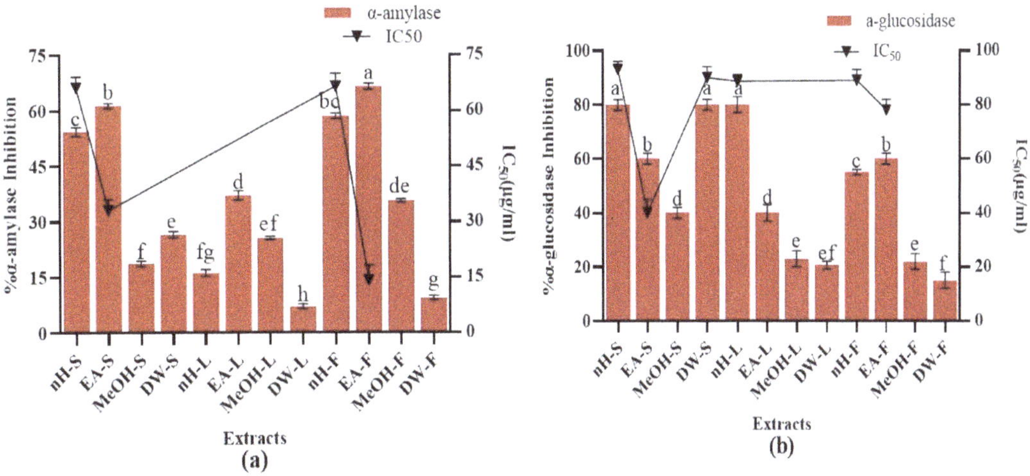

Figure 4. Demonstration of (**a**) α-amylase inhibition and (**b**) α-glucosidase inhibition potential of *P. roxburghii* crude extracts. Assays were conducted in a triplicate manner and the data have been demonstrated as mean ± standard deviation. Significantly different means ($p < 0.05$) are represented by different superscripts ($^{a-h}$).

A detailed study was executed for the first time to investigate α-amylase and α-glucosidase inhibition potential of different parts of *P. roxburghii*. Previous study demonstrated that plants containing tannins such as *Camelia sinensis Var.*, *Artocarpus heterophyllus* Lam., *Persea Americana* Mil., and *Syzygium polyanthum* Walp. are potent inhibitors of α-glucosidase [86]. A wide range of tannins such as ellagic acid, gallic acid, gallocatechin, ellagi- and gallo-tannins are present in *P. roxburghii* might be involved behind the α-glucosidase inhibition activity [49]. However, further studies are needed to prove this hypothesis.

3.3.4. Preliminary Toxicity Potential

Brine Shrimp Lethality Assay

Brine shrimp lethality assay is a preliminary method to investigate the cytotoxic effect of test samples [87]. Brine shrimp lethality potential was noticed to be in direct relation to the concentration of test extracts (Table 3). Out of the twelve extracts subjected for estimation of cytotoxic activity, 41.67% of the extracts ($LC_{50} < 50$ μg/mL) were considered as extremely cytotoxic while 33.33% of crude extracts ($LC_{50} \geq 50$ but ≤ 200 μg/mL) were classified as moderately cytotoxic. Other 25% of the extracts (>200 μg/mL) were categorized as weakly cytotoxic. Overall, DW-F extract was noticed as most cytotoxic extract showing LC_{50} 9.36 ± 0.91 μg/mL while LC_{50} of doxorubicin (positive control) was 5.93 μg/mL. Previously, brine shrimp lethality potential of methanolic seed extract was investigated and mild cytotoxicity potential was reported [88]. Undertaken study was carried out on various other parts of *P. roxburghii* and promising brine shrimp lethality potential of nH and DW extracts of fruit was revealed for the first time. During the present study, 100% of the test extracts exhibited LC_{50} values of less than 1000 μg/mL, referring to the existence of cytotoxic phytometabolites in the subject plant.

Table 3. Brine shrimp lethality and protein kinase inhibition potential of crude extracts of *P. roxburghii*.

Extract Codes	% Brine Shrimp Mortality					Protein Kinase Inhibition	
	200 (µg/mL)	100 (µg/mL)	50 (µg/mL)	25 (µg/mL)	LC_{50} (µg/mL)	Clear Zone (mm)	Bald Zone (mm)
nH-S	70 ± 10^{b}	40 ± 7.5^{d}	2 ± 11.5^{g}	20 ± 0^{e}	130.93 ± 0.56^{b}	–	–
EA-S	100 ± 0^{a}	100 ± 0^{a}	50 ± 5.7^{b}	30 ± 0^{d}	39.52 ± 0.42^{f}	–	–
MeOH-S	100 ± 0^{a}	40 ± 5.7^{d}	40 ± 0^{c}	30 ± 7.5^{d}	88.51 ± 0.59^{e}	–	–
DW-S	100 ± 0^{a}	50 ± 0^{c}	0 ± 0^{h}	0 ± 0^{g}	100 ± 0.67^{c}	–	–
nH-L	30 ± 0^{c}	30 ± 0^{e}	0 ± 0^{h}	0 ± 0^{g}	$>200 \pm 0.73^{a}$	–	–
EA-L	100 ± 5.7^{a}	100 ± 0^{a}	70 ± 0^{a}	60 ± 7.5^{b}	20 ± 1.16^{g}	–	–
MeOH-L	30 ± 11.5^{c}	20 ± 0^{f}	10 ± 0^{f}	10 ± 5.7	$>200 \pm 1.52^{a}$	–	–
DW-L	100 ± 0^{a}	50 ± 0^{c}	30 ± 5.7^{d}	20 ± 0^{e}	93.1 ± 0.36^{d}	–	–
nH-F	100 ± 0^{a}	100 ± 0^{a}	70 ± 0^{a}	70 ± 0^{a}	18.84 ± 0.49^{gh}	–	7 ± 0.97^{ab}
EA-F	100 ± 7.5^{a}	90 ± 7.5^{b}	50 ± 0^{b}	40 ± 0^{c}	35.5 ± 0.53^{g}	–	–
MeOH-F	30 ± 5.7^{c}	16 ± 0^{g}	16 ± 0^{e}	10 ± 0^{f}	$>200 \pm 1.21^{a}$	–	–
DW-F	100 ± 0^{a}	100 ± 0^{a}	0 ± 0^{h}	60 ± 0^{b}	9.36 ± 0.91^{i}	–	8 ± 0.4^{a}

– = not detected. Assays were conducted in a triplicate manner and the data have been demonstrated as mean ± standard deviation. The values with different superscripts ($^{a-i}$) depict significantly ($p < 0.05$) different mean values.

Protein Kinase Inhibition Assay

Protein kinase regulates multiple functions of the cell cycle including apoptosis, metabolism, growth, and differentiation. In the case of human cancers, irregularities of growth factor signaling pathways were notified. Hence, protein kinase inhibition can serve as a target for drug development by overcoming the irregularities in signaling pathways [89]. Protein kinase inhibition activity of the test extracts (100 µg/disc) was investigated as well (Table 3). DW and nH fruit extracts showed 8 ± 0.4 and 7 ± 0.97 mm bald phenotype zones, respectively while no clear zone was found. The remaining extracts were unable to express any protein kinase inhibition activity. The absence of ZOI in case of DMSO confirmed the non-toxic effect of negative control. Protein kinase is involved in the regulation of various processes of cell cycle, i.e., control of cycle growth, metabolism, differentiation, and apoptosis. Abnormalities of growth factor signaling pathways have been notified in various forms of human cancer. Hence, protein kinase inhibitors capable to repair the dysregulations in these signaling pathways can act as promising targets for anticancer drug development [89]. Previously, this plant has shown significant anticancer potential against different cancer cell lines. A combined herbal therapy comprised of *Punica granatum* and *P. roxburghii* showed strong cytotoxic activity against HepG2 hepatocellular carcinoma cell line in correlation with its antioxidant effect, hence proving the significant anticancer potential of the subject plant in the current study [90]. Not only the organic solvents extracts but the green silver nanoparticles synthesized using *P. roxburghii* leaves extracts depicted stupendous inhibitory activity against PANC-1, HCT-116 and MDA-MB-231 cell lines showing its potential to reduce the proliferation of cancerous cells [91]. During the present study, *P. roxburghii* DW extract showed mild protein kinase inhibition activity. The activity is supported by the results of a previous finding in which water extracts of *Sargassum oligocystum* depicted significant cytotoxic and antiproliferative activity against K562 and Daudi cell lines [92]. The linkage of protein kinase inhibitory activity and anticancer effects persuaded to draw the relation between previous findings and the results of the current studies. Although, a detailed study has been conducted for the first time to investigate the protein kinase inhibition potential of multiple parts of *P. roxburghii*.

4. Conclusions

Undertaken study demonstrates the polyphenolic analysis and pharmacological potential of crude extracts. Overall, leaf part showed maximum biological activities as compared to the stem and fruit parts of *P. roxburghii*. In case of leaf part DW and MeOH extracts were highly active whereas EA and nH extracts also showed some biological activities. The present study suggests that *P. roxburghii* crude extracts are a potential reservoir of phytoconstituents instigating the considerable antioxidant, antibacterial, antidiabetic, and cytotoxic compounds. These bioactive phytoconstituents could act as unique frameworks in a quest for innovative drugs. we recommend that subsequent studies should be carried out for the isolation of bioactive compounds from *P. roxburghii* (especially DW-L extract), which can be considered as potential candidates for the treatment of different ailments.

Author Contributions: A.N. executed all experimental work and prepared initial draft of the manuscript. M.Z.I.K. and M.A. compiled the data and were involved in the written work. N.A., M.K.O., A.A.-H. and W.H.A.-Q. were involved in H.A. analysis and interpretation of data. I.-u.-H. contributed in study design, supervised the execution of experiments and revised the manuscript. All authors have read and agreed to the published version of the manuscript.

Funding: This research was funded by King Saud University, Riyadh, Saudi Arabia under Researchers Support Project number (RSP-2021/374).

Institutional Review Board Statement: Not applicable.

Informed Consent Statement: Not applicable.

Data Availability Statement: Data sharing not applicable.

Acknowledgments: The authors would like to acknowledge Rizwana Aleem Qureshi, Department of Plant Sciences, Faculty of Biological Sciences, Quaid-i-Azam University Islamabad, Pakistan, for the identification of the plant, used in the current study. Authors also appreciate Researchers Support Project number (RSP-2021/374) King Saud University, Riyadh, Saudi Arabia for support.

Conflicts of Interest: Authors declare no conflict of interests.

Sample Availability: The crude extracts prepared for the current study are available from the authors.

References

1. Auddy, B.; Ferreira, M.; Blasina, F.; Lafon, L.; Arredondo, F.; Dajas, F.; Tripathi, P.; Seal, T.; Mukherjee, B. Screening of anti-oxidant activity of three Indian medicinal plants, traditionally used for the management of neurodegenerative diseases. *J. Ethnopharmacol.* **2003**, *84*, 131–138. [CrossRef]
2. Mustafa, G.; Arif, R.; Atta, A.; Sharif, S.; Jamil, A. Bioactive compounds from medicinal plants and their importance in drug discovery in Pakistan. *Matrix Sci. Pharma* **2017**, *1*, 17–26. [CrossRef]
3. Orhan, I.E. Pharmacognosy: Science of natural products in drug discovery. *Bioimpacts* **2014**, *4*, 109–110. [CrossRef] [PubMed]
4. Ahmed, H.; Irshad Khan, M.Z.; Waseem, D.; Nazli, A.; Waleed Baig, M. Phytochemical Analysis and antioxidant potential of *Ficus benghalensis* L. *J. Bioresour. Manag.* **2017**, *4*, 3–29.
5. Verma, S.; Singh, S. Current and future status of herbal medicines. *Vet. World* **2008**, *1*, 347–350. [CrossRef]
6. Sasidharan, S.; Chen, Y.; Saravanan, D.; Sundram, K.; Latha, L.Y. Extraction, isolation and characterization of bioactive compounds from plants' extracts. *Afr. J. Tradit. Complement. Altern. Med.* **2011**, *8*, 1–10. [CrossRef]
7. Yao, J.; Weng, Y.; Dickey, A.; Wang, K.Y. Plants as factories for human pharmaceuticals: Applications and challenges. *Int. J. Mol. Sci.* **2015**, *16*, 28549–28565. [CrossRef] [PubMed]
8. Cragg, G.M.; Newman, D.J. Natural products: A continuing source of novel drug leads. *Biochem. Biophys. Acta Gen. Subj.* **2013**, *1830*, 3670–3695. [CrossRef]
9. Simoens, C.; Vermorken, J.B.; Korst, A.E.; Pauwels, B.; De Pooter, C.M.; Pattyn, G.G.; Lambrechts, H.A.; Breillout, F.; Lardon, F. Cell cycle effects of vinflunine, the most recent promising Vinca alkaloid, and its interaction with radiation, in vitro. *Cancer Chemother. Pharmacol.* **2006**, *58*, 210–218. [CrossRef]
10. Mok, T.S.; Wu, Y.-L.; Thongprasert, S.; Yang, C.-H.; Chu, D.-T.; Saijo, N.; Sunpaweravong, P.; Han, B.; Margono, B.; Ichinose, Y. Gefitinib or carboplatin–paclitaxel in pulmonary adenocarcinoma. *N. Engl. J. Med.* **2009**, *361*, 947–957. [CrossRef]
11. Naeem, I.; Taskeen, A.; Mubeen, H.; Maimoona, A. Characterization of flavonols present in barks and needles of *Pinus wallichiana* and *Pinus roxburghii*. *Chem. Asian J.* **2010**, *22*, 41–47.
12. Shahwar, D.; Raza, M.A.; Saeed, A.; Riasat, M.; Chattha, F.I.; Javaid, M.; Ullah, S. Antioxidant potential of the extracts of *Putranjiva roxburghii*, *Conyza bonariensis*, *Woodfordia fruiticosa* and *Senecio chrysanthemoids*. *Afr. J. Biotechnol.* **2012**, *11*, 4288–4295.

13. Chaudhary, N.S.; Shee, C.; Islam, A.; Ahmad, F.; Yernool, D.; Kumar, P.; Sharma, A.K. Purification and characterization of a trypsin inhibitor from *Putranjiva roxburghii* seeds. *Phytochemistry* **2008**, *69*, 2120–2126. [CrossRef]
14. Tripathi, N.; Kumar, N. *Putranjiva roxburghii* oil—A potential herbal preservative for peanuts during storage. *J. Stored Prod. Res.* **2007**, *43*, 435–442. [CrossRef]
15. Vidhya, U.; Nishteswar, K. Putranjiva-a herb for pumsavana (male progeny facilitator)? *Int. J. Ayurveda Pharma Res.* **2015**, *3*, 11–16.
16. Mishra, S.; Kumar, S.; Darokar, M.P.; Shanker, K. Novel bioactive compound from the bark of *Putranjiva roxburghii* Wall. *Nat. Prod. Res.* **2019**, *35*, 1–3. [CrossRef]
17. Garg, H.; Mitra, C. *Putranjiva roxburghii* wall.—II: Triterpenes of the trunk bark. *Phytochemistry* **1968**, *7*, 2053–2055. [CrossRef]
18. Abhimanyu, K.K.; Ravindra, C.S.; Avanapu, R.S. A validated HPTLC method for the quantification of friedelin in *Putranjiva roxburghii* Wall extracts and in polyherbal formulations. *Bull. Fac. Pharm. Cairo Univ.* **2017**, *55*, 79–84. [CrossRef]
19. Garg, H.; Mitra, C. Roxburghonic acid—A friedelane triterpenoid keto acid of the leaf of *Putranjiva roxburghii*. *Phytochemistry* **1971**, *10*, 865–869. [CrossRef]
20. Sen, S.; Chakraborty, R.; De, B. Challenges and opportunities in the advancement of herbal medicine: India's position and role in a global context. *J. Herb. Med.* **2011**, *1*, 67–75. [CrossRef]
21. Fatima, H.; Khan, K.; Zia, M.; Ur-Rehman, T.; Mirza, B.; Haq, I.-U. Extraction optimization of medicinally important metabolites from *Datura innoxia* Mill.: An in vitro biological and phytochemical investigation. *BMC Complement. Altern. Med.* **2015**, *15*, 1–18. [CrossRef] [PubMed]
22. Khan, M.Z.I.; Zahra, S.S.; Ahmed, M.; Fatima, H.; Mirza, B.; Haq, I.-u.; Khan, S.U. Polyphenolic profiling of *Ipomoea carnea* Jacq. by HPLC-DAD and its implications in oxidative stress and cancer. *Nat. Prod. Res.* **2019**, *33*, 2099–2104. [CrossRef]
23. Bibi, G.; Ullah, N.; Mannan, A.; Mirza, B. Antitumor, cytotoxic and antioxidant potential of *Aster thomsonii* extracts. *Afr. J. Pharmacy Pharmacol.* **2011**, *5*, 252–258.
24. Ahmed, M.; Fatima, H.; Qasim, M.; Gul, B. Polarity directed optimization of phytochemical and *in vitro* biological potential of an indigenous folklore: *Quercus dilatata* Lindl. ex Royle. *BMC Complement. Altern. Med.* **2017**, *17*, 1–16. [CrossRef] [PubMed]
25. Zahra, S.S.; Ahmed, M.; Qasim, M.; Gul, B.; Zia, M.; Mirza, B.; Haq, I.-u. Polarity based characterization of biologically active extracts of *Ajuga bracteosa* Wall. ex Benth. and RP-HPLC analysis. *BMC Complement. Altern. Med.* **2017**, *17*, 443–456. [CrossRef] [PubMed]
26. Nasir, B.; Ahmad, M.; Zahra, S.S.; Fatima, H.; Ur-Rehman, T. Pharmacological evaluation of *Fumaria indica* (hausskn.) Pugsley; a traditionally important medicinal plant. *Pak. J. Bot.* **2017**, *49*, 119–132.
27. Nair, S.S.; Kavrekar, V.; Mishra, A. In vitro studies on alpha amylase and alpha glucosidase inhibitory activities of selected plant extracts. *Eur. J. Exp. Biol.* **2013**, *3*, 128–132.
28. Azwanida, N. A review on the extraction methods use in medicinal plants, principle, strength and limitation. *Med. Aromat. Plants* **2015**, *4*, 2167-0412.
29. Trusheva, B.; Trunkova, D.; Bankova, V. Different extraction methods of biologically active components from propolis: A preliminary study. *Chem. Cent. J.* **2007**, *1*, 1–4. [CrossRef]
30. Jha, A.K.; Prasad, K.; Prasad, K.; Kulkarni, A. Plant system: Nature's nanofactory. *Colloids Surf. B* **2009**, *73*, 219–223. [CrossRef]
31. Saraf, S. Applications of novel drug delivery system for herbal formulations. *Fitoterapia* **2010**, *81*, 680–689.
32. Keshav, P.; Goyal, D.K.; Kaur, S. GC–MS screening and antiparasitic action of *Putranjiva roxburghii* leaves against sensitive and resistant strains of *Leishmania donovani*. *J. Parasit. Dis.* **2021**, *45*, 1–12. [CrossRef] [PubMed]
33. Kumar, S.; Yadav, M.; Yadav, A.; Yadav, J. Impact of spatial and climatic conditions on phytochemical diversity and in vitro antioxidant activity of Indian *Aloe vera* (L.) Burm. f. *Afr. J. Bot.* **2017**, *111*, 50–59. [CrossRef]
34. Dai, J.; Mumper, R.J. Plant phenolics: Extraction, analysis and their antioxidant and anticancer properties. *Molecules* **2010**, *15*, 7313–7352. [CrossRef] [PubMed]
35. Li, A.-N.; Li, S.; Zhang, Y.-J.; Xu, X.-R.; Chen, Y.-M.; Li, H.-B. Resources and biological activities of natural polyphenols. *Nutrients* **2014**, *6*, 6020–6047. [CrossRef]
36. Parr, A.J.; Bolwell, G.P. Phenols in the plant and in man. The potential for possible nutritional enhancement of the diet by modifying the phenols content or profile. *J. Sci. Food Agric.* **2000**, *80*, 985–1012.
37. Croft, K.D. The Chemistry and biological effects of flavonoids and phenolic acids. *Ann. N. Y. Acad. Sci.* **1998**, *854*, 435–442. [CrossRef] [PubMed]
38. Cos, P.; Ying, L.; Calomme, M.; Hu, J.P.; Cimanga, K.; Van Poel, B.; Pieters, L.; Vlietinck, A.J.; Berghe, D.V. Structure−activity relationship and classification of flavonoids as inhibitors of xanthine oxidase and superoxide scavengers. *J. Nat. Prod.* **1998**, *61*, 71–76. [CrossRef]
39. Al-Snafi, A.E. Phenolics and flavonoids contents of medicinal plants, as natural ingredients for many therapeutic purposes-a review. *IOSR J. Pharm.* **2020**, *10*, 42–81.
40. Kubo, I.; Fujita, K.-I.; Nihei, K.-I.; Nihei, A. Antibacterial activity of akyl gallates against *Bacillus subtilis*. *J. Agric. Food Chem.* **2004**, *52*, 1072–1076. [CrossRef]
41. Chen, F.; Long, X.; Yu, M.; Liu, Z.; Liu, L.; Shao, H. Phenolics and antifungal activities analysis in industrial crop Jerusalem artichoke (*Helianthus tuberosus* L.) leaves. *Ind. Crops Prod.* **2013**, *47*, 339–345. [CrossRef]
42. Uozaki, M.; Yamasaki, H.; Katsuyama, Y.; Higuchi, M.; Higuti, T.; Koyama, A.H. Antiviral effect of octyl gallate against DNA and RNA viruses. *Antivir. Res.* **2007**, *73*, 85–91. [CrossRef]

43. You, B.R.; Moon, H.J.; Han, Y.H.; Park, W.H. Gallic acid inhibits the growth of HeLa cervical cancer cells via apoptosis and/or necrosis. *Food Chem. Toxicol.* **2010**, *48*, 1334–1340. [CrossRef]
44. Kumar, S.; Pandey, A.K. Chemistry and biological activities of flavonoids: An overview. *Sci. World J.* **2013**, *2013*, 1–16. [CrossRef] [PubMed]
45. Pereira, D.M.; Valentão, P.; Pereira, J.A.; Andrade, P.B. Phenolics: From chemistry to biology. *Molecules* **2009**, *14*, 2202–2211. [CrossRef]
46. Jucá, M.M.; Cysne Filho, F.M.S.; de Almeida, J.C.; Mesquita, D.d.S.; Barriga, J.R.d.M.; Dias, K.C.F.; Barbosa, T.M.; Vasconcelos, L.C.; Leal, L.K.A.M.; Ribeiro, J.E. Flavonoids: Biological activities and therapeutic potential. *Nat. Prod. Res.* **2020**, *34*, 692–705. [CrossRef] [PubMed]
47. Xu, S.-J.; Wang, X.; Wang, T.-Y.; Lin, Z.-Z.; Hu, Y.-J.; Huang, Z.-L.; Yang, X.-J.; Xu, P. Flavonoids from *Rosa roxburghii* Tratt prevent reactive oxygen species-mediated DNA damage in thymus cells both combined with and without PARP-1 expression after exposure to radiation in vivo. *Aging* **2020**, *12*, 16368–16389. [CrossRef]
48. Ye, S.; Shao, Q.; Zhang, A. *Anoectochilus roxburghii*: A review of its phytochemistry, pharmacology, and clinical applications. *J. Ethnopharmacol.* **2017**, *209*, 184–202. [CrossRef]
49. Gupta, M. A review of pharmacological properties, pharmacognosy and therapeutic actions of *Putranjiva roxburghii* Wall. (Putranjiva). *Int. J. Herb. Med.* **2016**, *4*, 104–108.
50. Chen, J.H.; Ho, C.-T. Antioxidant activities of caffeic acid and its related hydroxycinnamic acid compounds. *J. Agric. Food Chem.* **1997**, *45*, 2374–2378. [CrossRef]
51. Jiang, R.-W.; Lau, K.-M.; Hon, P.-M.; Mak, T.C.; Woo, K.-S.; Fung, K.-P. Chemistry and biological activities of caffeic acid derivatives from *Salvia miltiorrhiza*. *Curr. Med. Chem.* **2005**, *12*, 237–246. [CrossRef]
52. Espíndola, K.M.M.; Ferreira, R.G.; Narvaez, L.E.M.; Silva Rosario, A.C.R.; da Silva, A.H.M.; Silva, A.G.B.; Vieira, A.P.O.; Monteiro, M.C. Chemical and pharmacological aspects of caffeic acid and its activity in hepatocarcinoma. *Front. Oncol.* **2019**, *9*, 541–551. [CrossRef]
53. Wang, G.-F.; Shi, L.-P.; Ren, Y.-D.; Liu, Q.-F.; Liu, H.-F.; Zhang, R.-J.; Li, Z.; Zhu, F.-H.; He, P.-L.; Tang, W. Anti-hepatitis B virus activity of chlorogenic acid, quinic acid and caffeic acid in vivo and in vitro. *Antivir. Res.* **2009**, *83*, 186–190. [CrossRef] [PubMed]
54. Chao, P.-C.; Hsu, C.-C.; Yin, M.-C. Anti-inflammatory and anti-coagulatory activities of caffeic acid and ellagic acid in cardiac tissue of diabetic mice. *Nutr. Metab.* **2009**, *6*, 1–8. [CrossRef] [PubMed]
55. Kikuzaki, H.; Hisamoto, M.; Hirose, K.; Akiyama, K.; Taniguchi, H. Antioxidant properties of ferulic acid and its related compounds. *J. Agric. Food Chem.* **2002**, *50*, 2161–2168. [CrossRef]
56. Ogiwara, T.; Satoh, K.; Kadoma, Y.; Murakami, Y.; Unten, S.; Atsumi, T.; Sakagami, H.; Fujisawa, S. Radical scavenging activity and cytotoxicity of ferulic acid. *Anticancer Res.* **2002**, *22*, 2711–2717.
57. Kim, J.K.; Park, S.U. A recent overview on the biological and pharmacological activities of ferulic acid. *EXCLI J.* **2019**, *18*, 132–138.
58. Ou, S.; Kwok, K.C. Ferulic acid: Pharmaceutical functions, preparation and applications in foods. *J. Sci. Food Agric.* **2004**, *84*, 1261–1269. [CrossRef]
59. Kiliç, I.; Yeşiloğlu, Y. Spectroscopic studies on the antioxidant activity of p-coumaric acid. *Biomol. Spectrosc.* **2013**, *115*, 719–724. [CrossRef] [PubMed]
60. Lou, Z.; Wang, H.; Rao, S.; Sun, J.; Ma, C.; Li, J. p-Coumaric acid kills bacteria through dual damage mechanisms. *Food Control* **2012**, *25*, 550–554. [CrossRef]
61. Arruda, C.; Ribeiro, V.P.; Mejia, J.A.A.; Almeida, M.O.; Goulart, M.O.; Candido, A.C.B.B.; dos Santos, R.A.; Magalhaes, L.G.; Martins, C.H.G.; Bastos, J.K. Green propolis: Cytotoxic and leishmanicidal activities of artepillin C, p-coumaric acid, and their degradation products. *Rev. Bras. Farmacogn.* **2020**, *30*, 169–176. [CrossRef]
62. Itoh, A.; Isoda, K.; Kondoh, M.; Kawase, M.; Watari, A.; Kobayashi, M.; Tamesada, M.; Yagi, K. Hepatoprotective effect of syringic acid and vanillic acid on CCl_4-induced liver injury. *Biol. Pharm. Bull.* **2010**, *33*, 983–987. [CrossRef] [PubMed]
63. Tai, A.; Sawano, T.; Ito, H. Antioxidative properties of vanillic acid esters in multiple antioxidant assays. *Biosci. Biotechnol. Biochem.* **2012**, *76*, 314–318. [CrossRef] [PubMed]
64. Srinivasulu, C.; Ramgopal, M.; Ramanjaneyulu, G.; Anuradha, C.; Kumar, C.S. Syringic acid (SA)—A review of its occurrence, biosynthesis, pharmacological and industrial importance. *Biomed. Pharmacother.* **2018**, *108*, 547–557. [CrossRef]
65. Fernández, I.S.; Cuevas, P.; Angulo, J.; López-Navajas, P.; Canales-Mayordomo, Á.; González-Corrochano, R.; Lozano, R.M.; Valverde, S.; Jiménez-Barbero, J.; Romero, A. Gentisic acid, a compound associated with plant defense and a metabolite of aspirin, heads a new class of in vivo fibroblast growth factor inhibitors. *J. Biol. Chem.* **2010**, *285*, 11714–11729. [CrossRef] [PubMed]
66. Abedi, F.; Razavi, B.M.; Hosseinzadeh, H. A review on gentisic acid as a plant derived phenolic acid and metabolite of aspirin: Comprehensive pharmacology, toxicology, and some pharmaceutical aspects. *Phytother. Res.* **2020**, *34*, 729–741. [CrossRef] [PubMed]
67. Yang, C.S.; Lee, M.-J.; Chen, L. Human salivary tea catechin levels and catechin esterase activities: Implication in human cancer prevention studies. *Biomark. Prev.* **1999**, *8*, 83–89.
68. Veluri, R.; Weir, T.L.; Bais, H.P.; Stermitz, F.R.; Vivanco, J.M. Phytotoxic and antimicrobial activities of catechin derivatives. *J. Agric. Food Chem.* **2004**, *52*, 1077–1082. [CrossRef] [PubMed]

69. Venkatakrishnan, K.; Chiu, H.-F.; Cheng, J.-C.; Chang, Y.-H.; Lu, Y.-Y.; Han, Y.-C.; Shen, Y.-C.; Tsai, K.-S.; Wang, C.-K. Comparative studies on the hypolipidemic, antioxidant and hepatoprotective activities of catechin-enriched green and oolong tea in a double-blind clinical trial. *Food Funct.* **2018**, *9*, 1205–1213. [CrossRef]
70. Ghayur, M.N.; Khan, H.; Gilani, A.H. Antispasmodic, bronchodilator and vasodilator activities of (+)-catechin, a naturally occurring flavonoid. *Arch. Pharm. Res.* **2007**, *30*, 970–975. [CrossRef]
71. Dong, X.; Fu, J.; Yin, X.; Cao, S.; Li, X.; Lin, L.; Huyiligeqi; Ni, J. Emodin: A review of its pharmacology, toxicity and pharmacokinetics. *Phytother. Res.* **2016**, *30*, 1207–1218. [CrossRef]
72. Lin, C.-C.; Chang, C.-H.; Yang, J.-J.; Namba, T.; Hattori, M. Hepatoprotective effects of emodin from *Ventilago leiocarpa*. *J. Ethnopharmacol.* **1996**, *52*, 107–111. [CrossRef]
73. Wang, R.; He, R.; Li, Z.; Lin, X.; Wang, L. HPLC-Q-Orbitrap-MS/MS phenolic profiles and biological activities of extracts from roxburgh rose (*Rosa roxburghii* Tratt.) leaves. *Arab. J. Chem.* **2021**, *14*, 103257. [CrossRef]
74. Kedare, S.B.; Singh, R. Genesis and development of DPPH method of antioxidant assay. *J. Food Sci. Technol.* **2011**, *48*, 412–422. [CrossRef]
75. Mishra, K.; Ojha, H.; Chaudhury, N.K. Estimation of antiradical properties of antioxidants using DPPH assay: A critical review and results. *Food Chem.* **2012**, *130*, 1036–1043. [CrossRef]
76. Abdel-Hameed, E.-S.S. Total phenolic contents and free radical scavenging activity of certain Egyptian Ficus species leaf samples. *Food Chem.* **2009**, *114*, 1271–1277. [CrossRef]
77. Perumal Samy, R.; Gopalakrishnakone, P. Therapeutic potential of plants as anti-microbials for drug discovery. *Evid. Based Complement. Alternat. Med.* **2010**, *7*, 283–294. [CrossRef] [PubMed]
78. Arif, T.; Bhosale, J.; Kumar, N.; Mandal, T.; Bendre, R.; Lavekar, G.; Dabur, R. Natural products–antifungal agents derived from plants. *J. Asian Nat. Prod. Res.* **2009**, *11*, 621–638. [CrossRef]
79. Sher, A. Antimicrobial activity of natural products from medicinal plants. *Gomal J. Med. Sci.* **2009**, *7*, 1–17.
80. Ali, H.; Houghton, P.; Soumyanath, A. α-Amylase inhibitory activity of some Malaysian plants used to treat diabetes; with particular reference to *Phyllanthus amarus*. *J. Ethnopharmacol.* **2006**, *107*, 449–455. [CrossRef]
81. Ramkumar, K.M.; Thayumanavan, B.; Palvannan, T.; Rajaguru, P. Inhibitory effect of *Gymnema montanum* leaves on α-glucosidase activity and α-amylase activity and their relationship with polyphenolic content. *Med. Chem. Res.* **2010**, *19*, 948–961. [CrossRef]
82. Mai, T.T.; Thu, N.N.; Tien, P.G.; Van Chuyen, N. Alpha-glucosidase inhibitory and antioxidant activities of Vietnamese edible plants and their relationships with polyphenol contents. *J. Nutr. Sci Vitaminol.* **2007**, *53*, 267–276. [CrossRef] [PubMed]
83. Benalla, W.; Bellahcen, S.; Bnouham, M. Antidiabetic medicinal plants as a source of alpha glucosidase inhibitors. *Curr. Diabetes Rev.* **2010**, *6*, 247–254. [CrossRef]
84. Derosa, G.; Maffioli, P. α-Glucosidase inhibitors and their use in clinical practice. *Arch. Med. Sci. AMS* **2012**, *8*, 899–906. [CrossRef] [PubMed]
85. Jo, S.; Ka, E.; Lee, H. Comparison of antioxidant potential and rat intestinal α-glucosidases inhibitory activities of quercetin, rutin, and isoquercetin. *Int. J. Appl. Res. Nat. Prod.* **2009**, *2*, 52–60.
86. Elya, B.; Handayani, R.; Sauriasari, R.; Hasyyati, U.S.; Permana, I.T.; Permatasari, Y.I. Antidiabetic activity and phytochemical screening of extracts from Indonesian plants by inhibition of alpha amylase, alpha glucosidase and dipeptidyl peptidase IV. *Pak. J. Biol. Sci.* **2015**, *18*, 279. [CrossRef]
87. Sarah, Q.S.; Anny, F.C.; Mir, M. Brine shrimp lethality assay. *Bangladesh J. Pharmacol.* **2017**, *12*, 186–189. [CrossRef]
88. Raghavendra, H.; Prashith, K.T.; Valleesha, N.; Sudharshan, S.; Chinmaya, A. Screening for cytotoxic activity of methanol extract of *Putranjiva roxburghii* Wall (Euphorbiaceae) seeds. *Pharmacogn. J.* **2010**, *2*, 335–337. [CrossRef]
89. Fabbro, D.; Ruetz, S.; Buchdunger, E.; Cowan-Jacob, S.W.; Fendrich, G.; Liebetanz, J.; Mestan, J.; O'Reilly, T.; Traxler, P.; Chaudhuri, B. Protein kinases as targets for anticancer agents: From inhibitors to useful drugs. *Pharmacol. Ther.* **2002**, *93*, 79–98. [CrossRef]
90. Kaur, P.; Mehta, R.G.; Singh, B.; Arora, S. Development of aqueous-based multi-herbal combination using principal component analysis and its functional significance in HepG2 cells. *BMC Complement. Altern. Med.* **2019**, *19*, 1–17. [CrossRef]
91. Balkrishna, A.; Sharma, V.K.; Das, S.K.; Mishra, N.; Bisht, L.; Joshi, A.; Sharma, N. Characterization and anti-cancerous effect of *Putranjiva roxburghii* seed extract mediated silver nanoparticles on human colon (HCT-116), pancreatic (PANC-1) and breast (MDA-MB 231) cancer cell lines: A comparative study. *Int. J. Nanomed.* **2020**, *15*, 573. [CrossRef] [PubMed]
92. Zandi, K.; Ahmadzadeh, S.; Tajbakhsh, S.; Rastian, Z.; Yousefi, F.; Farshadpour, F.; Sartavi, K. Anticancer activity of Sargassum oligocystum water extract against human cancer cell lines. *Eur. Rev. Med. Pharmacol. Sci.* **2010**, *14*, 669–673. [PubMed]

molecules

Article

Profiling of Antifungal Activities and In Silico Studies of Natural Polyphenols from Some Plants

Beenish Khanzada [1,2], Nosheen Akhtar [3,*] Mohammad K. Okla [4], Saud A. Alamri [4], Abdulrahman Al-Hashimi [4], Muhammad Waleed Baig [5], Samina Rubnawaz [2], Hamada AbdElgawad [6], Abdurahman H. Hirad [4], Ihsan-Ul Haq [5] and Bushra Mirza [2,*]

1 Institute of Biochemistry, University of Sindh, Jamshoro 76080, Pakistan; beenish@usindh.edu.pk
2 Department of Biochemistry, Quaid-i-Azam University, Islamabad 45320, Pakistan; samina.r.nawaz@gmail.com
3 Department of Biological Sciences, National University of Medical Sciences, Rawalpindi 46000, Pakistan
4 Botany and Microbiology Department, College of Science, King Saud University, Riyadh 11451, Saudi Arabia; malokia@ksu.edu.sa (M.K.O.); saualamri@ksu.edu.sa (S.A.A.); aalhashimi@ksu.edu.sa (A.A.-H.); hiirad1@gmail.com (A.H.H.)
5 Department of Pharmacy, Quaid-i-Azam University, Islamabad 45320, Pakistan; mwbg7@yahoo.com (M.W.B.); ihaq@qau.edu.pk (I.-U.H.)
6 Integrated Molecular Plant Physiology Research, Department of Biology, University of Antwerp, 2020 Antwerpen, Belgium; hamada.abdelgawad@uantwerpen.be
* Correspondence: nosheenakhtar@numspak.edu.pk (N.A.); bushramirza@qau.edu.pk (B.M.); Tel.: +92-3125648441 (N.A.)

Citation: Khanzada, B.; Akhtar, N.; Okla, M.K.; Alamri, S.A.; Al-Hashimi, A.; Baig, M.W.; Rubnawaz, S.; AbdElgawad, H.; Hirad, A.H.; Haq, I.-U.; et al. Profiling of Antifungal Activities and In Silico Studies of Natural Polyphenols from Some Plants. *Molecules* **2021**, *26*, 7164. https://doi.org/10.3390/molecules26237164

Academic Editors: Ana Paula Duarte, Ângelo Luís and Eugenia Gallardo

Received: 27 October 2021
Accepted: 22 November 2021
Published: 26 November 2021

Abstract: A worldwide increase in the incidence of fungal infections, emergence of new fungal strains, and antifungal resistance to commercially available antibiotics indicate the need to investigate new treatment options for fungal diseases. Therefore, the interest in exploring the antifungal activity of medicinal plants has now been increased to discover phyto-therapeutics in replacement to conventional antifungal drugs. The study was conducted to explore and identify the mechanism of action of antifungal agents of edible plants, including *Cinnamomum zeylanicum, Cinnamomum tamala, Amomum subulatum, Trigonella foenumgraecum, Mentha piperita, Coriandrum sativum, Lactuca sativa,* and *Brassica oleraceae var. italica.* The antifungal potential was assessed via the disc diffusion method and, subsequently, the extracts were assessed for phytochemicals and total antioxidant activity. Potent polyphenols were detected using high-performance liquid chromatography (HPLC) and antifungal mechanism of action was evaluated in silico. *Cinnamomum zeylanicum* exhibited antifungal activity against all the tested strains while all plant extracts showed antifungal activity against *Fusarium solani.* Rutin, kaempferol, and quercetin were identified as common polyphenols. In silico studies showed that rutin displayed the greatest affinity with binding pocket of fungal 14-alpha demethylase and nucleoside diphosphokinase with the binding affinity (K_d, −9.4 and −8.9, respectively), as compared to terbinafine. Results indicated that *Cinnamomum zeylanicum* and *Cinnamomum tamala* exert their antifungal effect possibly due to kaempferol and rutin, respectively, or possibly by inhibition of nucleoside diphosphokinase (NDK) and 14-alpha demethylase (CYP51), while *Amomum subulatum* and *Trigonella foenum graecum* might exhibit antifungal potential due to quercetin. Overall, the study demonstrates that plant-derived products have a high potential to control fungal infections.

Keywords: antifungal; edible plants; molecular docking; antioxidant activities; polyphenols

1. Introduction

Fungal infections are continuously on the rise and one of the major causes of morbidity and mortality, particularly in immune-compromised patients [1,2]. Among the most predominant fungal infections are those caused by *Candida, Aspergillus, Fusarium species* (spp.), and *Mucor* spp. [2]. A fungus, such as *Fusarium* spp., produces mycotoxins that can cause mycotoxicosis in humans upon ingestion of food, having a colony of this fungus. *Fusarium solani* induces skin lesions and onychomycosis [3]. *Aspergillus* spp. mediated aspergillosis is increasing among patients undergoing chemotherapy and in patients with low immunity [4]. Moreover, *Mucor* spp. mediated mucormycosis is also widely reported [5]. For the treatment of fungal infections, four types of antibiotics are often offered, i.e., azole (Fluconazole), polyenes (Amphotericin B), echinocandins (caspofungin), flucytosine (5 fluorocytosine) [6], all offering a different mode of action. Terbinafine inhibits squalene epoxidase (fungal cell wall enzyme involved in ergosterol biosynthesis pathway) while azoles mostly inhibit 14-alpha demethylase [7,8]. Despite increased awareness and improved treatment strategies, drug resistance among fungal pathogens is enduring to develop, leading serious threat to public health and healthcare systems, worldwide. Dependence on antifungal antibiotics and their recurrent doses might lead to the development of resistance, as recently reported by Doung et al. in the case of itraconazole resistant isolates of *Aspergillus flavus* [9]. Moreover, many antifungal agents have been reported for complications in host tissue [8]. For instance, Aspergillosis is mostly treated with azoles but hepatotoxicity and visual disturbance are seen as side effects [10]. Hence, there is an urgent need to find novel agents to treat fungal infections with greater antifungal activity and fewer side effects.

To protect themselves against pathogens, plants produce and exude a myriad of secondary metabolites, which play important roles as defense mechanisms against various infections. The approach of traditional medicinal plant therapy is preferable due to fewer side effects and better efficacy [11]. The use of phytochemicals, either alone or in combination with conventional drugs, could be a better solution for fungal infections, due to reduced toxicity and minimum environmental impact [12]. In this connection, alcoholic extracts of medicinal plants could offer a better extraction, as some of the alcoholic extracts have been found to act as a more functional antifungal drug than conventional antibiotics [13]. Antifungal activity of aqueous extracts has also been observed and reported extensively [13]. Among phytochemicals, polyphenols have been reported to have antifungal effects against various fungal pathogens either by plasma membrane disruption, inhibition of cell wall and DNA/RNA/protein synthesis, and mitochondrial dysfunction [6]. For instance, quercetin, myricetin, and naringenin exhibit antifungal effects against *Candida albicans* and *Saccharomyces cerevisiae* [14]. Elagic acid and caffeic acid are recently reported to inhibit *Candida auris* by modifying fungal cell wall [15]. Another polyphenol, curcumin has shown a synergistic effect with itraconazole, co-trimoxazole, and amphotericin B by enhancing the reactive oxygen species production [16]. Nowadays, in silico tools are being used as an efficient method to predict the anti-microbial efficacy of phytobiotics [17]. In this regard, target identification for specific polyphenol should be focused for developing plant based drugs. For instance, superoxide dismutase, catalase, and isocitrate lyase have been proposed as molecular targets for curcumin via in silico studies [16]. Likewise, other plant polyphenols should be assessed for targeting specific fungal enzymes.

Hence, assessment of the medicinal plants for antifungal potential should be prioritized, and the mechanism of action should be focused to discover novel phyto-therapeutic agents. In this contribution, we selected eight plants, including *Cinnamomum zeylanicum (C. zeylanicum), Cinnamomum tamala (C. tamala), Amomum subulatum (A. subulatum), Trigonella foenumgraecum (T. foenumgraecum), Mentha piperita (M. piperita), Coriandrum sativum (C. sativum), Lactuca sativa (L. sativa),* and *Brassica oleraceae var. italica (B. oleraceae)*) and evaluated their antifungal potential. These common plants were selected due to their minimal/nontoxicity, significant antioxidant and antimicrobial properties, and their frequent use in routine diet. Active phytochemicals of plants were quantified using high-performance liquid chromatography (HPLC). Furthermore, in silico studies were performed to comprehend the underlying antifungal mechanism of detected polyphenols.

2. Results

2.1. Percent Extract Recovery

Percent extract yield for selected herbs and spices was calculated, as shown in Table 1. Among spices, *C. zeylanicum* has the highest percent yield, while *M. piperita* presented a high percent recovery among herbs.

Table 1. Percent extract recovery of plants using methanol for extraction. Extraction was executed in triplicate and the data are presented as mean ± standard deviation (SD).

Plant Name	Percent Yield (%)
C. zeylanicum	26.51 ± 2.22
C. tamala	18.79 ± 1.22
A. subulatum	9.45 ± 1.11
T. foenumgraecum	5.075 ± 0.52
M. piperita	10.275 ± 1.11
C. sativum	5.995 ± 0.91
L. sativa	6.37 ± 0.81
B. oleraceae	7.025 ± 1.02

2.2. Antifungal Activity

The antifungal potential of extracts was assessed against four pathogenic fungal strains. Results showed that all studied plant extracts inhibited the growth of *Fusarium solani*. *C. zeylanicum*) inhibited the growth of all fungal strains with a maximum zone of inhibition (13.6 mm) against *Fusarium solani*. *A. subulatum* exhibited antifungal activity against *Fusarium solani*, *Aspergillus niger*, and *Aspergillus flavus*. Overall, extracts of *T. foenumgraecum*, *C. zeylanicum*, *C. tamala*, and *A. subulatum* showed greater antifungal activity against *Fusarium solani* than herbal extracts, i.e., *C. sativum*, *L. sativa*, *B. oleraceae*, and *M. piperita* (7.6 to 10 mm), as shown in Table 2. No other extract inhibited the growth of *Aspergillus flavus* except *C. zeylanicum* and *A. subulatum*. *M. piperita* and *T. foenumgraecum* exhibited antifungal activity against *Fusarium solani* and *Aspergillus niger* while *B. oleraceae* possessed antifungal potential against *Fusarium solani* and *Mucor spp*. *C. sativum* and *L. sativa* exhibited antifungal potential only against *Fusarium solani* (Table 2). Terbinafine, the standard antifungal drug, showed significant antifungal potential against all the fungal strains.

Table 2. Antifungal activity against tested strains at 100 µg/disc concentration. Experiments were executed in triplicate and the data are presented as mean ± standard deviation (SD).

-	Zone of Inhibition (mm)			
-	*Fusarium solani*	*Aspergillus niger*	*Aspergillus flavus*	*Mucor* Spp.
C. zeylanicum	13.0 ± 0.5	10.0 ± 0.1	8.1 ± 0.1	10.0 ± 0.1
C. tamala	11.3 ± 1.0	-	-	7.1 ± 0.1
A. subulatum	13.6 ± 0.1	7.1 ± 0.1	5.0 ± 0.0	-
T. foenumgraecum	9.3 ± 0.0	7.1 ± 0.2	-	-
M. piperita	10.0 ± 0.1	6.0 ± 0.1	-	-
C. sativum	7.6 ± 0.3	-	-	-
L. sativa	9.3 ± 0.4	-	-	-
B. oleraceae	8.3 ± 0.5	-	-	9.2 ± 0.0
Terbinafine	20 ± 0.9	22 ± 1.03	23 ± 1.0	22 ± 1.0

Cinnamomum zeylanicum (C. zeylanicum), Cinnamomum tamala (C. zamala), Amomum subulatum (A. subulatum), Trigonella foenum graecum (T. foenumgraecum), Mentha piperita (M. piperita), Coriandrum sativum (C.sativum), and *Brasicca oleraceae var. italica (B. oleraceae).*

2.3. *Antioxidant Activities*

2.3.1. Total Antioxidant Capacity and Total Reducing Power

Results of TAC showed a wide range of values exhibited by plant extracts, expressed as the number of ascorbic acid equivalents, from 5.58 to 48.4 mg vit. C eq/g extract as shown in Table 3. *C. tamala* showed maximum TAC followed by *C. zeylanicum* and *M. piperita*. *C. sativum* displayed the lowest TAC. Furthermore, total reducing power results indicated a wide range of reducing power of extracts, ranging from 8.7 to 65.04 µg vit. C eq./mg Extract. The maximum TRP values were shown by *M. piperita* followed by *C. zeylanicum* and *C. tamala* (Table 3).

Table 3. Antioxidant potential and total phenolic and flavonoid content of ethanol extracts of selected plants. Experiments were executed in triplicate and the data are presented as mean ± standard deviation (SD).

S.No	Plant Name	Total Reducing Power (Vit C Equiv mg/g Extract)	Total Antioxidant Capacity (Vit C Equiv mg/g Extract)	DPPH Free Radical Scavanging (%)		Total Phenolic Content (mg GAE/g Extract)	Total Flavonoid Content (mg QE/g Extract)
-	-	-	-	% Scavanging at 1000 ppm (%)	IC50 (mg/mL)	-	-
1	*C. zeylanicum*	63.08 ± 0.22	36.75 ± 0.63	78.82 ± 2.52	25.4	82.42 ± 5.62	23.66 ± 0.13
2	*C. tamala*	57.72 ± 0.41	48.47 ± 0.91	82.93 ± 2.11	8.681	176.51 ± 1.52	13.69 ± 0.32
3	*A. subulatum*	8.48 ± 2.42	29.97 ± 0.71	84.73 ± 3.51	283.4	24.42 ± 0.07	2.45 ± 0.42
4	*T. foenumgraecum*	9.28 ± 2.31	21.20 ± 0.82	75 ± 3.32	485.9	25.51 ± 0.34	28.79 ± 0.38
5	*M. piperita*	65.04 ± 1.11	32.73 ± 0.82	83.42 ± 3.52	124.9	136.22 ± 6.41	26.78 ± 1.11
6	*C. sativum*	15.72 ± 0.82	5.58 ± 2.12	83.52 ± 4.52	250.9	25.17 ± 0.82	11.56 ± 0.57
7	*L. sativa*	23.22 ± 0.21	14.56 ± 1.06	80.77 ± 5.52	221.6	28.63 ± 3.22	15.28 ± 1.29
8	*B. oleraceae*	8.72 ± 2.85	21.89 ± 1.11	60.99 ± 2.53	759.4	17.17 ± 3.22	2.47 ± 1.11

2.3.2. DPPH Scavenging Assay

A percentage inhibition value was used to express the DPPH free radical scavenging activities of extracts. The inhibition values ranged from 60.9 to 84.7% for ethanol extract of selected plants at 1000 µg/mL concentration, as shown in Table 3. Free radical scavenging activity was observed in the *A. subulatum*, followed by *M. piperita* and *C. sativum*. Results revealed that plants showing the lowest IC_{50} values were *C. tamala*, *C. zeylanicum*, and *M. piperita*.

2.4. *Phytochemical Investigation*

2.4.1. Total Phenolic and Flavonoid Content

The highest amount of TPC was quantified in *C. tamala* (176.5 ± 1.5 mg GAE/g Extract). Among ethanol extract of herbs, *M. piperita* showed the highest TPC (136.2 ± 6.4 mg GAE/g Extract), as presented in Table 3. Phenolic content declined in the following order *C. tamala* > *M. piperita* > *C. zeylanicum* > *A. subulatum* > *L. sativa Lactuca sativa* > *T. foenumgraecum* > *B. oleraceae* > *C. sativum*. Among spices, *T. foenumgraecum* represented the highest TFC while *M. piperita* showed the highest TFC among herbs (Table 3). The lowest TFC was detected in ethanolic extract of *A. subulatum* and *B. oleraceae*. The descending order of TFC among the extracts was *T. foenumgraecum* > *M. piperita* > *C. zeylanicum* > *L. sativum* > *C.tamala* > *C. sativum* > *B. oleraceae* > *A. subulatum*

2.4.2. Reversed-Phase HPLC Analysis

Quantitative polyphenol detection was carried out using HPLC (using a total of 18 standards), by comparing chromatograms of samples with that of standards (Table 4). A total of 13 polyphenols were detected in studied plant extracts, i.e., rutin, kaempferol, vanillic acid, quercetin, apigenin, ferulic acid, catechin, gentisic acid, syringic acid, plumbagin, caffeic acid, coumaric acid, and emodin. Rutin was found to be the most common polyphenol followed by vanillic acid, syringic acid, caffeic acid, quercetin and emodin. The highest amount of rutin (8.34 ± 0.26 μg/mg extract) and vanillic acid (2.35 ± 0.04 μg/mg extract) was found in *C. tamala*. While the maximum amount of ferulic acid (3.01 ± 0.26 μg/mg extract) and caffeic acid (0.43 ± 0.04 μg/mg extract) were quantified in *M. piperita*. Quercetin was found to be highest in *T. foenumgraecum* followed by *A. subulatum*. The representative chromatograms are shown in Figure 1.

Table 4. Identification and quantification of polyphenols (µg/mg Extract) in ethanol extracts of selected plants by HPLC. Experiments were executed in triplicate and the data are presented as mean ± standard deviation (SD).

S.N	Plant	Rutin (R)	Vanillic Acid (VA)	Quercetin (Q)	Ferulic Acid (FA)	Syringic Acid (SA)	Kaempferol (K)	Plumbagin (PL)	Apigenin (A)	Catechin (C)	Luteolin (L)	Emodin (E)	Caffeic Acid (CF)	Coumaric Acid (CA)	Gentisic Acid (GS)
1	*C. zeylanicum*	-	0.21 ± 0.02	-	-	-	0.63 ± 0.02	-	-	-	-	-		-	-
2	*C. tamala*	8.34 ± 0.26	2.35 ± 0.04	-	-	-	-	0.11 ± 0.04	-	-	-	0.54 ± 0.03		-	-
3	*A. subulatum*	-	0.21 ± 0.01	0.85 ± 0.03	-	0.2 ± 0.04	-	-	-	-	-	-		-	-
4	*T. foenumgraecum*	6.32 ± 0.03	-	1.35 ± 0.04	-	0.24 ± 0.06	-	-	0.53 ± 0.03	-	-	0.05 ± 0.002		–	-
5	*M. piperita*	4.3 ± 0.05	-	-	3.01 ± 0.26	0.22 ± 0.07	-	-	-	-	-	-	0.43 ± 0.04	-	-
6	*C. sativum*	3.44 ± 0.06	-	-	-	-	-	-	-	-	-	-	0.27 ± 0.02	-	1.01 ± 0.22
7	*L. sativa*	3.07 ± 0.11	-	-	-	-	-	-	-	0.55 ± 0.03	-	-	0.37 ± 0.04	-	-
8	*B. oleraceae*	0.84 ± 0.03	-	-	-	-	-	-	-	-	-	-	-	-	-

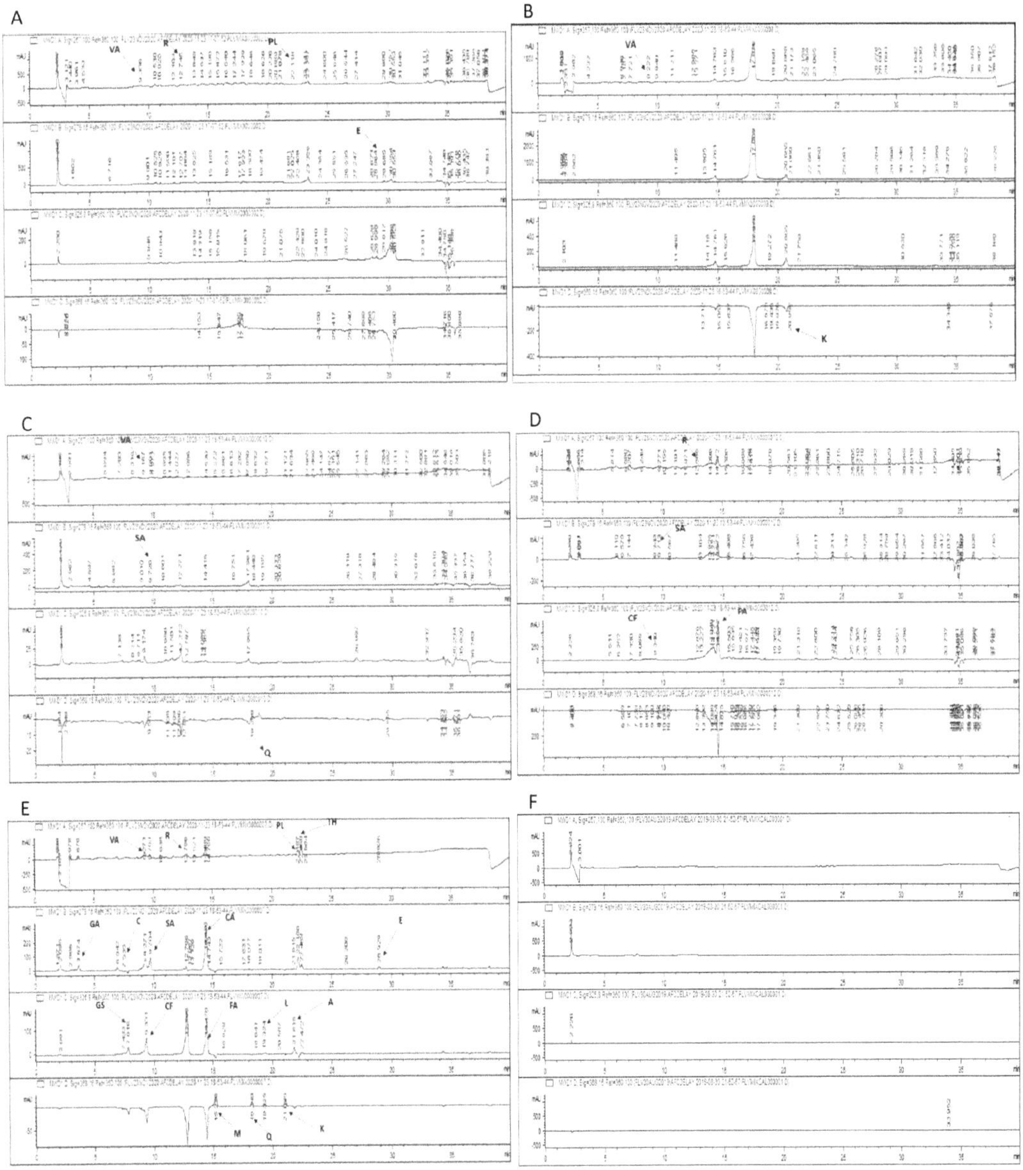

Figure 1. HPLC profile of (**A**) Ethanolic Extract of C.tamala; (**B**) Ethanolic extract of C.zeylanicum; (**C**) Ethanolic extract of A.subulatum;. (**D**) Ethanolic Extract of M.piperita; (**E**) Standard polyphenols; and (**F**) Blank. R; Rutin, VA; Vanillic acid, Q; Quercetin, FA; Ferulic acid, SA; Syringic acid, K; Kaempferol, PL; Plumbagion, C; Catechin, L; Luteolin, E; Emodin, CF; Caffeic acid, CA; Coumaric acid, and GS; Gentisic acid.

2.5. Molecular Docking Analysis

To explore the underlying mechanism of antifungal activity, the docking interactions of common polyphenols were analyzed against two important fungal enzymes. Among all polyphenols, rutin showed the greatest affinity for 14-alpha demethylase (CYP51) and nucleoside diphosphokinase (NDK) with the lowest K_d (dissociation constant) values of −9.4 and −8.9, respectively (Table 5). The lower the K_d values, the better the binding affinity of a ligand would be with its target. Results showed that rutin interacted with fungal 14-alpha demethylase by forming hydrogen bonds with tyrosine 90 (Figure 2a), while, quercetin formed two hydrogen bonds with histidine 415 and cystine 417 (Figure 2b). Binding affinity was estimated to be −9.4 and −8 for rutin and quercetin, respectively (Table 5). Binding affinity for interaction with 14-alpha demethylase, were found to be increased in the following order: rutin > catechin > quercetin > kaempferol > vanillic acid > ferulic acid. Rutin and kaempferol showed the highest binding affinity for fungal nucleoside diphosphokinase with K_d −8.9 and −8.2, respectively. Rutin and kaempferol both interacted with nucleoside diphosphokinase by forming hydrogen bonds with arginine C:19 and glutamine D:147 (Figure 2d,e), while rutin also showed interaction with arginine C:19, glutamate: 30, glycine A: 20, aspergine: 21, and serine D: 27 (Figure 2d).

Table 5. Binding affinities of polyphenols interaction with fungal proteins in terms of K_d values using molecular docking analysis.

Polyphenol	14-alphaDemethylase (CYP51)	Nucleoside Diphosphokinase (NDK)
Rutin	−9.4	−8.9
Quercetin	−8	−7.8
Kaempferol	−7.9	−8.2
Vanillic acid	−5.7	−5.6
Ferulic acid	−6.1	−5.9
Catechin	−8.1	−7.7

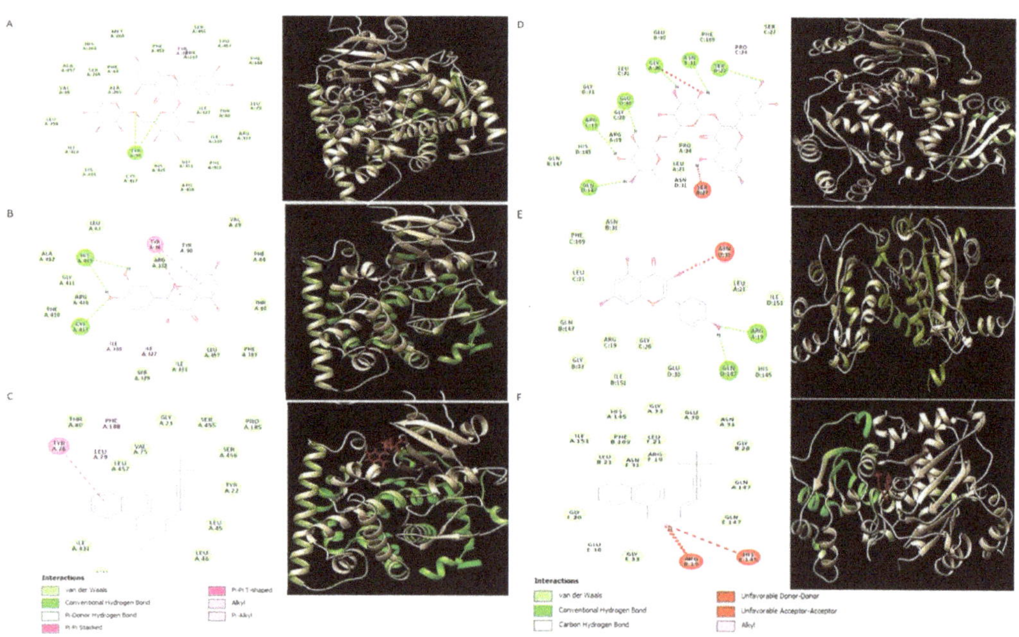

Figure 2. Graphical representation of polyphenols binding modes in fungal proteins. (**A**) Rutin with 14 alpha demethylase (CYP51). (**B**) Quercetin with CYP51. (**C**) Terbinafine with CYP51. (**D**) Rutin with nucleoside diphospho kinase (NDK). (**E**) Kaempferol with NDK. (**F**) Terbinafine with NDK.

3. Discussion

Increasing incidence of drug-resistant fungi, the emergence of new fungal strains, and the toxicity profile of antifungal drugs have led to the use of medicinal plants as potential antifungal means. Further, understanding the mechanism by which plant antifungal agents, i.e., polyphenols, interact with crucial fungal proteins should be of prime importance in identifying the molecular targets [18]. Therefore, in the present research, crude ethanol extracts of selected plants, i.e., *C. zeylanicum, C. tamala, A. subulatum, T. foenumgraecum, M. piperita, C. sativum, L. sativa*, and *B. oleraceae var.italica* were prepared and screened for their antifungal activities, and polyphenols were identified in each plant. The polyphenols are strong antioxidants that have an important role in controlling microbial diseases. The respective polyphenols were further assessed for interaction with fungal proteins, via in silico analysis. Major findings of the current research are: (i) *C. zeylanicum, C. tamala*, and *A. subulatum* displayed significant antifungal activity among studied plants; (ii) rutin, kaempferol, and quercetin were identified as the most common antifungal polyphenols in selected plants; (iii) rutin displayed the greatest affinity with binding pocket of fungal 14-alpha demethylase (CYP51) and nucleoside diphosphokinase (NDK) with the lowest K_d values as compared to standard drug terbinafine; (iv) *C. zeylanicum* is predicted to possess antifungal activity due to kaempferol (inhibition of CYP51 and NDK; (v) *A. subulatum* and *T. foenumgraecum* antifungal effect are likely due to quercetin; and (vi): rutin was the major polyphenol found in *C. tamala* and all herbs, which might be responsible for antifungal activity by either competitive or allosteric inhibition of both studied fungal enzymes.

As per results of the current study, *C. zeylanicum* showed the highest percent yield in comparison to other plant extracts. Different plants exhibited different yields, perhaps due to differences in their phytochemical composition and their variable solubility [19]. Furthermore, ethanolic extracts *of C. zeylanicum, C. tamala*, and *A. subulatum* exhibited significant antifungal activities against different studied strains. A recent report has documented the antifungal efficacy of *C. zeylanicum* bark powder, its water suspensions, and its essential oils against *Fusarium oxysporum*. [20]. The *C. zeylanicum* methanol, *n*-hexane, and aqueous extracts are also reported to exhibit inhibitory effects against *Alternaria solani* [21]. To the best of our knowledge, there is no report presenting the antifungal effects of ethanol extract of *C. zeylanicum and A. subulatum* against *Fusarium solani*, which we observed in our study. We found that *C. zeylanicum* and *A. subulatum* ethanolic extracts only moderately inhibited the growth of both *Aspergillus* species. In a previous report, methanolic extract of *A. subulatum* exhibited a considerably higher zone of inhibition (i.e., 19 mm) against *Aspergillus niger* [22]. This might be due to different experimental conditions and different solvent system used. *T. foenumgraecum* is reported to inhibit the growth of fungal strains and exhibited inhibition effects against *Fusarium oxysporum* and *Fusarium. Oxysporum* [23]. In our study, the *T. foenumgraecum* has shown inhibition potential against *Fusarium solani* and *Aspergillus niger*. Extract of *M. piperita* inhibited the growth of two fungal strains, i.e., *Aspergillus niger* and *Fusarium solani. M. piperita* extract has been already shown to cause inhibitory effects against radial fungal growth and production of aflatoxin by *Aspergillus* species [24]. Moreover, the *B. oleraceae var.italica* in our study inhibited the growth of *Fusarium solani* and *Mucor spp*.

Next, our results revealed that plants having antifungal activities exhibited excellent antioxidant potential, evaluated by free radical scavenging assay, total antioxidant capacity, and total reducing power activity. The reducing power of *C. zeylanicum* leaf and bark extracts might be due to the di- and monohydroxyl substitutions in the aromatic ring, which possess potent hydrogen donating abilities as described by Shimada et al. [25]. Plants with high phenolic content are mostly targeted by pharmacists to treat infections [26]. It indicates that polyphenols are the actual contributor to the antioxidant and antifungal activity of the studied plants. In the current investigation, potent antifungal activities of *C. tamala, C. zeylanicum, A. subulatum*, and *M. piperita* were possibly observed due to their high phenolic content. The significant antioxidant potential of these plants might be the reason for their antifungal efficacy, as plant flavonoids are reported to inhibit biofilm formation by stimulating membrane disturbances, which reduces the fungal cell size and causes leakage of intracellular components [27]. Phenols are also oxidized in response to infection into their respective quinones which can further inactivate the fungal enzymes. Polyphenols have ortho para directing groups (which tend to donate electrons) which may contribute to their antioxidant as well as antifungal properties [28]. Hence, the findings of the current investigation support the previous views that, due to antioxidant compounds such as polyphenols, the plant extracts exhibit pronounced efficacy as an antimycotoxin [29,30].

To detect the common antifungal polyphenols from plants under study, we performed HPLC analysis. Kaempferol was detected in *C. zeylanicum*. In a previous study, HPLC chromatogram showed the presence of quercetin and kaempferol in cinnamon extract [31]. Polyphenol has been reported as justifiable resources for the control of fungal biofilms [32]. Kaempferol and quercetin seemed to be the contributor to the antifungal activity of *C. zeylanicum* and *A. subulatum* ethanol extracts, respectively. In addition, rutin might be involved in the antifungal action of all other test extracts. Rutin and quercetin are reported to enhance the antifungal activity of amphotericin B [33]. Quercetin has also been reported to exhibit individual or synergic antifungal properties with fluconazole (an inhibitor of fungal fatty acid synthase) [34]. Many studies presented the rutin as the potent agent for exhibiting the antifungal responses of the extracts [35]. Although phenolic compounds exhibit variable mechanisms for antimicrobial pharmacology, many of them act by promoting damage to the function of the cell membrane or cell wall [32,36]

To explore the underlying mechanisms inhibiting the fungal growth, in silico modeling was carried out which further confirmed the inhibition of fungal enzymes by the polyphenols detected in the extracts. We evaluated the inhibition potential of detected polyphenols against two fungal enzymes, i.e., CYP51 and NDK. The CYP51 belongs to the cytochrome P450 monooxygenase superfamily and mediates a crucial step of the synthesis of ergosterol, which is a fungal-specific sterol. The NDK catalyzes the reversible exchange of the γ-phosphate between nucleoside triphosphate (NTP) and nucleoside diphosphate (NDP) [37,38]. Among different reported antifungal mechanisms, rutin, and quercetin, found in our plant extracts, seem to inhibit fungal strains via inhibition of CYP51 enzyme, same as that of azole drugs. In a previous study, rutin was among the seven plant molecules showing excellent binding energy against CYP51 [39]. In the same study, quercetin formed five hydrogen bonds with HIS468, GLY307, THR311, LYS143, and TYR143, having binding energy −7.54 kcal/mol [39]. In our study, quercetin exhibited the binding energy of −8 kcal/mol to CYP51, which is in line with the previously mentioned finding. Many studies used terbinafine as a standard antifungal agent due to its efficacy against broad spectrum of pathogenic fungi [40]. As terbinafin also works by inhibition of ergosterol biosynthesis (via inhibition of squalene epoxidase) [41], in the present study, it was used as a control to compare the antifungal potential of plant polyphenols. Interestingly the K_d values of rutin (−9.4) were much lower than that of terbinafine (−8.9) which clearly indicates high inhibition potential of rutin for fungal CYP51 as compared to terbinafine. It may be as terbinafine and rutin both inhibit fungal ergostrol biosynthesis pathway but both work by inhibiting different enzymes of same pathway. In another report, fluconazole and ketozole showed binding affinity values of −7.6 and −10, respectively, with fungal alpha demethylase [42,43]. It indicates that rutin can inhibit CYP51 either alone or in combination with the above-mentioned drugs. Kaempferol revealed its antifungal potential by low K_d values against fungal NDK, which is reported for regulation of the spore development and pathogenicity [44]. Quercetin and rutin also showed a significant binding affinity with NDK, which attributes to their dual mechanism of action. The antifungal efficacy of these polyphenols also supports the structure-activity relationship analysis, recently revealed [28]. According to this study, phenolic compounds that do not support a conjugated π system have a low electrophilicity index, and compounds having a low electrophilicity index have good antifungal activities.

Other detected polyphenols such as vanillic acid, ferulic acid, catechin, and caffeic acid also have antifungal properties [45] but they might use an antifungal mechanism other than inhibiting these two fungal enzymes tested in the present study.

4. Materials and Methods

4.1. *Plant Collection*

Spices (*C. zeylanicum, C. tamala, A. subulatum,* and *T. foenumgraecum*) and herbs (*M. piperita, C. sativum, L. sativa,* and *B. oleraceae*) were collected during September 2020 from the local market of Bharakahu, Islamabad, Pakistan. The specimens were identified by Dr. Muhammad Zafar, Department of Plant Sciences, Quaid-i-Azam University, Islamabad.

4.2. *Extract Preparation*

The herbs were washed thoroughly under running tap water and dried under shade for three weeks. Dried spices and herbs were subjected to a fine powder and stored at room temperature in air-tight containers.Then, 200 g fine powder of each plant was macerated for five days, using analytical grade ethanol (1000 mL), in air-tight glass bottles. Maceration was performed to increase the contact between plant material and solvent (Ethanol) and to soften plant's cell wall so that plant phytochemicals soluble in ethanol may be released. Sonication (after maceration) was added to further enhance the disruption of the plant cell wall and facilitate the release of phytochemicals. The extracts were filtered using Whatman #1 filter paper (Sigma, USA). Extracts were concentrated by vacuum evaporation in a rotary evaporator (Buchi, Switzerland) and dried in a vacuum hood at 40 °C, and stored at 15 °C till further use. The extracts were filtered using Whatman #1 filter paper. Extracts were concentrated by vacuum evaporation in a rotary evaporator (Buchi, Switzerland) and dried in a vacuum hood at 40 °C, and stored at 15 °C till further use.

The extracts, once dried, were weighed to find out % recovery by the following formula:

$$\text{Extract recovery}\left(\%\frac{w}{w}\right) = \frac{\text{A}}{200} \times 100 \quad (1)$$

where A = weight of dry extract.

4.3. *Antifungal Activity*

The antifungal activity of extracts was evaluated by the agar disc diffusion method, performed in triplicate [19]. The spores of test fungal strains Fusarium solani (FCBP-0291), Mucor species (FCBP-0300), Aspergillus niger (FCBP-0198), and Aspergillus flavus (FCBP-0064) were procured from the department of Pharmacy, Quaid-i-Azam University, Islamabad, Pakistan. These strains were collected in solution (0.02% v/v tween 20 in H_2O) and the turbidity was adjusted according to McFarland 0.5 turbidity standard. Then, fungal strain (100 μL) was streaked on sabouraud dextrose agar. Discs of filter paper, impregnated with test extract (5 μL; 20 mg/mL), positive control (Terbinafine, 20 μg/mL), and negative control (DMSO) were employed on agar plates and incubated (24–48 h; 28 °C). Afterward, the average diameter (mm) of the growth inhibition zone around all the discs was measured and recorded.

4.4. *Antioxidant Activities*

4.4.1. Total Reducing Power (TRP)

The reducing power assay was performed as described by Ahmed et al., with some modifications [19]. Briefly, test extract (50 μL; 20 mg/mL in DMSO) was mixed with 475 μL of phosphate buffer (0.2 mol/L, pH 6.6) and potassium ferricyanide (1% w/v in H_2O). The mixture was incubated (20 min; 50 °C) and then 500 μL of trichloroacetic acid was added. Afterward, centrifuged at room temperature (10 min). The upper layer (500 μL) was mixed with eqal volumn of distilled water and $FeCl_3$ (100 μL; 0.1% w/v in H_2O). Then, 200 μL was takenfrom this was taken and the absorbance of the was measured at 645nm using microplate reader (Biotech, Elx-800, St. Louis, MO, USA). The assay was performed in triplicate. The chemicals and standards used in the assays were purchased from Sigma-Aldrich (St. Louis, MO, USA).

4.4.2. Total Antioxidant Capacity (TAC)

Total antioxidant capacity was determined by the phosphomolybdenum method [46]. An aliquot of 10 µL extract (20 mg/mL in DMSO) was added to 190 µL of reagent solution (0.6 M sulphuric acid, 28 mM sodium phosphate, and 4 mM ammonium molybdate solution in H_2O) in 96 well plate. After incubation for 90 min at 95 °C in the water bath, the reaction mixture was cooled at room temperature and absorbance was taken at 630 nm using microplate reader (Biotech, Elx-800, USA). Both TRP and TAC were expressed as the number of mg of vitimin C (vit. C) equivalents per gram of extract (vit. C eq/g). The chemicals and standards used in the assay were purchased from Sigma-Aldrich.

4.4.3. DPPH (2,2-Diphenyl1–1-picryl-hydrazyl radical) Free Radical Scavenging

Discoloration of purple-colored DPPH was used to measure the free radical scavenging activities of extracts [19,46]. Here, 190 µL of DPPH solution (in methanol) was added in each well (96 well plate) followed by the addition of 10 µL of four different dilutions (125, 250, 500, and 1000 µg/mL) of each extract. Samples were incubated in the dark at 37 °C for 15 min and absorbance was measured at 517 nm using microplate reader (Biotech, Elx-800, St. Louis, MO, USA). The assay was performed in triplicate and ascorbic acid was used as a positive control in each experiment. Percentage free radical scavenging was derived from the given formula:

$$\% \text{ Scavenging} = \text{Absorbance of Negative control} - \text{Absorbance of extract}/\text{Absorbance of negative control} \times 100 \quad (2)$$

The inhibitory concentration at which the test sample showed 50% scavenging (IC_{50}) was determined using Graph Pad Prism 5 software.

4.5. Phytochemical Analysis

4.5.1. Total Phenolic Content (TPC)

TPC was measured by the Folin–Ciocalteu method as reported previously [47]. Briefly, 10 µL (20 mg/mL in DMSO) of each extract was mixed with Folin–Ciocalteu reagent (10 fold diluted with water). After 5 min of incubation at room temperature, 98 µL of 6% sodium bicarbonate was added. The resulting mixture was kept at 25 °C for 90 min and absorbance was measured at 630 nm using microtitre plate reader. The standard curve equation for gallic acid was obtained (y = 0.0732x − 0.0205), with R^2 value of 0.999. The assay was performed in triplicate and TPC was determined as mg of gallic acid equivalent (GAE) per gram of extract (mg GAE/g Extract). The standard gallic acid and other chemicals used in this assay were purchased from Sigma-Aldrich.

4.5.2. Total Flavonoid Content (TFC)

The TFC of the extracts was determined by aluminium chloride colorimetric assay, with modifications [19]. Briefly, from each extract stock (10 µL; 20 mg/mL in DMSO), aluminium chloride (10 µL; 10% *w*/*v* in H_2O), 1.0 M potassium acetate (10 µL), and distilled water (170 µL) were mixed and incubated at room temperature (30 min). The absorbance of the mixture was calculated at 405 nm.

The resultant TFC was expressed in microgram equivalents of quercetin per milligram Extract (µg QE/mg Extract). The standard curve equations for quercetin (Sigma-Aldrich) (y = 0.01x − 0.004) and the R^2 value of 0.996 were obtained.

4.5.3. High-Performance Liquid Chromatography (HPLC) Analysis

Polyphenols were detected and quantified using HPLC analysis of medicinal plants. HPLC system, Agilent Chem station Rev. B.02–01-SR1 (260), equipped with a Zorbex-C8 analytical column (4.6 × 250 nm, 5 µm particle size) in combination with a diode array detector (DAD; Agilent technologies, Waldbronn, Germany) was used, following previously reported method [18]. All HPLC standards were purchased from Sigma Aldrich. Wavelengths used to detect the standards were 257 nm for rutin, vanillic acid, and plumbagin, 279 nm for catechin, syringic acid, coumaric acid, and emodin, 325 nm for caffeic acid, apigenin, luteolin, gentisic acid, and ferulic acid, while quercetin and kaempferol were analyzed at 368 nm. Different polyphenols were identified and quantified by comparing the UV absorption spectra and the retention time of samples was compared with those of standards and the results were expressed as µg/mg extract.

4.6. Molecular Docking

Molecular docking was performed to predict the possible antifungal mechanisms of polyphenols, detected via HPLC analysis. Protein databank (PDB) was used to obtain 3D structures of fungal enzymes 14-alpha-demethylase and nucleoside diphosphokinase with PDB IDs 5FRB and 6K3H, respectively. A 3D structure CYP51B protein was modeled using Swiss Prot. Pubchem was used to obtain structures of ligands (i.e.: rutin, quercetin, apigenin, and kaempferol with PubChem IDs 5280805, 5280343, 5280443, and 5280863, respectively). Ligands and heteroatoms were removed and proteins were optimized and minimized using UCSF Chimera software to obtain structurally correct protein. Docking was performed using the Pyrx-virtual screening tool, binding affinities were saved as CSV files and Discovery studio was used to visualize the final protein–ligand interactions [48].

4.7. Statistical Analysis

Statistical analysis was carried out using SPSS statistics and MS Excel. Replication of each experiment was conducted thrice.

5. Conclusions

This study revealed new scientific insights about the antifungal activity of natural polyphenols and their mechanism of action. *C. zeylanicum, C. tamala,* and *A. subulatum* represent good sources of such antifungal polyphenols. Bio-formulation involving extracts from more than one plant may be necessary for effective antifungal application. Our study suggested that high-phenolic content of plant extracts is responsible for their antioxidant and fungal-inhibitory activity. These tested plants may be used for developing new, safer, and effective fungicides. Moreover, detected polyphenols can be used either alone or synergistically with the fungal antibiotics to reduce their toxic effects and to increase the antifungal efficacy.

Author Contributions: Conceptualization, B.K. and N.A.; methodology, B.K.; software, M.W.B., N.A; validation, B.K., A.A.-H. and S.A.A.; formal analysis, S.R.; investigation, B.K. and I.-U.H.; resources, B.M.; data curation, H.A. and I.-U.H.; writing—original draft preparation, B.K, N.A.; writing—review and editing, N.A.; visualization, A.H.H.; supervision, B.M. and N.A.; project administration, M.K.O.; and funding acquisition, M.K.O. All authors have read and agreed to the published version of the manuscript.

Funding: The research was funded by Project number (RSP-2021/374) King Saud University, Riyadh, Saudi Arabia.

Institutional Review Board Statement: Not applicable.

Informed Consent Statement: Not applicable.

Data Availability Statement: The datasets used and/or analyzed during the current study are available from the corresponding author on reasonable request.

Acknowledgments: The authors extend their appreciation to the Researchers Supporting Project number (RSP-2021/374) King Saud University, Riyadh, Saudi Arabia. Authors are also thankful to Muhammad Zafar, Department of Plant Sciences, Quaid-i-Azam University, Islamabad for the identification of plant specimens and Sikander Azam, Department of Bioinformatics, Quaid-i-Azam University, Islamabad for help in docking studies.

Conflicts of Interest: The authors declare no conflict of interest.

Sample Availability: Samples of plant material are available from the authors.

References

1. Bongomin, F.; Gago, S.; Oladele, R.O.; Denning, D.W. Global and multi-national prevalence of fungal diseases—estimate precision. *J. Fungi* **2017**, *3*, 57. [CrossRef] [PubMed]
2. Ganesan, K.; Guo, S.; Fayyaz, S.; Zhang, G.; Xu, B. Targeting programmed fusobacterium nucleatum Fap2 for colorectal cancer therapy. *Cancers* **2019**, *11*, 1592. [CrossRef]
3. Thomas, B.; Audonneau, N.C.; Machouart, M.; Debourgogne, A. Fusarium infections: Epidemiological aspects over 10 years in a university hospital in France. *J. Infect. Public Health* **2020**, *13*, 1089–1093. [CrossRef] [PubMed]
4. Blaize, M.; Mayaux, J.; Nabet, C.; Lampros, A.; Marcelin, A.-G.; Thellier, M.; Piarroux, R.; Demoule, A.; Fekkar, A. Fatal invasive aspergillosis and coronavirus disease in an immunocompetent patient. *Emerg. Infect. Dis.* **2020**, *26*, 1636. [CrossRef] [PubMed]
5. Venkatesh, D.; Dandagi, S.; Chandrappa, P.R.; Hema, K. Mucormycosis in immunocompetent patient resulting in extensive maxillary sequestration. *J. Oral Maxillofac. Pathol. JOMFP* **2018**, *22*, S112. [CrossRef]
6. Aboody, M.S.A.; Mickymaray, S. Anti-fungal efficacy and mechanisms of flavonoids. *Antibiotics* **2020**, *9*, 45. [CrossRef]
7. Taghipour, S.; Shamsizadeh, F.; Pchelin, I.M.; Rezaei-Matehhkolaei, A.; Mahmoudabadi, A.Z.; Valadan, R.; Ansari, S.; Katiraee, F.; Pakshir, K.; Zomorodian, K. Emergence of terbinafine resistant Trichophyton mentagrophytes in Iran, harboring mutations in the squalene epoxidase (SQLE) gene. *Infect. Drug Resist.* **2020**, *13*, 845. [CrossRef]
8. Mickymaray, S.; Alturaiki, W. Antifungal efficacy of marine macroalgae against fungal isolates from bronchial asthmatic cases. *Molecules* **2018**, *23*, 3032. [CrossRef] [PubMed]
9. Duong, T.M.N.; Nguyen, P.T.; Le, T.V.; Nguyen, H.L.P.; Nguyen, B.N.T.; Nguyen, B.P.T.; Nguyen, T.A.; Chen, S.C.-A.; Barrs, V.R.; Halliday, C.L. Drug-Resistant Aspergillus flavus Is Highly Prevalent in the Environment of Vietnam: A new challenge for the management of aspergillosis? *J. Fungi* **2020**, *6*, 296. [CrossRef] [PubMed]
10. Schauwvlieghe, A.F.; Buil, J.B.; Verweij, P.E.; Hoek, R.A.; Cornelissen, J.J.; Blijlevens, N.M.; Henriet, S.S.; Rijnders, B.J.; Brüggemann, R.J. High-dose posaconazole for azole-resistant aspergillosis and other difficult to treat mould infections. *Mycoses* **2020**, *63*, 122–130. [CrossRef]
11. Giordani, C.; Simonetti, G.; Natsagdorj, D.; Choijamts, G.; Ghirga, F.; Calcaterra, A.; Quaglio, D.; De Angelis, G.; Toniolo, C.; Pasqua, G. Antifungal activity of Mongolian medicinal plant extracts. *Nat. Prod. Res.* **2020**, *34*, 449–455. [CrossRef]
12. de Andrade Monteiro, C.; Ribeiro Alves dos Santos, J. Phytochemicals and their antifungal potential against pathogenic yeasts. *Phytochem. Hum. Health* **2019**, 1–31. [CrossRef]
13. Keikha, N.; Shafaghat, M.; Mousavia, S.M.; Moudi, M.; Keshavarzi, F. Antifungal effects of ethanolic and aqueous extracts of Vitex agnus-castus against vaginal isolates of Candida albicans. *Curr. Med. Mycol.* **2018**, *4*, 1. [CrossRef]
14. Li, K.; Xing, S.; Wang, M.; Peng, Y.; Dong, Y.; Li, X. Anticomplement and antimicrobial activities of flavonoids from Entada phaseoloides. *Nat. Prod. Commun.* **2012**, *7*, 1934578X1200700715. [CrossRef]
15. Sangta, J.; Wongkaew, M.; Tangpao, T.; Withee, P.; Haituk, S.; Arjin, C.; Sringarm, K.; Hongsibsong, S.; Sutan, K.; Pusadee, T. Recovery of Polyphenolic Fraction from Arabica Coffee Pulp and Its Antifungal Applications. *Plants* **2021**, *10*, 1422. [CrossRef]
16. Rocha, O.B.; do Carmo Silva, L.; de Carvalho Júnior, M.A.B.; de Oliveira, A.A.; de Almeida Soares, C.M.; Pereira, M. In vitro and in silico analysis reveals antifungal activity and potential targets of curcumin on *Paracoccidioides* spp. *Braz. J. Microbiol.* **2021**, *52*, 1–15. [CrossRef]
17. Rampone, S.; Pagliarulo, C.; Marena, C.; Orsillo, A.; Iannaccone, M.; Trionfo, C.; Sateriale, D.; Paolucci, M. In silico analysis of the antimicrobial activity of phytochemicals: Towards a technological breakthrough. *Comput. Methods Programs Biomed.* **2021**, *200*, 105820. [CrossRef]
18. Dahiya, P.; Purkayastha, S. Phytochemical screening and antimicrobial activity of some medicinal plants against multi-drug resistant bacteria from clinical isolates. *Indian J. Pharm. Sci.* **2012**, *74*, 443. [PubMed]
19. Ahmed, M.; Fatima, H.; Qasim, M.; Gul, B. Polarity directed optimization of phytochemical and in vitro biological potential of an indigenous folklore: Quercus dilatata Lindl. ex Royle. *Bmc Complementary Altern. Med.* **2017**, *17*, 1–16. [CrossRef]
20. Kowalska, J.; Tyburski, J.; Krzymińska, J.; Jakubowska, M. Cinnamon powder: An in vitro and in vivo evaluation of antifungal and plant growth promoting activity. *Eur. J. Plant Pathol.* **2020**, *156*, 237–243. [CrossRef]
21. Yeole, G.; Teli, N.; Kotkar, H.; Mendki, P. Cinnamomum zeylanicum extracts and their formulations control early blight of tomato. *J. Biopestic.* **2014**, *7*, 110.
22. Agnihotri, S.; Wakode, S. Antimicrobial activity of essential oil and various extracts of fruits of greater cardamom. *Indian J. Pharm. Sci.* **2010**, *72*, 657. [CrossRef]
23. Omezzine, F.; Bouaziz, M.; Daami-Remadi, M.; Simmonds, M.S.J.; Haouala, R. Chemical composition and antifungal activity of Trigonella foenum-graecum L. varied with plant ploidy level and developmental stage. *Arab. J. Chem.* **2017**, *10*, S3622–S3631. [CrossRef]
24. Skrinjar, M.; Mandic, A.; Misan, A.; Sakac, M.; Saric, L.C.; Zec, M. Effect of Mint (*Mentha piperita* L.) and Caraway (*Carum carvi* L.) on the growth of some toxigenic Aspergillus species and Aflatoxin B1 production. *Matica Srp. Proc. Nat. Sci.* **2009**, *116*, 131–139. [CrossRef]

25. Abeysekera, W.; Premakumara, G.; Ratnasooriya, W. In vitro antioxidant properties of leaf and bark extracts of ceylon cinnamon (*Cinnamomum zeylanicum* Blume). *Trop. Agric. Res.* **2013**, *24*, 128–138.
26. Prasad, K.N.; Yang, B.; Dong, X.; Jiang, G.; Zhang, H.; Xie, H.; Jiang, Y. Flavonoid contents and antioxidant activities from Cinnamomum species. *Innov. Food Sci. Emerg. Technol.* **2009**, *10*, 627–632. [CrossRef]
27. Lee, H.; Woo, E.-R.; Lee, D.G. Apigenin induces cell shrinkage in Candida albicans by membrane perturbation. *Fems Yeast Res.* **2018**, *18*, foy003. [CrossRef]
28. Appell, M.; Tu, Y.-S.; Compton, D.L.; Evans, K.O.; Wang, L.C. Quantitative structure-activity relationship study for prediction of antifungal properties of phenolic compounds. *Struct. Chem.* **2020**, *31*, 1621–1630. [CrossRef]
29. Samapundo, S.; De Meulenaer, B.; Osei-Nimoh, D.; Lamboni, Y.; Debevere, J.; Devlieghere, F. Can phenolic compounds be used for the protection of corn from fungal invasion and mycotoxin contamination during storage? *Food Microbiol.* **2007**, *24*, 465–473. [CrossRef]
30. Palumbo, J.D.; O'Keeffe, T.L.; Mahoney, N.E. Inhibition of ochratoxin A production and growth of Aspergillus species by phenolic antioxidant compounds. *Mycopathologia* **2007**, *164*, 241–248. [CrossRef]
31. Hussain, Z.; Khan, J.A.; Arshad, M.I.; Muhammad, F.; Abbas, R.Z.; Moreno, M.; Moreno Murillo, B. Comparative characterization of cinnamon, cinnamaldehyde and kaempferol for phytochemical, antioxidant and pharmacological properties using acetaminophen-induced oxidative stress mouse model. *Boletín Latinoam. Caribe Plantas Med. Aromáticas* **2021**, *20*, 339–350. [CrossRef]
32. Rocha, M.F.G.; Sales, J.A.; da Rocha, M.G.; Galdino, L.M.; de Aguiar, L.; Pereira-Neto, W.d.A.; de Aguiar Cordeiro, R.; Castelo-Branco, D.d.S.C.M.; Sidrim, J.J.C.; Brilhante, R.S.N. Antifungal effects of the flavonoids kaempferol and quercetin: A possible alternative for the control of fungal biofilms. *Biofouling* **2019**, *35*, 320–328. [CrossRef]
33. Oliveira, V.; Carraro, E.; Auler, M.; Khalil, N. Quercetin and rutin as potential agents antifungal against Cryptococcus spp. *Braz. J. Biol.* **2016**, *76*, 1029–1034. [CrossRef]
34. Bitencourt, T.A.; Komoto, T.T.; Massaroto, B.G.; Miranda, C.E.S.; Beleboni, R.O.; Marins, M.; Fachin, A.L. Trans-chalcone and quercetin down-regulate fatty acid synthase gene expression and reduce ergosterol content in the human pathogenic dermatophyte Trichophyton rubrum. *BMC Complementary Altern. Med.* **2013**, *13*, 1–6. [CrossRef]
35. Pinho, F.V.; da Cruz, L.C.; Rodrigues, N.R.; Waczuk, E.P.; Souza, C.E.; Coutinho, H.D. Phytochemical Composition, Antifungal and Antioxidant Activity of Duguetia furfuracea A. *St.-Hill.* **2016**, *2016*, 7821051.
36. Ngolong Ngea, G.L.; Qian, X.; Yang, Q.; Dhanasekaran, S.; Ianiri, G.; Ballester, A.R.; Zhang, X.; Castoria, R.; Zhang, H. Securing fruit production: Opportunities from the elucidation of the molecular mechanisms of postharvest fungal infections. *Compr. Rev. Food Sci. Food Saf.* **2021**, *20*, 2508–2533. [CrossRef]
37. Nguyen, S.; Jovcevski, B.; Pukala, T.L.; Bruning, J.B. Nucleoside selectivity of Aspergillus fumigatus nucleoside-diphosphate kinase. *FEBS J.* **2021**, *288*, 2398–2417. [CrossRef]
38. Monk, B.C.; Sagatova, A.A.; Hosseini, P.; Ruma, Y.N.; Wilson, R.K.; Keniya, M.V. Fungal Lanosterol 14α-demethylase: A target for next-generation antifungal design. *Biochim. Biophys. Acta (BBA)—Proteins Proteom.* **2020**, *1868*, 140206. [CrossRef]
39. Jadhav, A.K.; Khan, P.K.; Karuppayil, S.M. Phytochemicals as potential inhibitors of lanosterol 14 α-demethylase (CYP51) enzyme: An in silico study on sixty molecules. *Int. J. Appl. Pharm.* **2020**, 18–30. [CrossRef]
40. Pérez, A. Terbinafine: Broad new spectrum of indications in several subcutaneous and systemic and parasitic diseases. *Mycoses* **1999**, *42*, 111–114. [CrossRef]
41. Monod, M.; Feuermann, M.; Yamada, T. Terbinafine and Itraconazole Resistance in Dermatophytes. In *Dermatophytes and Dermatophytoses*; Springer Nature: Cham, Switzerland, 2021; pp. 415–429. [CrossRef]
42. Mhatre, S.; Patravale, V. Drug repurposing of triazoles against mucormycosis using molecular docking: A short communication. *Comput. Biol. Med.* **2021**, *136*, 104722. [CrossRef]
43. Godamudunage, M.P.; Lampe, J.N.; Scott, E.E. Comparison of Cytochrome P450 3A4 and 3A7 with Azole Inhibitors. *FASEB J.* **2017**, *31*, 669.5.
44. Wang, Y.; Wang, S.; Nie, X.; Yang, K.; Xu, P.; Wang, X.; Liu, M.; Yang, Y.; Chen, Z.; Wang, S. Molecular and structural basis of nucleoside diphosphate kinase–mediated regulation of spore and sclerotia development in the fungus Aspergillus flavus. *J. Biol. Chem.* **2019**, *294*, 12415–12431. [CrossRef]
45. de Jesús Joaquín-Ramos, A.; López-Palestina, C.U.; Pinedo-Espinoza, J.M.; Altamirano-Romo, S.E.; Santiago-Saenz, Y.O.; Aguirre-Mancilla, C.L.; Gutiérrez-Tlahque, J. Phenolic compounds, antioxidant properties and antifungal activity of jarilla (Barkleyanthus salicifolius [Kunth] H. Rob & Brettell). *Chil. J. Agric. Res.* **2020**, *80*, 352–360.
46. Younus, I.; Ismail, H.; Rizvi, C.B.; Dilshad, E.; Saba, K.; Mirza, B.; Tahir, M. Antioxidant, analgesic and anti-inflammatory activities of in vitro and field-grown Iceberg lettuce extracts. *J. Pharm. Pharmacogn. Res.* **2019**, *7*, 343–355.
47. Foo, S.C.; Yusoff, F.M.; Ismail, M.; Basri, M.; Yau, S.K.; Khong, N.M.; Chan, K.W.; Ebrahimi, M. Antioxidant capacities of fucoxanthin-producing algae as influenced by their carotenoid and phenolic contents. *J. Biotechnol.* **2017**, *241*, 175–183. [CrossRef]
48. Baig, M.W.; Nasir, B.; Waseem, D.; Majid, M.; Khan, M.Z.I.; Haq, I.-U. Withametelin: A biologically active withanolide in cancer, inflammation, pain and depression. *Saudi Pharm. J.* **2020**, *28*, 1526–1537. [CrossRef]

molecules

Article

Nutraceutical Content and Genetic Diversity Share a Common Pattern in New Pomegranate Genotypes

Carmen Arlotta, Valeria Toscano, Claudia Genovese *, Pietro Calderaro, Giuseppe Diego Puglia and Salvatore Antonino Raccuia

Institute for Agricultural and Forest Systems in the Mediterranean, National Research Council (ISAFOM-CNR), 95128 Catania, Italy; arlottacarmen@gmail.com (C.A.); valeria.toscano@cnr.it (V.T.); pietro.calderaro@cnr.it (P.C.); giuseppediego.puglia@cnr.it (G.D.P.); salvatoreantonino.raccuia@cnr.it (S.A.R.)

* Correspondence: claudia.genovese@cnr.it; Tel.: +39-095-613-9951

Abstract: The nutraceutical value of pomegranate in the treatment of many diseases is well-documented and is linked to its richness in phenolic compounds. This study aims to evaluate the nutraceutical and genetic diversity of novel pomegranate genotypes (G1–G5) in comparison to leading commercial pomegranate varieties, i.e., 'Wonderful', 'Primosole', 'Dente di Cavallo' and 'Valenciana'. Morphometric measurements were carried out on fruits, accompanied by chemical characterization (total phenolic content, antioxidant activity, carbohydrates and minerals) and the development of four new polymorphic SSR markers involved in the flavonoid pathway. The cultivars displayed a marked variability in the weight and shape of the fruits, as well as in the weight of the arils and juice yield. The highest level of total phenolic content and antioxidant activity was found in 'Wonderful' and G4, while the lowest was in 'Dente di Cavallo'. Furthermore, the results showed that pomegranate juice is an excellent source of minerals, especially potassium, which plays a key role in organ functioning. The new flavonoid-related markers effectively differentiated the cultivars with the same diversity pattern as morpho-chemical characterization, so the SSRs developed in the present study can be used as a rapid tool for the identification of pomegranate cultivars with relevant nutraceutical traits, such as the new genotypes investigated.

Keywords: *Punica granatum* L.; bio-agronomic traits; genetic markers; antioxidant activity; total phenolic content; carbohydrates; HPAEC-PAD; minerals; potassium; nutraceutical value

Citation: Arlotta, C.; Toscano, V.; Genovese, C.; Calderaro, P.; Puglia, G.D.; Raccuia, S.A. Nutraceutical Content and Genetic Diversity Share a Common Pattern in New Pomegranate Genotypes. *Molecules* **2022**, *27*, 389. https://doi.org/10.3390/molecules27020389

Academic Editors: Ana Paula Duarte, Ângelo Luís and Eugenia Gallardo

Received: 22 December 2021
Accepted: 5 January 2022
Published: 8 January 2022

1. Introduction

Punica granatum L. belongs to the *Punicaceae* family, and it is an ancient and appreciated fruit crop. Pomegranate is a fruit tree widely grown in several countries, especially those with Mediterranean-like climates, characterized by high exposure to sunlight, mild winters with minimal temperatures not lower than −12 °C and dry, hot summers without rain during the last stages of the fruit development. Under such conditions, the fruit can develop to its best size and with an optimal colour and sugar accumulation without the danger of splitting [1]. Pomegranates have traditionally been used for the production of fresh juice from the arils, the edible parts of the fruit, but, recently, there has been a great and increasing demand for industrial processing to obtain bottled juice, jams, oil, supplements and anti-ageing creams [2,3]. The juice represents, on average, 30–40% of the total fruit weight and is a good source of minerals (potassium, phosphorus, calcium, iron and magnesium), as well as glucose, fructose [4] and fibres. Moreover, pomegranate juice contains organic acids (citric acid, malic acid), vitamin C, vitamin E, coenzyme Q and polyphenols, such as ellagitannins (punicalagin) and anthocyanins [5]. The juice concentration, the content of sugars, the colour, the polyphenol content and the quality of the product depend on the variety and degree of ripeness of the fruits. Climatic and environmental conditions can influence these traits. Pomegranate varieties differ in their taste, ranging from sweet to sour, and this is related to the phenols, organic acids and sugars contained in the fruit. The most accepted varieties are those that have fruits with a more acidic flavour, rich in phenols with important health properties [6].

In recent years, there has been an increase in pomegranate fruit production. Pomegranate berries are considered a functional product of great benefit for the human diet due to the high nutraceutical content present in their juice. Hydrolysable tannins, anthocyanins and minerals contained in them have useful health properties [2], such as cancer prevention and therapy [7] and on chronic inflammatory diseases and cardiovascular diseases [8]. The beneficial effects and the antioxidant activity of the juice and fruit, in general, are attributed to the phenolic content that acts against free radicals, molecules highly harmful to the cells themselves. Moreover, today, the market demand comes from not only consumers but also from many foods and nutraceutical and cosmetics companies, and, for this reason, it has become increasingly important to characterize its different varieties to obtain high-quality products with economic interests.

Moreover, in the last few years in Italy, especially in the southern regions, there has been a rapid expansion of the areas cultivated with pomegranate that increased from 7 ha in 2008 to 1303 ha in 2020 [9]. In Sicily, pomegranate represents a minor fruit tree species even if it was cultivated since Arab domination, but it maintained great importance in private gardens and orchards until recent times. Thanks to uncontrolled sexual propagation that occurred from its introduction, nowadays, many local genotypes can be found in Sicily, likely well adapted to different climate conditions from the sea up to 800 m above sea level [10]. Recent studies were aimed to unveil the mechanism for pomegranate flavonoids production and accumulation [11–13]. At the same time, several investigations were carried out on genetic variability screening through molecular markers, i.e., simple-sequence repeat (SSR), allowing varietal identification [14–18]. However, all these studies relied only on neutral genes, which are not linked to market valuable traits. Therefore, the development of non-neutral markers associated with relevant traits can be highly useful for the rapid identification of new potential valuable varieties and future breeding programs. To make the most of the potential beneficial effects of new pomegranate genotypes, it is necessary to carry out wide-ranging research work of characterizing the various ecotypes.

Many studies have been published on the evaluation of the pomegranate germplasm using morphological, chemical and genetic variability, but an integrative approach integrating all these data would be beneficial for cultivar selection and improvement. In this study, an integrated morphological, chemical and molecular approach was used to characterize new pomegranate genotypes focusing on relevant nutraceutical traits, such as minerals and phenolic content.

2. Results and Discussion

2.1. Morphological Characteristics of Fruits

Morphological characteristics play an important role in consumer and market choice. Fruit size, aril yield and aril size are key traits for the fresh market and breeding programs. In this work, significant morphometric differences among the studied genotypes were found. The average fruit weight found in this study was 413.96 g, with high variability among genotypes, ranging from 255.87 g in G4 to a maximum of 825.82 g observed in WD (Table 1). Variation in fruit weight could depend on the cultivar and ecological condition [3]. These data showed that, as regards VL and WD, the average fruit weight was greater than reported in the literature [10,19], while the data of PS and DC were lower than reported by La Malfa et al. [10].

Table 1. Morphologic characteristics of fruit and peel.

Genotype	Fruit Weight (g)	Fruit Circumference (mm)	Fruit Length (mm)	Fruit Diameter (mm)	Fruit Shape (FL/FD)	Septum Number	Peel Weight (g)	Peel Yield (%)	Dry Peel Yield (%)
G1	285.72 ± 28.61 ^{cd}	271.40 ± 8.65 ^{e}	71.40 ± 2.71 ^{d}	82.60 ± 2.41 ^{de}	0.87 ± 0.02 ^{a}	6.60 ± 0.55 ^{b}	106.78 ± 10.39 ^{d}	37.52 ± 3.53 ^{c}	28.67 ± 3.49 ^{cd}
G2	385.92 ± 29.97 ^{c}	303.20 ± 10.24 ^{c}	79.64 ± 1.68 ^{c}	98.08 ± 8.55 ^{c}	0.82 ± 0.07 ^{a}	6.60 ± 0.55 ^{b}	162.15 ± 17.11 ^{c}	41.10 ± 2.22 ^{bc}	27.53 ± 1.76 ^{d}
G3	304.07 ± 34.41 ^{cd}	276.00 ± 10.84 ^{de}	74.40 ± 2.31 ^{cd}	83.20 ± 2.78 ^{de}	0.90 ± 0.03 ^{a}	5.80 ± 0.45 ^{b}	111.92 ± 11.63 ^{d}	37.06 ± 4.74 ^{c}	29.49 ± 1.87 ^{cd}
G4	255.87 ± 23.26 ^{d}	263.00 ± 7.88 ^{e}	70.24 ± 2.57 ^{d}	80.64 ± 3.74 ^{e}	0.88 ± 0.04 ^{a}	7.00 ± 1.00 ^{b}	121.95 ± 22.29 ^{d}	47.76 ± 8.64 ^{ab}	33.02 ± 1.69 ^{ab}
G5	632.42 ± 105.24 ^{b}	362.00 ± 21.68 ^{b}	90.74 ± 3.00 ^{b}	106.44 ± 6.09 ^{b}	0.86 ± 0.04 ^{a}	8.20 ± 0.45 ^{a}	295.72 ± 40.29 ^{b}	47.27 ± 6.11 ^{ab}	19.75 ± 1.21 ^{e}
PS	297.85 ± 54.84 ^{cd}	276.00 ± 16.74 ^{cd}	67.60 ± 2.41 ^{d}	82.40 ± 3.58 ^{de}	0.83 ± 0.02 ^{a}	6.20 ± 0.45 ^{b}	117.74 ± 16.10 ^{d}	40.07 ± 6.06 ^{bc}	33.96 ± 2.38 ^{ab}
DC	359.42 ± 46.66 ^{cd}	295.00 ± 11.73 ^{cd}	79.56 ± 5.49 ^{c}	90.62 ± 4.65 ^{d}	0.88 ± 0.07 ^{a}	7.00 ± 0.00 ^{b}	127.79 ± 23.83 ^{d}	35.49 ± 4.41 ^{c}	36.05 ± 0.97 ^{a}
VL	378.51 ± 28.24 ^{c}	297.60 ± 3.37 ^{cd}	80.98 ± 6.05 ^{c}	90.00 ± 2.47 ^{d}	0.91 ± 0.08 ^{a}	7.00 ± 0.00 ^{b}	161.06 ± 24.53 ^{c}	42.49 ± 4.97 ^{bc}	31.28 ± 2.91 ^{bc}
WD	825.83 ± 142.80 ^{a}	386.00 ± 19.62 ^{a}	105.05 ± 8.61 ^{a}	114.65 ± 5.18 ^{a}	0.92 ± 0.06 ^{a}	6.75 ± 1.50 ^{b}	430.34 ± 70.61 ^{a}	52.24 ± 3.02 ^{a}	20.97 ± 1.54 ^{e}

Data are expressed as mean ± SD (n = 3). Significant differences ($p \leq 0.05$) are indicated by different letters (from ^{a} to ^{e}). FL, fruit length; FD, fruit diameter; PS, 'Primosole'; DC, 'Dente di Cavallo'; VL, 'Valenciana'; WD, 'Wonderful'.

Moreover, the fruit circumference showed significant differences among genotypes, whose average was 303.35 mm. WD and G5 showed the highest values (386 mm and 362 mm, respectively), while the lowest value was observed in G4 (263 mm). Accordingly, WD and G5 displayed also the highest fruit length and diameter values, while the lowest values were found in G4 and PS, as shown in Table 1. The average fruit size of WD grown in Sicily was greater than the WD cultivated in California, as previously reported [19]. This could probably be due to the different climate, temperature and humidity conditions in the two environments. According to the fruit weight, WD also showed the highest weight of peel and arils (430.34 g and 395.48 g, respectively) (Tables 1 and 2).

Table 2. Morphologic characteristics of arils and seeds.

Genotype	Number of Arils	Arils Weight (g)	Weight of 100 Arils (g)	Arils Yield (%)	Dry Arils Yield (%)	Seeds Weight (g)	Seeds Yield (%)
G1	464 ± 79.80 ^{bc}	178.94 ± 25.61 ^{b}	39.98 ± 1.96 ^{b}	62.48 ± 3.54 ^{ab}	16.35 ± 0.83 ^{d}	54.59 ± 10.80 ^{c}	19.04 ± 2.58 ^{ab}
G2	568 ± 50.23 ^{bc}	223.78 ± 17.42 ^{b}	43.00 ± 4.32 ^{b}	58.02 ± 2.23 ^{abcd}	16.73 ± 0.76 ^{cd}	80.42 ± 3.89 ^{b}	20.90 ± 1.21 ^{a}
G3	479 ± 107.12 ^{bc}	192.15 ± 31.89 ^{b}	41.84 ± 3.46 ^{b}	62.95 ± 4.74 ^{ab}	17.17 ± 0.87 ^{cd}	53.15 ± 3.78 ^{c}	17.59 ± 1.58 ^{abc}
G4	372 ± 87.48 ^{c}	133.92 ± 26.77 ^{c}	38.76 ± 4.3 ^{b}	52.24 ± 8.64 ^{bcd}	17.97 ± 0.26 ^{bc}	42.68 ± 8.07 ^{c}	16.59 ± 1.86 ^{bc}
G5	674 ± 146.24 ^{b}	336.70 ± 85.89 ^{a}	50.87 ± 1.86 ^{a}	52.73 ± 6.12 ^{cd}	16.93 ± 0.56 ^{cd}	123.63 ± 22.48 ^{a}	19.54 ± 1.34 ^{ab}
PS	510 ± 158.46 ^{bc}	180.11 ± 45.77 ^{b}	38.49 ± 5.93 ^{b}	59.94 ± 6.06 ^{abc}	18.99 ± 1.27 ^{b}	46.29 ± 9.00 ^{c}	15.66 ± 2.85 ^{bc}
DC	550 ± 78.25 ^{bc}	231.63 ± 31.62 ^{b}	43.85 ± 2.74 ^{b}	64.52 ± 4.41 ^{a}	18.03 ± 0.45 ^{bc}	57.32 ± 5.81 ^{c}	16.02 ± 1.17 ^{bc}
VL	543 ± 104.58 ^{bc}	217.45 ± 22.10 ^{b}	43.01 ± 6.53 ^{b}	57.52 ± 4.97 ^{abcd}	17.29 ± 0.74 ^{cd}	50.60 ± 14.03 ^{c}	13.44 ± 4.16 ^{c}
WD	1028 ± 219.96 ^{a}	395.48 ± 80.39 ^{a}	38.60 ± 3.42 ^{b}	47.77 ± 3.02 ^{d}	23.08 ± 0.51 ^{a}	118.51 ± 34.84 ^{a}	14.19 ± 2.16 ^{c}

Data are expressed as mean ± SD (n = 3). Significant differences ($p \leq 0.05$) are indicated by different letters (from ^{a} to ^{d}). PS, 'Primosole'; DC, 'Dente di Cavallo'; VL, 'Valenciana'; WD, 'Wonderful'.

The latter result was in agreement with the value reported for the WD cultivar grown in Spain [20] but substantially differs from what was obtained for the same cultivar grown in America [19], probably, once again, due to dissimilar environmental growth conditions. The value reported for the arils weight of our VL (217.45 g) was lower than reported by Alcaraz-Mármol [20].

Because the ready-to-eat arils are attracting increasing international demand due to their health and nutraceutical characteristics, the arils' features determine the economic value of the fruit and are important parameters for growers, market and industry [21]. The genotypes studied showed a large variability on arils traits: DC showed the highest arils yield (64.52%), while the lowest was observed in WD (47.77%), but, also, two Sicilian genotypes, G1 and G3, displayed high arils yield (62.48% and 62.95%, respectively). Furthermore, the G5 genotype has been shown to have the biggest arils, with a weight of 100 arils equal to 50.87 g; PS and WD have been shown to have the lowest (38.49 g and 38.60 g, respectively).

Moreover, the peel thickness is one important trait of market selection; fruits with thin skin and, therefore, with a lower peel yield, are more suitable for processing, while those with a thicker skin and higher peel yield are more suitable for transport and storage and can, therefore, be used for fresh consumption [3]. Regarding the peel, contrary to what was observed for arils yield, the WD genotype showed the highest peel yield (52.24%), while

DC was the lowest (35.49%), showing that arils and peels yields are inversely correlated. However, the dry arils yield was highest in WD (23.08%), while dry peel yield was found to be highest in DC (36.05%), indicating that the percentage of humidity in these fruits, for the parameters analysed, is lower than the other cultivars examined.

2.2. *Juice Characteristics*

The juice yield is a very important trait from an industrial point of view to obtain bottled juice. According to the highest aril yield, DC also showed the highest juice yield, with 48.50% (Table 3); conversely, G5 and WD showed the lower value, with 33.18% and 33.58%, respectively. In general, the G1–G5 genotypes showed values of yield in arils and in juice higher or similar compared to the international genotypes (VL and WD). This classifies G1–G5 accessions as good products for the market and juice production.

Table 3. Physico-chemical characteristics and colour coordinates (L, a, b, C and h°) of pomegranate juice.

Genotype	Juice Weight (g)	Juice Yield (%)	pH	Total Soluble Solids (° Brix)	Lightness (L)	Colour a	Colour b	Chroma (C)	Hue (h°)
G1	124.34 ± 16.87 ^{cd}	43.43 ± 2.47 ^{abc}	3.63 ± 0.08 ^{ab}	16.06 ± 0.19 ^{bc}	17.70 ± 0.93 ^{b}	6.84 ± 0.67 ^{bc}	−1.62 ± 0.43 ^{c}	7.05 ± 0.58 ^{bc}	−13.51 ± 4.39 ^{c}
G2	143.35 ± 15.72 ^{cd}	37.12 ± 2.57 ^{bcd}	3.76 ± 0.10 ^{a}	15.76 ± 0.63 ^{c}	17.77 ± 0.80 ^{b}	5.86 ± 0.60 ^{c}	−1.70 ± 0.19 ^{c}	6.11 ± 0.58 ^{bc}	−16.27 ± 2.19 ^{c}
G3	139.00 ± 30.61 ^{cd}	45.35 ± 6.10 ^{ab}	3.77 ± 0.08 ^{a}	16.06 ± 0.58 ^{bc}	22.40 ± 4.05 ^{a}	10.49 ± 3.42 ^{a}	4.49 ± 1.47 ^{a}	11.21 ± 3.02 ^{a}	25.06 ± 8.27 ^{a}
G4	91.23 ± 20.76 ^{d}	35.64 ± 7.47 ^{cd}	3.54 ± 0.06 ^{b}	17.00 ± 0.37 ^{ab}	18.69 ± 1.33 ^{ab}	10.93 ± 1.47 ^{a}	2.52 ± 3.52 ^{ab}	11.51 ± 2.51 ^{a}	10.82 ± 13.15 ^{b}
G5	213.06 ± 64.13 ^{b}	33.18 ± 5.34 ^{d}	3.15 ± 0.05 ^{c}	16.08 ± 0.51 ^{bc}	21.06 ± 5.51 ^{ab}	10.68 ± 3.37 ^{a}	4.36 ± 1.94 ^{a}	11.40 ± 4.07 ^{a}	19.36 ± 7.84 ^{ab}
PS	133.81 ± 40.00 ^{cd}	44.27 ± 5.39 ^{abc}	3.72 ± 0.07 ^{a}	17.52 ± 0.85 ^{a}	18.57 ± 0.27 ^{ab}	8.75 ± 1.10 ^{ab}	−1.33 ± 0.09 ^{b}	8.81 ± 1.09 ^{ab}	−8.60 ± 1.20 ^{c}
DC	174.31 ± 26.58 ^{bc}	48.50 ± 3.78 ^{a}	3.64 ± 0.03 ^{ab}	17.54 ± 0.21 ^{a}	18.07 ± 1.78 ^{b}	6.66 ± 1.03 ^{bc}	−0.83 ± 1.01 ^{c}	6.78 ± 0.95 ^{bc}	−7.70 ± 9.23 ^{c}
VL	166.85 ± 20.15 ^{bc}	44.08 ± 4.11 ^{abc}	3.67 ± 0.10 ^{ab}	17.58 ± 0.60 ^{a}	19.59 ± 1.50 ^{ab}	4.88 ± 1.34 ^{c}	−1.59 ± 0.64 ^{c}	5.19 ± 1.19 ^{c}	−19.30 ± 9.41 ^{c}
WD	276.98 ± 51.18 ^{a}	33.58 ± 2.57 ^{d}	3.14 ± 0.20 ^{c}	17.75 ± 0.86 ^{a}	13.59 ± 0.74 ^{c}	7.08 ± 1.67 ^{bc}	1.30 ± 0.37 ^{bc}	7.20 ± 1.71 ^{bc}	10.32 ± 1.15 ^{b}

Data are expressed as mean ± SD (n = 3). Significant differences ($p \leq 0.05$) are indicated by different letters (from ^{a} to ^{d}). PS, 'Primosole'; DC, 'Dente di Cavallo'; VL, 'Valenciana'; WD, 'Wonderful'.

Determination of total soluble solids (TSS or °Brix) is important to establish the organoleptic quality of the juice. The range of °Brix values found was from 15.76 in G2 to 17.75 in WD; all the varieties tested had a °Brix value higher than the minimum threshold generally required for commercial use (>12%). Our values are similar to those found in previous studies on different Apulian (from 13.60 to 18.00), Spanish (from 15.10 to 17.70) and Californian (from 14.90 to 16.80) pomegranate genotypes [17,19,20]. The value obtained for WD (17.75) was higher than that reported by the same previous studies (16.90, 17.20 and 16.80), but this may be due to environmental conditions and harvesting time [17,19,20].

The highest pH value was measured for G2, G3 and PS (3.76, 3.77 and 3.72, respectively), while the lowest was for G5 (3.15) and WD (3.14). These pH values were in agreement with Ferrara [17] and Beaulieu [22] for WD (2.93 and 3.05) and with Todaro [23] for PS, DC and VL (3.67, 3.88 and 3.71, respectively), growth in the experimental farm of the Catania University (Italy, Sicily), but they were lower than reported by Alcaraz-Mármol [20] for WD (3.89) and VL (5.90).

Observed as well were differences in juice colour among the genotypes (Table 3); G3 showed the highest values of lightness of colour (L) (22.40), whereas WD showed the lowest (13.47, 13.59 and 14.63, respectively). The highest values of colour a (tending to red colour) were observed in G4 (10.93), while G3 (4.49) and G5 (4.36) showed the greatest value of colour b, which means that the juice colour tends to be yellow. As regards value C, it was observed that G3, G4 and G5 were the genotypes with the most intensity/purity of the colour, with values of 11.21, 11.51 and 11.40, respectively. In addition, for the value of h°, significant differences were found between cultivars; a very high value was found in G3 (25.06). According to the CIELAB colour parameters, cultivars with values of colour, a positive and colour b negative have pigmentation from red to blue (G1, G2, PS, DC and VL), while cultivars with values of colour a and colour b positives have pigmentation from red to yellow (G3, G4, G5, and WD). Moreover, VL showed the lowest value of colour a, C and h°, which means that its juice tends towards a clear pink colour.

2.3. Total Phenolic Content and Antioxidant Activity

Total phenolic content (TPC), expressed as gallic acid equivalent (GAE), ranged from 645.59 (in DC) to 1447.22 mg L^{-1} (in WD) (Figure 1A). The average value of TPC was 884.83 mg L^{-1}, showing to be significantly influenced by the genotype ($p \leq 0.05$). In particular, among the tested genotypes, the G4 showed the highest phenolic content (1203.77 mg L^{-1}), comparable to the WD (1447.22 mg L^{-1}), which is the most studied international genotype. Moreover, the genotypes with lower values were DC, G5 and VL. The values obtained were similar to those reported by Fanali [24] for the six old Italian pomegranate varieties collected in the experimental farm of Tuscia University (range from 500 to 1400 mg GAE L^{-1}). In comparison with previous literature data, Sicilian genotypes showed TPC lower than some Croatian [25] (from 1985.6 to 2948.7 mg GAE L^{-1}) and Iranian cultivars et al. [26] (from 2957.9 to 9853.2 mg GAE L^{-1}).

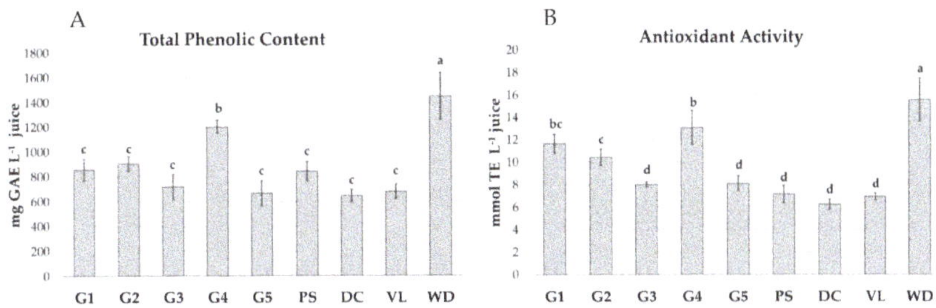

Figure 1. Total phenolic content (**A**) and antioxidant activity (**B**) in pomegranate juices. Each value is the mean of three replicates and is reported with standard deviation. Value bars indicated with different letters (from a to d) are significantly different ($p \leq 0.05$).

The antioxidant activity (AA), expressed as Trolox equivalent (TE), followed the same trend of the TPC, with the mean value of 9.69 mmol L^{-1} varying significantly ($p \leq 0.05$) between 6.24 in DC and 15.56 in WD (Figure 1B). WD showed the highest AA, followed by G4. The WD and VL values in this study were similar to those reported by Mena [27] (15.30 and 7.01, respectively) for the same cultivars grown in Spain (commercially available accessions).

The AA and TPC levels were positively and significantly correlated ($r = 0.974$).

2.4. Quantitative Determination of Carbohydrates by HPAE-PAD

The main carbohydrates detected in juices were glucose and fructose (Table 4).

Table 4. Carbohydrate content (g L^{-1}) of pomegranate juices.

Genotype	Glucose (G)	Fructose (F)	Total	Ratio G/F
G1	45.33 ± 0.79^{de}	33.54 ± 0.39^{de}	78.87	1.35
G2	64.16 ± 0.09^{a}	68.90 ± 0.24^{a}	133.06	0.93
G3	43.18 ± 1.20^{ef}	32.03 ± 0.80^{e}	75.20	1.35
G4	46.21 ± 2.29^{d}	34.44 ± 1.54^{d}	80.65	1.34
G5	36.83 ± 0.96^{g}	27.84 ± 0.63^{g}	64.67	1.32
PS	48.69 ± 0.28^{c}	35.51 ± 0.15^{d}	84.20	1.37
DC	52.81 ± 1.16^{b}	38.36 ± 1.36^{c}	91.17	1.38
VL	40.96 ± 2.62^{f}	29.89 ± 1.72^{f}	70.85	1.37
WD	62.54 ± 0.23^{a}	62.50 ± 0.04^{b}	125.04	1.00

Data are expressed as mean ± SD (n = 3). Significant differences ($p \leq 0.05$) are indicated by different letters (from ^{a} to ^{g}). PS, 'Primosole'; DC, 'Dente di Cavallo'; VL, 'Valenciana'; WD, 'Wonderful'.

The glucose concentration was mainly greater than fructose, with the ratio glucose/fructose taking values from 0.93 in G2 to 1.38 in DC. Similar profiles were previously described for other cultivars [28,29] (range from 0.96 to 1.12). In contrast, it is reported that the Spanish genotypes have almost always higher levels of fructose (from 55.4 to 82.4 g L^{-1}) than glucose (from 55.3 to 78.0 g L^{-1}) [30], as reported for WD (from 77.8 to 95.6 g L^{-1} for fructose and from 55.4 to 89.4 g L^{-1} for glucose) and VL (from 75.8 to 97.4 g L^{-1} for fructose and from 44.4 to 81.1 g L^{-1} for glucose) grown in Spain [20,27].

In this work, G2 showed the highest carbohydrate content, followed by WD. In particular, the highest amount of glucose was detected in G2 and in WD with 64.16 g L^{-1} and 62.54 g L^{-1}, respectively, the lowest in G5 with 36.83 g L^{-1}. A similar trend existed for fructose with G2, which showed also the highest amount of fructose (68.90 g L^{-1}), while G5 was the lowest (27.84 g L^{-1}). The sugar profile contributes to potential health benefits and determines the sensory attributes of pomegranate; juice red pomegranate cultivars usually have a sourer taste than pink-white genotypes. Figure 2 shows the chromatogram of one of the PJs (DC).

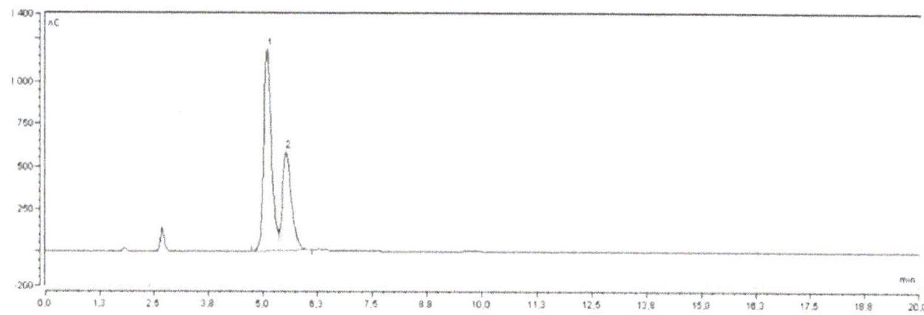

Figure 2. HPAE-PAD chromatogram of pomegranate juice (DC). Glucose (1) and fructose (2).

2.5. *Quantitative Determination of Minerals by IC*

The minerals in pomegranate juice, shown in Table 5, varied significantly among the genotypes. The highest content of macro-elements present in all the samples of juices was potassium (1816.73 mg L^{-1}), followed by chlorides (416.90 mg L^{-1}) and phosphates (367.04 mg L^{-1}), according to Al-Maiman and Ahmad [31], who reported potassium as the highest among the mineral elements in pomegranate juice.

Table 5. Minerals (mg L^{-1}) of pomegranate juices.

Parameter	G1	G2	G3	G4	G5	PS	DC	VL	WD	Mean
Anions (A-)										
Fluorides	42.97 ± 1.52^{e}	41.80 ± 0.72^{e}	48.27 ± 1.33^{c}	45.60 ± 1.52^{d}	41.67 ± 0.29^{e}	39.03 ± 1.22^{f}	59.90 ± 0.61^{a}	42.37 ± 0.42^{e}	51.33 ± 0.75^{b}	**45.88**
Chlorides	365.17 ± 7.88^{f}	296.07 ± 2.02^{h}	317.63 ± 2.87^{g}	380.83 ± 3.86^{e}	285.70 ± 1.04^{h}	489.63 ± 7.94^{c}	637.10 ± 12.00^{a}	422.00 ± 1.67^{d}	557.97 ± 6.33^{b}	**416.90**
Phosphates	262.80 ± 5.95^{h}	333.40 ± 3.95^{d}	310.87 ± 7.49^{f}	294.47 ± 4.91^{g}	317.17 ± 2.72^{e}	387.67 ± 0.61^{c}	463.87 ± 8.28^{b}	387.80 ± 0.36^{c}	545.30 ± 8.31^{a}	**367.04**
Sulphates	69.53 ± 0.40^{i}	83.70 ± 1.51^{g}	113.20 ± 1.45^{d}	90.93 ± 2.01^{f}	139.97 ± 1.40^{b}	126.60 ± 0.44^{c}	206.73 ± 3.23^{a}	93.97 ± 1.21^{e}	79.60 ± 1.47^{h}	**111.58**
Total A-	**740.47**	**754.97**	**789.97**	**811.83**	**784.51**	**1042.93**	**1367.60**	**946.14**	**1234.20**	
Cations (C+)										
Sodium	9.21 ± 0.92^{e}	13.33 ± 0.49^{c}	13.58 ± 0.24^{c}	12.81 ± 0.31^{cd}	11.70 ± 1.05^{d}	14.08 ± 0.14^{c}	18.53 ± 1.69^{a}	16.09 ± 0.88^{b}	18.84 ± 0.86^{a}	**14.24**
Potassium	1710.42 ± 19.95^{d}	1819.33 ± 13.83^{c}	1679.56 ± 68.23^{d}	1287.99 ± 70.45^{e}	2056.63 ± 19.41^{b}	1668.04 ± 0.56^{d}	2142.66 ± 16.43^{b}	1749.37 ± 28.68^{d}	2236.56 ± 51.49^{a}	**1816.73**
Magnesium	54.32 ± 1.26^{c}	57.57 ± 0.54^{b}	72.08 ± 3.02^{a}	53.98 ± 2.81^{c}	69.58 ± 0.61^{a}	49.00 ± 1.02^{d}	70.60 ± 0.73^{a}	48.75 ± 0.93^{d}	57.84 ± 1.85^{b}	**59.30**
Calcium	8.66 ± 1.69^{c}	7.67 ± 1.06^{c}	6.98 ± 0.40^{c}	8.05 ± 0.69^{c}	30.38 ± 1.20^{b}	7.29 ± 0.53^{c}	9.95 ± 1.18^{c}	8.15 ± 1.60^{c}	36.93 ± 1.20^{a}	**13.78**
Total C+	**1782.61**	**1897.90**	**1772.20**	**1362.83**	**2168.29**	**1738.41**	**2241.74**	**1822.36**	**2350.17**	
Total Minerals	**2523.08**	**2652.87**	**2562.17**	**2174.66**	**2952.80**	**2781.34**	**3609.34**	**2768.50**	**3584.37**	

Data are expressed as mean ± SD (n = 3). Significant differences ($p \leq 0.05$) are indicated by different letters (from [a] to [i]). PS, 'Primosole'; DC, 'Dente di Cavallo'; VL, 'Valenciana'; WD, 'Wonderful'.

Concerning anions (Figure 3A), the highest values were found in the genotype DC (1367.6 mg L^{-1}), followed by WD (1234.20 mg L^{-1}). Lower values were generally found in the Sicilian genotypes, especially in G1 (740.47 mg L^{-1}). The relative order of concentration of anions was $Cl^{-} > PO_4^{3-} > SO_4^{2-} > F^{-}$. The content of chlorides, sulphates and fluorides was higher in DC, whereas the content of phosphates was higher in WD.

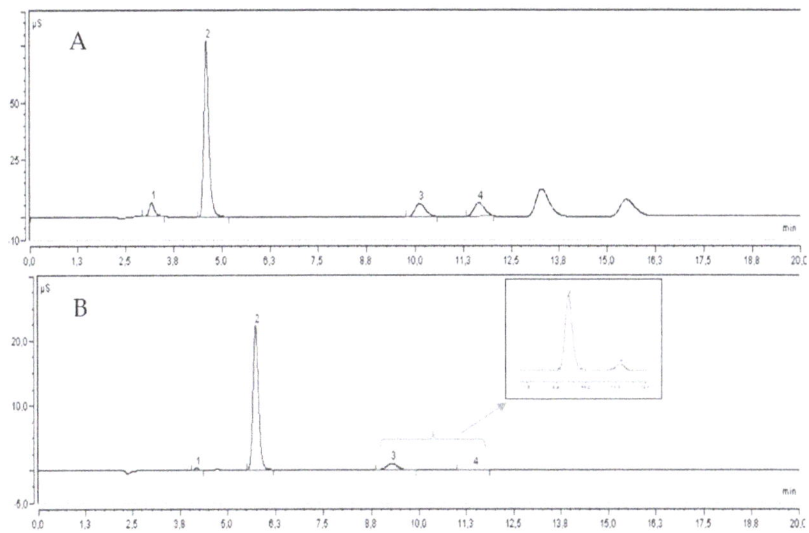

Figure 3. IC chromatogram of pomegranate juice (DC). Fluorides (1), chlorides (2), phosphates (3) and sulphates (4) (**A**). Sodium (1), potassium (2), magnesium (3) and calcium (4) (**B**).

Regarding cations (Figure 3B), the greatest results were generally found in WD (2350.17 mg L^{-1}) and DC (2241.74 mg L^{-1}); the lowest value was in G4 (1362.84 mg L^{-1}). The order of concentration of cations was $K^+ > Mg^{2+} > Na^+ > Ca^{2+}$, similar to that found by Al-Maiman and Ahmad [31] in 'Taifi' cultivars, except for the magnesium, which was lower. Potassium, sodium and calcium were predominant in WD, whereas magnesium was in G3. Overall, among the genotypes analysed, DC and WD showed the highest concentration of minerals, 3609.34 and 3584.37 mg L^{-1}, respectively. The pomegranate juice appears to be a good source of nutrients, and variation in mineral composition could originate from the pomegranate genotypes as well as agro-climatic conditions, handling practices and manufacturing conditions.

2.6. Molecular Analyses

Molecular analyses performed using SSRs produced a total of 96 alleles, with the maximum number of alleles per locus ranging from four to fourteen and a length of the amplified bands between 130 bp and 367 bp (Figure 4).

The polymorphism information content (PIC) was used to measure genetic diversity. High, medium or low loci polymorphism is in accordance with PIC > 0.5, 0.5 > PIC > 0.25 and PIC < 0.25, respectively [32]. A high PIC for all the markers used in this study was observed, on average 0.753, ranging from 0.469 of pg4 to 0.891 of pg14 (Table 6), higher than the previous analysis with the same microsatellites [32,33]. Among the new microsatellite markers developed in this study, MYBmp04 and MYBmp01 displayed a very high variability among the studied genotypes, with PIC values of 0.875 and 0.775, respectively (Table 6).

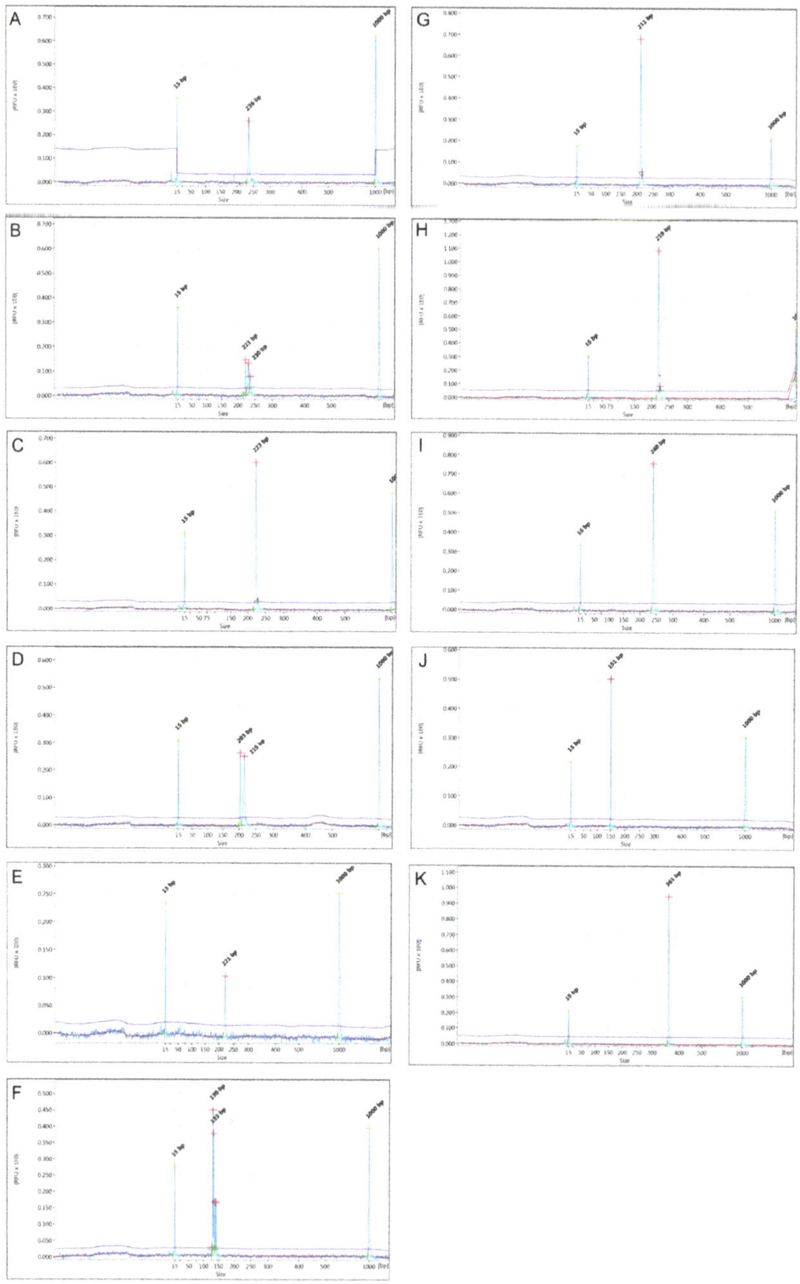

Figure 4. Capillary electrophoresis run of the PCR amplification products of 'Wonderful' (WD) variety with literature derived ones Pg4 (**A**), Pg10 (**B**), Pg14 (**C**), Pg21 (**D**), Pg22 I (**E**), Pg17 (**F**), Pom047 (**G**) and new developed markers MYBmp01 (**H**), MYBmp02 (**I**), MYBmp03 (**J**), MYBmp04 (**K**) and MYB derived markers were always homozygous, while Pg17, Pg21 and Pg10 were heterozygous. Peaks at 15 and 1000 bp correspond to the alignment marker used in the analysis.

Table 6. Microsatellite allele data obtained using 11 polymorphic microsatellite loci in the tested pomegranate genotypes.

Locus	Major Allele Frequency	Genotypes Number	Alleles Number	Ho	He	Nei	I	PIC
Pg4	0.667	4	4	0.000	0.521	0.510	0.961	0.469
Pg10(a)	0.458	7	7	0.542	0.704	0.689	1.392	0.642
Pg14	0.167	14	14	0.000	0.918	0.899	2.463	0.891
Pg21	0.250	14	11	0.458	0.887	0.869	2.205	0.862
Pg22	0.250	10	10	0.000	0.876	0.858	2.109	0.843
Pg17	0.313	13	9	0.417	0.836	0.819	1.902	0.797
Pom047	0.208	10	10	0.000	0.894	0.875	2.178	0.862
MYBmp01	0.292	8	8	0.000	0.819	0.802	1.794	0.775
MYBmp02	0.458	7	7	0.000	0.723	0.708	1.516	0.671
MYBmp03	0.542	5	5	0.000	0.653	0.639	1.258	0.597
MYBmp04	0.208	11	11	0.000	0.904	0.885	2.279	0.875
Average	**0.347**	**9.364**	**8.727**	**0.129**	**0.794**	**0.778**	**1.823**	**0.753**

Ho, observed heterozygosity; He, expected heterozygosity; Nei, Nei's genetic diversity; I, Shannon's information index; PIC, polymorphism information content.

Although to a lesser degree, MYBmp02 and MYBmp03 markers showed relevant variability and all of them could be recommended for further genetic analyses aimed at detecting molecular diversity in the pomegranate germplasm collection. The Shannon index (I) and heterozygosity (He) values were consistent with the PIC trend. The Nei's genetic diversity and Shannon's information index in Sicilian pomegranates in the present study were higher in comparison to Indian pomegranates based on ISSR markers reported by Narzary [34]. It can be shown that the SSR marker is a trustworthy technique for assessing genetic diversity in pomegranate genotypes. The average number of alleles was 8.727, which is significantly higher with respect to what was reported for Iranian pomegranate genotypes analysed with chloroplast SSRs [35] and even with respect to previous analyses with nuclear microsatellites as well [33,36]. This set of four new MYB-related SSRs, along with the other seven SSRs, might be useful for population genetic analyses, such as genotyping and linkage mapping.

The cluster UPGMA (unweighted pair group method with arithmetic mean) analysis (Figure 5) performed with all 11 microsatellites clustered the tested genotypes into three main groups and showed that there is a genetic differentiation among the international genotypes.

The first one was composed of DC, G1, G2 and VL, while the second one included G3, G4, PS and WD. On the other hand, G5 was distinctly separated, forming an outgroup, from all the other tested varieties. DC and G1 clustered together; also, G4 and G3, WD and PS, in general, are closely related, while VL showed association with G1–G5 genotypes, in particular with G1, G2 and DC. The affinity between VL and most of the G1–G5 genotypes could be due to a common genetic origin since they have been domesticated within the Mediterranean basin; similarly, the clustering of PS together with WD could be explained by a common genetic matrix between the two, as shown by Parvaresh [32], referring to the "Palermo" genotype. This different clusterisation of the two international commercial varieties could have arisen from the highly different phenolic compounds composition between WD and VL, as reported along with this study.

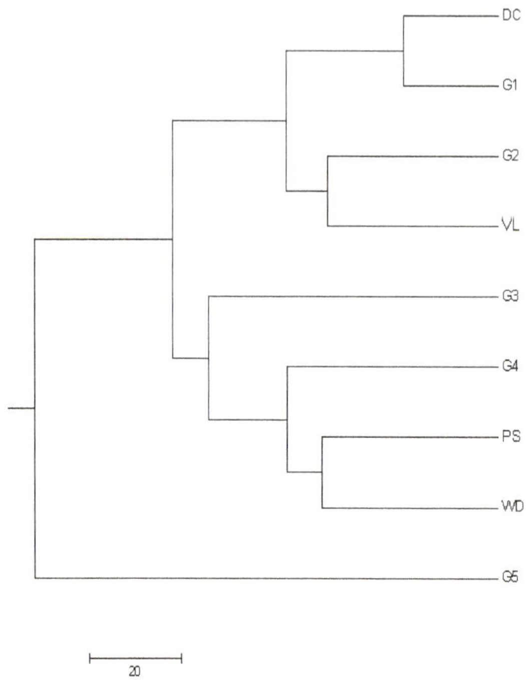

Figure 5. UPGMA dendrogram of 9 pomegranate genotypes based on the nuclear microsatellite (SSR) markers described in Table 6.

2.7. Principal Component Analysis

The selection of the most valuable fruit characteristics, including fruit, aril and juice weight, shape (circumference), juice °Brix, juice pH, juice colour, minerals (as cations and anions) and antioxidant activity, allowed to identify a specific differentiation pattern in which G1–G4, 'Valenciana' and 'Primosole' are closely related, while DC, G5 and WD were apart from the group and between each other (Figure 6A).

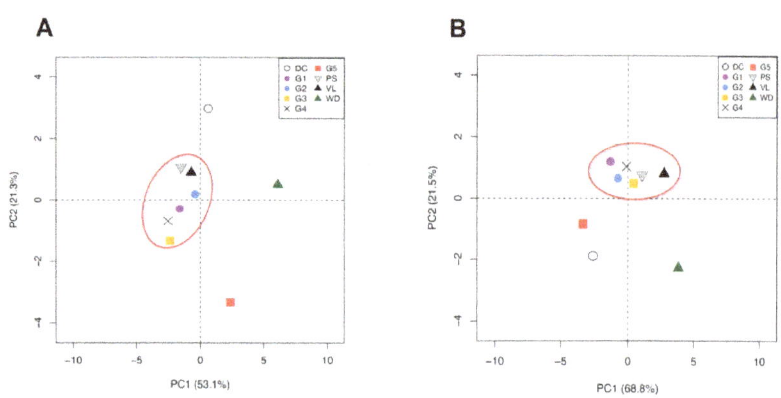

Figure 6. Principal component analysis of the sampled genotypes using morphological and physicochemical characters (**A**) and four non-neutral SSRs (**B**). Represented values are the mean of 5 individual measurements.

The total variability of morphological and chemical factors is described by four factors, with the first two principal factors explaining 74.4%. The first factor accounted for 53.1%, while the second one contributed 21.3%. The PCA on molecular data (Figure 6B) revealed the same clustering pattern to the morpho-chemical one; in fact, the G1–G4 genotypes were grouped and G5, WD and DC were more isolated. The total genetic variability is described by four factors, with the first two principal factors explaining 90.3%. Differently from the result of UPGMA elaboration, the use of microsatellites derived from MYB related genes allowed a more efficient separation of genotypes reproducing the morpho-chemical characteristics. The Mantel test showed that flavonoid-associated SSR Euclidean distances were significantly correlated with morpho-chemical traits among the sampled genotypes ($r = 0.62$, $p < 0.05$). This finding could represent a valid tool to rapidly analyse the technological characteristics, such as fruit shape and weight, minerals content, juice features and antioxidant properties, of new pomegranate genotypes using genetic markers.

3. Materials and Methods

3.1. Plant Material

The sampling was carried out on fruits and leaves of five pomegranate Sicilian genotypes, from 'Genotype 1′ to 'Genotype 5′ (from G1 to G5), two Sicilian commercial cultivars 'Primosole'(PS) and 'Dente di Cavallo' (DC) and two international world-leading commercial varieties 'Valenciana'(VL) and 'Wonderful'(WD), collected in Sicily in October 2016 (Figure 7). All cultivars were harvested from a germplasm collection of CNR-ISAFOM, the section of Catania (Italy), grown under the same environmental conditions and with the same applied agronomic practices. For each accession, five fruits at the marketing ripeness stage were randomly collected.

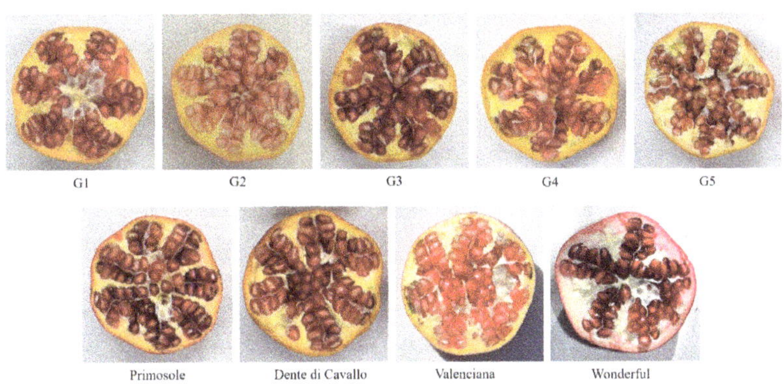

Figure 7. Cross section of pomegranate genotypes studied in this work.

3.2. Chemicals

Folin-Ciocalteau, sodium carbonate, 2,2-diphenyl-1-picrylhydrazyl (DPPH), methanol, 6-hydroxy-2,5,7,8-tetramethylchroman-2-carboxylic acid (Trolox), fructose, glucose, sodium hydroxide 50% and standard solutions for IC 1000 mg L^{-1} of sodium, potassium, magnesium, calcium, fluoride, chloride, sulphate and phosphate were purchased from Sigma-Aldrich (St. Louis, MO, USA). Methanesulphonic acid, sodium carbonate 0.5 M and sodium bicarbonate 0.5 M were obtained from Thermo Scientific (Waltham, MA, USA). Gallic acid was provided by Extrasynthese (Genay, France). Water Type I reagent grade was produced using a Milli-Q water purification system (Millipore, MA, USA).

3.3. Bio-Agronomic Traits and Physico-Chemical Analysis

Five mature fruits per genotype were analysed. Arils were manually separated from the peel and morphometric measurements were carried out on different fruit organs: peel, arils and seeds. Eighteen quantitative fruit morphological traits were examined, including fruit weight (g), fruit diameter (mm), fruit length (mm), fruit circumference (mm), number of septa, number of arils per fruit, arils weight (g), the weight of 100 arils per fruit (g), seeds weight (g), peel weight (g), juice weight (g). From these data, the peel yield (%), arils yield (%), juice yield (%) and seeds yield (%) were calculated. Furthermore, peel and arils were dried for 48 h at 105 °C and dry peel and arils yield (%) were determined. The pomegranate juice (PJ) was obtained by manual squeezing of arils through sterile gauze. The PJ was separated into aliquots and some juice was immediately frozen at −20 °C for downstream analyses. On the fresh juice, the following analyses were carried out: total soluble solids (TSS), pH and colour. The TSS was determined with a digital refractometer DBR 45 (Giorgio Bormac S.r.l., Carpi, Italy) and results were reported as °Brix at 20 °C. The pH measurements were performed using a digital pH meter XS Instruments mod. PC510 at 20 °C. The colour was measured with a Chroma Meter CR-400 (Konica Minolta, Osaka, Japan) using the CIELAB colour system [37]. Three colour measurements were made (L, a, b) and chroma (C) and hue (h°) units of colour space were detected.

3.4. Total Phenolic Content

TPC was determined by the method of Dewanto [38], with some modifications. The PJ, appropriately diluted with ultrapure water, was centrifuged at 5000 rpm for 4 min and 125 μL of supernatant was mixed with 625 μL of Folin–Ciocalteu reagent 5-fold-diluted with ultrapure water. After 6 min, 1.25 mL of 7% Na_2CO_3 aqueous solution and 1 mL of ultrapure water were added. The mixture was shaken and placed in the dark at room temperature for 1 h. After incubation, the total content of phenolic compounds was measured at 760 nm using the BioSpectrometer UV/Vis spectrophotometer (Eppendorf, Hamburg, Germany). Gallic acid standard solution (25–200 mg L^{-1}) was used for the calibration curve (R^2 = 0.9943). All measurements were performed in triplicate. The results were expressed as mg of gallic acid equivalent per litre of juice (mg GAE L^{-1}).

3.5. Antioxidant Activity

AA of PJ was determined using the DPPH method described by Brand-Williams [39] with some modifications. PJ diluted in the ratio of 1:100 with methanol was centrifuged at 5000 rpm for 4 min. Afterwards, 100 μL of supernatant was mixed with 2 mL of 0.1 mM DPPH in methanol. After incubating at room temperature for 30 min in the dark, the absorbance of the mixture was measured at 517 nm. Trolox was used as a reference (calibration range 10–200 μmol L^{-1}; R^2 = 0.9992). All samples were analysed in triplicates and the results are expressed as mmol Trolox per litre of juice (mmol TE L^{-1}).

3.6. Quantitative Determination of Carbohydrates by HPAE-PAD

The quantification of the main carbohydrates (fructose and glucose) present in PJ was determined by the method of Corradini [40]. PJ, diluted in the ratio 1:1000 with deionized water and filtered (0.45 μm) was analysed using a High-Performance Anion-Exchange chromatography with Pulsed Amperometric Detection (HPAE-PAD), Thermo Scientific Dionex ICS3000 (Sunnyvale, CA, USA), consisting of a quaternary gradient inert pump, a pulsed amperometric detector and AS40 automated sampler. The separation was carried out on a Dionex CarboPac PA10 analytical column (250 × 4 mm). The acquisition of all the chromatograms was performed with Chromeleon Chromatography Management System. All experiments were carried out at 30 °C under isocratic elution using NaOH 100 mM with a flow rate of 0.8 mL min^{-1}. All analyses were performed in triplicate for each agronomic sample, quantified by calibration curve (range 0.5–100 mg L^{-1}; R^2 = 0.9982 for glucose and R^2 = 0.9995 for fructose) and the results are reported in mg L^{-1}. Relative standard deviations (RSD %) of peak retention times were <0.8%.

3.7. Quantitative Determination of Minerals by IC

The mineral content was determined using the ion chromatography (IC) method [41]. The most important inorganic anionic and cationic constituents of PJ were analysed: F^-, Cl^-, $PO_4{}^{3-}$ and $SO_4{}^{2-}$ for anions, Na^+, K^+, Mg^{2+} and Ca^{2+} for cations.

PJ was diluted in the ratio 1:100 with deionized water, filtered with a syringe filter 0.45 µm and subjected to analysis. All chromatographic analyses were performed by a Thermo Scientific Dionex ICS3000 ion chromatography (Sunnyvale, CA, USA) composed of an isocratic pump, a cationic or anionic suppressor, a conductance detector equipped with a temperature compensated conductivity cell, an injection valve with a 25 µL loop and a column thermostat compartment. The ion separation was carried out with two ion-exchange columns: anions were separated on a Dionex IonPac AS22 column (250 × 4 mm) with an IonPac AG22 guard column (50 × 4 mm) and cations were determined using an IonPac CS12A column (250 × 4 mm) equipped with IonPac CG12A guard column (50 × 4 mm).

An aqueous solution containing 20 mM methanesulfonic acid was used for the elution of cations. The mobile phase containing 4.5 mM sodium carbonate and 1.4 mM sodium bicarbonate was used for the elution of anions. Flow rate of 1.0 mL min^{-1} and 1.2 mL min^{-1} was used for the separation of cations and anions, respectively, maintaining the column temperature at 30 °C during analysis. All analyses were performed in triplicate for each agronomic sample, quantified by calibration curve (range 0.5–100 mg L^{-1}; $R^2 > 0.999$) and the results are reported in mg L^{-1} of juice. Relative standard deviations (RSD %) of peak retention times ranged from 0.7% to 2.1%.

3.8. Molecular Analyses

Genomic DNA was extracted from 100 mg of leaf pomegranate using the CTAB method described by Ebrahimi [33]. Samples quantity and quality were assessed by checking them on a 1% agarose gel and by measuring with an Eppendorf BioSpectrometer® (Eppendorf AG, Hamburg, Germany) their absorbance at 260 nm and 280 nm to assess DNA purity.

Sixty-four loci characterized in previous studies [16,32,33,42–44] were considered for the analysis of genetic relationships among cultivars. Out of 64 loci, 7 markers were selected based on heterozygosity, the number of alleles, allele size and amplification reproducibility, preferring tetra- and tri-nucleotides compared to di-nucleotides. Furthermore, 4 new gene-derived primer pairs were designed using putative *Punica granatum* MYB gene sequences (GenBank Acc. num. HM056531.1 and MT495437.1) involved in the biosynthesis of flavonoids, as reported in Arlotta [12] (Table 7).

DNA was diluted to 30 ng μL^{-1} for PCR amplification. PCR assays were performed in a reaction mixture of 25 μL^{-1} including: 90 ng of genomic DNA, 0.25 µM of each primer, 200 µM dNTPs, 1 U of Q5 High-Fidelity DNA polymerase (New England Biolabs, Ipswich, MA, USA) and 4 μL^{-1} of 5x Q5 Reaction Buffer. DNA amplifications were performed in a thermocycler with the following cycling program: an initial denaturation step of 98 °C for 30 s, followed by 30 cycles of 98 °C for 10 s, at optimal annealing temperature for 30 s, 72 °C for 30 s and a final extension at 72 °C for 2 min. The amplification results were analysed by capillary electrophoresis using QIAxcel High Resolution Gel Cartridge (Qiagen, Hilden, Germany).

GeneALEx version 6.5 [45] was used to obtain a pairwise population matrix calculated through Nei's genetic diversity, while expected heterozygosity (He), observed heterozygosity (Ho), Shannon index (I), the number of alleles and UPGMA dendrogram were calculated along with the number of alleles, major allele frequency, number of genotypes and PIC value for each primer.

Table 7. Primer sets used to investigate genetic diversity among pomegranate cultivars and the annealing temperature (Ta) chosen for each primer.

Locus	Repeat Motif	Primer Sequence (5′-3′)	Ta °C	Reference
Pg4	(TC)12 TT(TC)20	F: CTGATGTAATGGCTGAGCAAA R: GCACTTGAACAAAGAGAATGC	63	Ebrahimi et al. 2010
Pg10(a)	(AG)9 GG(AG)14	F: TGCTAGACAGAACTGGGAGAAC R: AGAGAGTGGGGTTTCCATTG	63	Ebrahimi et al. 2010
Pg14	(AG)32	F: GCACATTTCTTCCACCTTCC R: GGTTACAATGCACAGAGTCCAC	62	Ebrahimi et al. 2010
Pg21	(AG)7	F: CAAGACAGAAGCACCATCCA R: TCTCCCAAATCAGACCAACC	62	Ebrahimi et al. 2010
Pg22	(ACAT)3 (AT)3 (AG)22 (AT)3	F: CCCCGCACTTAGAATCTATTA R: TCCAGTTCCAATCGACAGAC	56	Ebrahimi et al. 2010
Pg17	(TCA)14	F: CATCAGACTACGATGGCACT R: GCATAATAGCCTTCAATTTACA	57	Parvaresh et al. 2010
Pom047	(CT)24	F: GCCTATCTCGTGATCACATC R: AATGGGAGCGGACTAACTAT	57	Rania et al. 2012
MYBmp01	(CT)9	F: GATGAAGATGACAAAACACCCC R: TGGGAGCTAGACAGAGTGACAA	60	Present study
MYBmp02	(GA)12	F: TCCTCAAGCAGACCCAGAAA R: TGCTGTTCTTGTTACGCCTT	62	Present study
MYBmp03	(AGC)4	F: AGGCGTAACAAGAACAGCAA R: AGCAACAGTCTTCCACCTCC	62	Present study
MYBmp04	(GAG)4	F: CTCGCTTGTCTTGCTAAAGGAT R: CGAGGAACTTATTGACCCACTC	57	Present study

3.9. Statistical Analysis

All analyses were carried out in triplicate and data were expressed as mean ± standard deviation. Pearson correlation coefficient (r) in bivariate linear correlation followed by Student's test was used to compare phenolic content and antioxidant activity. Differences between means at the 95% ($p \leq 0.05$) confidence level were considered statistically significant.

All data were submitted to Bartlett's test for the homogeneity of variance and the data that were not homogeneous were logarithmically transformed. All the homogeneous data were analysed using analysis of variance (ANOVA) by CoSTAT program.

Two different principal component analyses (PCA) were performed to evaluate the potentiality of identified genetic markers. In the first PCA, we used morphological and chemical traits: arils, juice weight and fruit weight, circumference, juice °Brix, juice pH, colour L, chroma, TPC, AA, minerals (as cations and anions) and sugars. Another PCA was performed using only SSRs associated with flavonoid production. A Mantel test was performed to measure the correlation between the two matrices from each PCA as Euclidean distance matrices based on 9999 replications. Data were analysed using the R environment for statistical computing (R Development Core Team, 2021).

4. Conclusions

This study evaluated new pomegranate genotypes by a multidisciplinary approach using leading commercial varieties as a reference. A significant variability has been observed for the qualitative and chemical traits among the pomegranate genotypes. The SSR markers used in this study were suitable for the assessment of variability in the pomegranate germplasm for the preservation of this species. To the best of our knowledge, this research is the first study that described a common diversity pattern between the relevant bio-agronomic traits and genetic markers in *P. granatum*. The use of this set of *MYB*-related markers might represent a promising tool for the rapid genetic characterization of pomegranate accessions focused on marketable traits. Furthermore, all the microsatellite markers evaluated in the present study confirmed the relationships observed with morphological and chemical characteristics and can be used as genetic tools for breeding purposes.

As regards the nutraceutical content, the results showed that pomegranate juice is an excellent source of minerals that are essential for human health. Consuming a fruit per day, indeed, may cover the daily requirement of many minerals, especially potassium, most present in WD, DC and G5 among the genotypes investigated. Moreover, G4 presented the highest total phenolic content and antioxidant activity, even higher than the well-known marketable 'Dente di Cavallo', 'Valenciana' and 'Primosole'. The G1–G3 and G5 genotypes presented a value of total phenolic content and antioxidant activity comparable to the commercial ones, representing a valid alternative to the most common cultivars for nutraceutical purposes.

Author Contributions: Conceptualization, S.A.R.; methodology, C.A., V.T., C.G., G.D.P. and S.A.R.; validation, C.A., V.T., C.G., G.D.P. and S.A.R.; formal analysis, C.A., V.T., C.G. and G.D.P.; investigation, C.A., V.T., C.G., G.D.P. and P.C.; resources, S.A.R. and P.C.; data curation, C.A., V.T., C.G. and G.D.P.; writing—original draft preparation, C.A., V.T., C.G. and G.D.P.; writing—review and editing, C.G., V.T. and G.D.P.; supervision, S.A.R.; funding acquisition, S.A.R. All authors have read and agreed to the published version of the manuscript.

Funding: This research received no external funding.

Informed Consent Statement: Not applicable.

Data Availability Statement: Not applicable.

Acknowledgments: The authors kindly thank Helena Domenica Pappalardo and Giusi D'Amante for helping in fruit harvesting and processing.

Conflicts of Interest: The authors declare no conflict of interest.

Sample Availability: Samples of the compounds are not available from the authors.

References

1. Holland, D.; Hatib, K.; Bar-Ya'akov, I. Pomegranate: Botany, Horticulture, Breeding. In *Horticultural Reviews*; Janick, J., Ed.; John Wiley & Sons Inc.: Hoboken, NJ, USA, 2009; Volume 35, pp. 127–191.
2. Khadivi, A.; Ayenehkar, D.; Kazemi, M.; Khaleghi, A. Phenotypic and pomological characterization of a pomegranate (*Punica granatum* L.) germplasm collection and identification of the promising selections. *Sci. Hortic.* **2018**, *238*, 234–245. [CrossRef]
3. Melgarejo-Sánchez, P.; Martínez, J.J.; Legua, P.; Martínez, R.; Hernández, F.; Melgarejo, P. Quality, antioxidant activity and total phenols of six Spanish pomegranates clones. *Sci. Hortic.* **2015**, *182*, 65–72. [CrossRef]
4. Tezcan, F.; Gültekin-Özgüven, M.; Diken, T.; Özçelik, B.; Erim, F.B. Antioxidant activity and total phenolic, organic acid and sugar content in commercial pomegranate juices. *Food Chem.* **2009**, *115*, 873–877. [CrossRef]
5. Topalović, A.; Knežević, M.; Gačnik, S.; Mikulic-Petkovsek, M. Detailed chemical composition of juice from autochthonous pomegranate genotypes (*Punica granatum* L.) grown in different locations in Montenegro. *Food Chem.* **2020**, *330*, 127261. [CrossRef] [PubMed]
6. Cristofori, V.; Caruso, D.; Latini, G.; Dell'Agli, M.; Cammilli, C.; Rugini, E.; Bignami, C.; Muleo, R. Fruit quality of Italian pomegranate (*Punica granatum* L.) autochthonous varieties. *Eur. Food Res. Technol.* **2011**, *232*, 397–403. [CrossRef]

7. Tibullo, D.; Caporarello, N.; Giallongo, C.; Anfuso, C.; Genovese, C.; Arlotta, C.; Puglisi, F.; Parrinello, N.; Bramanti, V.; Romano, A.; et al. Antiproliferative and Antiangiogenic Effects of Punica granatum Juice (PGJ) in Multiple Myeloma (MM). *Nutrients* **2016**, *8*, 611. [CrossRef]
8. Vlachojannis, C.; Erne, P.; Schoenenberger, A.W.; Chrubasik-Hausmann, S. A critical evaluation of the clinical evidence for pomegranate preparations in the prevention and treatment of cardiovascular diseases. *Phytother Res.* **2015**, *29*, 501–508. [CrossRef]
9. ISTAT. Available online: http://dati.istat.it/index.aspx?queryid=33705 (accessed on 15 March 2021).
10. La Malfa, S.; Gentile, A.; Domina, F.; Tribulato, E. Primosole: A new selection from Sicilian pomegranate germplasm. *Acta Hortic.* **2009**, *818*, 125–132. [CrossRef]
11. Ben-Simhon, Z.; Judeinstein, S.; Nadler-Hassar, T.; Trainin, T.; Bar-Ya'akov, I.; Borochov-Neori, H.; Holland, D. A pomegranate (*Punica granatum* L.) WD40-repeat gene is a functional homologue of Arabidopsis TTG1 and is involved in the regulation of anthocyanin biosynthesis during pomegranate fruit development. *Planta* **2011**, *234*, 865–881. [CrossRef]
12. Arlotta, C.; Puglia, G.D.; Genovese, C.; Toscano, V.; Karlova, R.; Beekwilder, J.; De Vos, R.C.H.; Raccuia, S.A. MYB5-like and bHLH influence flavonoid composition in pomegranate. *Plant Sci.* **2020**, *298*, 110563. [CrossRef]
13. Trainin, T.; Harel-Beja, R.; Bar-Ya'akov, I.; Ben-Simhon, Z.; Yahalomi, R.; Borochov-Neori, H.; Ophir, R.; Sherman, A.; Doron-Faigenboim, A.; Holland, D. Fine Mapping of the "black" Peel Color in Pomegranate (*Punica granatum* L.) Strongly Suggests That a Mutation in the Anthocyanidin Reductase (ANR) Gene Is Responsible for the Trait. *Front. Plant Sci.* **2021**, *12*, 265. [CrossRef]
14. Hasnaoui, N.; Buonamici, A.; Sebastiani, F.; Mars, M.; Zhang, D.; Vendramin, G.G. Molecular genetic diversity of Punica granatum L. (pomegranate) as revealed by microsatellite DNA markers (SSR). *Gene* **2012**, *493*, 105–112. [CrossRef]
15. Ono, N.N.; Britton, M.T.; Fass, J.N.; Nicolet, C.M.; Lin, D.; Tian, L. Exploring the Transcriptome Landscape of Pomegranate Fruit Peel for Natural Product Biosynthetic Gene and SSR Marker Discovery(F). *J. Integr Plant Biol* **2011**, *53*, 800–813. [CrossRef]
16. Currò, S.; Caruso, M.; Distefano, G.; Gentile, A.; La Malfa, S. New microsatellite loci for pomegranate, Punica granatum (Lythraceae). *Am. J. Bot.* **2010**, *97*, e58–e60. [CrossRef]
17. Ferrara, G.; Giancaspro, A.; Mazzeo, A.; Giove, S.L.; Matarrese, A.M.S.; Pacucci, C.; Punzi, R.; Trani, A.; Gambacorta, G.; Blanco, A.; et al. Characterization of pomegranate (*Punica granatum* L.) genotypes collected in Puglia region, Southeastern Italy. *Sci. Hortic.* **2014**, *178*, 70–78. [CrossRef]
18. Tarantino, A.; Difonzo, G.; Disciglio, G.; Frabboni, L.; Paradiso, V.M.; Gambacorta, G.; Caponio, F. Fresh pomegranate juices from cultivars and local ecotypes grown in southeastern Italy: Comparison of physicochemical properties, antioxidant activity and bioactive compounds. *J. Sci. Food Agric.* 2021. [CrossRef]
19. Chater, J.M.; Merhaut, D.J.; Jia, Z.; Mauk, P.A.; Preece, J.E. Fruit quality traits of ten California-grown pomegranate cultivars harvested over three months. *Sci. Hortic.* **2018**, *237*, 11–19. [CrossRef]
20. Alcaraz-Mármol, F.; Nuncio-Jáuregui, N.; García-Sánchez, F.; Martínez-Nicolás, J.J.; Hernández, F. Characterization of twenty pomegranate (*Punica granatum* L.) cultivars grown in Spain: Aptitudes for fresh consumption and processing. *Sci. Hortic.* **2017**, *219*, 152–160. [CrossRef]
21. Passafiume, R.; Perrone, A.; Sortino, G.; Gianguzzi, G.; Saletta, F.; Gentile, C.; Farina, V. Chemical–physical characteristics, polyphenolic content and total antioxidant activity of three Italian-grown pomegranate cultivars. *NFS J.* **2019**, *16*, 9–14. [CrossRef]
22. Beaulieu, J.C.; Lloyd, S.W.; Preece, J.E.; Moersfelder, J.W.; Stein-Chisholm, R.E.; Obando-Ulloa, J.M. Physicochemical properties and aroma volatile profiles in a diverse collection of California-grown pomegranate (*Punica granatum* L.) germplasm. *Food Chem.* **2015**, *181*, 354–364. [CrossRef] [PubMed]
23. Todaro, A.; Cavallaro, R.; La Malfa, S.; Continella, A.; Gentile, A.; Fischer, U.; Spagna, G. Anthocyanin profile and antioxidant activity of freshly squeezed pomegranate (*Punica granatum* L.) juices of sicilian and spanish provenances. *Ital. J. Food Sci.* **2016**, *28*, 464–479. [CrossRef]
24. Fanali, C.; Belluomo, M.G.; Cirilli, M.; Cristofori, V.; Zecchini, M.; Cacciola, F.; Russo, M.; Muleo, R.; Dugo, L. Antioxidant activity evaluation and HPLC-photodiode array/MS polyphenols analysis of pomegranate juice from selected Italian cultivars: A comparative study. *Electrophoresis* **2016**, *37*, 1947–1955. [CrossRef] [PubMed]
25. Radunić, M.; Jukić Špika, M.; Goreta Ban, S.; Gadže, J.; Díaz-Pérez, J.C.; MacLean, D. Physical and chemical properties of pomegranate fruit accessions from Croatia. *Food Chem.* **2015**, *177*, 53–60. [CrossRef] [PubMed]
26. Tehranifar, A.; Zarei, M.; Nemati, Z.; Esfandiyari, B.; Vazifeshenas, M.R. Investigation of physico-chemical properties and antioxidant activity of twenty Iranian pomegranate (*Punica granatum* L.) cultivars. *Sci. Hortic.* **2010**, *126*, 180–185. [CrossRef]
27. Mena, P.; García-Viguera, C.; Navarro-Rico, J.; Moreno, D.A.; Bartual, J.; Saura, D.; Martí, N. Phytochemical characterisation for industrial use of pomegranate (*Punica granatum* L.) cultivars grown in Spain. *J. Sci. Food Agric.* **2011**, *91*, 1893–1906. [CrossRef]
28. Ozgen, M.; Durgaç, C.; Serçe, S.; Kaya, C. Chemical and antioxidant properties of pomegranate cultivars grown in the Mediterranean region of Turkey. *Food Chem.* **2008**, *111*, 703–706. [CrossRef]
29. Cam, M.; Hisil, Y.; Durmaz, G. Characterisation of pomegranate juices from ten cultivars grown in Turkey. *Int J. Food Prop.* **2009**, *12*, 388–395. [CrossRef]
30. Melgarejo, P.; Salazar, D.M.; Artés, F. Organic acids and sugars composition of harvested pomegranate fruits. *Eur. Food Res. Technol.* **2000**, *211*, 185–190. [CrossRef]
31. Al-Maiman, S.A.; Ahmad, D. Changes in physical and chemical properties during pomegranate (*Punica granatum* L.) fruit maturation. *Food Chem.* **2002**, *76*, 437–441. [CrossRef]

32. Parvaresh, M.; Talebi, M.; Sayed-Tabatabaei, B.E. Molecular diversity and genetic relationship of pomegranate (*Punica granatum* L.) genotypes using microsatellite markers. *Sci. Hortic.* **2012**, *138*, 244–252. [CrossRef]
33. Ebrahimi, S.; Seyed, T.B.; Sharif, N.B. Microsatellite isolation and characterization in pomegranate (*Punica granatum* L.). *Iran. J. Biotechnol.* **2010**, *8*, 156–163.
34. Narzary, D.; Rana, T.S.; Ranade, S.A. Molecular analyses of genetic diversity in Indian pomegranates using RAPD, DAMD and ISSR. *Fruit Veg. Cereal Sci. Biotechnol* **2010**, *4*, 126–143.
35. Norouzi, M.; Talebi, M.; Sayed-Tabatabaei, B.E. Chloroplast microsatellite diversity and population genetic structure of Iranian pomegranate (*Punica granatum* L.) genotypes. *Sci. Hortic.* **2012**, *137*, 114–120. [CrossRef]
36. Soriano, J.M.; Zuriaga, E.; Rubio, P.; Llácer, G.; Infante, R.; Badenes, M.L. Development and characterization of microsatellite markers in pomegranate (*Punica granatum* L.). *Mol. Breed.* **2011**, *27*, 119–128. [CrossRef]
37. Moss, J.R.; Otten, L. A relationship between colour development and moisture content during roasting of peanuts. *Can. Inst. Food Technol. J.* **1989**, *22*, 34–39. [CrossRef]
38. Dewanto, V.; Wu, X.; Adom, K.K.; Liu, R.H. Thermal processing enhances the nutritional value of tomatoes by increasing total antioxidant activity. *J. Agric. Food Chem.* **2002**, *50*, 3010–3014. [CrossRef] [PubMed]
39. Brand-Williams, W.; Cuvelier, M.E.; Berset, C. Use of a free radical method to evaluate antioxidant activity. *LWT* **1995**, *28*, 25–30. [CrossRef]
40. Corradini, C.; Cavazza, A.; Bignardi, C. High-performance anion-exchange chromatography coupled with pulsed electrochemical detection as a powerful tool to evaluate carbohydrates of food Interest: Principles and Applications. *Int. J. Carbohydr. Chem.* **2012**, 1–13. [CrossRef]
41. LaCourse, W.R. Ion chromatography in food analysis. In *Handbook of Food Analysis Instruments*; Otles, S., Ed.; CRC Press: Boca Raton, FL, USA, 2008; Chapter 9; pp. 161–196. [CrossRef]
42. Pirseyedi, S.M.; Valizadehghan, S.; Mardi, M.; Ghaffari, M.R.; Mahmoodi, P.; Zahravi, M.; Zeinalabedin, M.; Nekoui, S.M.K. Isolation and characterization of novel microsatellite markers in pomegranate (*Punica granatum* L.). *Int. J. Mol. Sci.* **2010**, *11*, 2010–2016. [CrossRef]
43. Jian, Z.H.; Liu, X.S.; Hu, J.B.; Chen, Y.H.; Feng, J.C. Mining microsatellite markers from public expressed sequence tag sequences for genetic diversity analysis in pomegranate. *J. Genet.* **2012**, *91*, 353–358. [CrossRef]
44. Rania, J.; Salwa, Z.; Najib, H.; Amal, B.D.; Messaoud, M.; Amel, S.H. Microsatellite polymorphism in Tunisian pomegranates (*Punica granatum* L.): Cultivar genotyping and identification. *Biochem. Syst. Ecol.* **2012**, *44*, 27–35. [CrossRef]
45. Peakall, R.; Smouse, P.E. GenAlEx 6.5: Genetic analysis in Excel. Population genetic software for teaching and research–an update. *Bioinformatics* **2012**, *28*, 2537–2539. [CrossRef] [PubMed]

molecules

MDPI

Article

Protective Effects of Green Tea Supplementation against Lead-Induced Neurotoxicity in Mice

Areej Al-Qahtani [1], Jamaan Ajarem [2], Mohammad K. Okla [1,*], Samina Rubnawaz [3,*], Saud A. Alamri [1], Wahidah H. Al-Qahtani [4], Ahmad R. Al-Himaidi [2], Hamada Abd Elgawad [5], Nosheen Akhtar [6], Saleh N. Maodaa [2] and Mostafa A. Abdel-Maksoud [2]

1 Department of Botany and Microbiology, College of Science, King Saud University, Riyadh 11451, Saudi Arabia; aamalkahtanee@kau.edu.sa (A.A.-Q.); saualamri@ksu.edu.sa (S.A.A.)
2 Department of Zoology, College of Science, King Saud University, Riyadh 11451, Saudi Arabia; jaarem@ksu.edu.sa (J.A.); ahimaidi@ksu.edu.sa (A.R.A.-H.); maodaa_28@yahoo.com (S.N.M.); harrany@gmail.com (M.A.A.-M.)
3 Department of Biochemistry, Faculty of Biological Sciences, Quaid I Azam University, Islamabad 45320, Pakistan
4 Department of Food Sciences & Nutrition, College of Food & Agriculture Sciences, King Saud University, Riyadh 11451, Saudi Arabia; wahida@ksu.edu.sa
5 Integrated Molecular Plant Physiology Research, Department of Biology, University of Antwerp, 2020 Antwerpen, Belgium; hamada.abdelgawad@uantwerpen.be
6 Department of Biological Sciences, National University of Medical Sciences, Rawalpindi 46000, Pakistan; nosheenakhtar@numspak.edu.pk
* Correspondence: malokla@ksu.edu.sa (M.K.O.); samina.r.nawaz@gmail.com (S.R.)

Citation: Al-Qahtani, A.; Ajarem, J.; Okla, M.K.; Rubnawaz, S.; Alamri, S.A.; Al-Qahtani, W.H.; Al-Himaidi, A.R.; Elgawad, H.A.; Akhtar, N.; Maodaa, S.N.; et al. Protective Effects of Green Tea Supplementation against Lead-Induced Neurotoxicity in Mice. *Molecules* **2022**, 27, 993. https://doi.org/10.3390/molecules27030993

Academic Editor: Luciana Mosca

Received: 21 October 2021
Accepted: 28 January 2022
Published: 1 February 2022

Abstract: The use of natural products as therapeutic agents is rapidly growing recently. In the current study, we investigated the protective effects of green tea supplementation on lead-induced toxicity in mice. Forty albino mice were divided into four groups as follows: A: control group; B: green tea receiving group; C: lead-intoxicated group; and D: lead-intoxicated group supplemented with green tea. At the end of the experiment, the animals were tested for neurobehavioral and biochemical alterations. Green tea was analyzed through Gas Chromatography–Mass Spectrometry (GC/MS) analysis. We found that supplementation with green tea ameliorated the lead-associated increase in body weight and blood glucose. Green tea supplementation also changed the blood picture that was affected due to lead toxicity and ameliorated lead-induced dyslipidemia. The group of mice that were supplemented with green tea has shown positive alterations in locomotory, anxiety, memory, and learning behaviors. The GC/MS analysis revealed many active ingredients among which the two most abundant were caffeine and 1,2-benzenedicarboxylic acid, mono(2-ethylhexyl) ester. We concluded that green tea supplementation has several positive effects on the lead-induced neurotoxicity in mice and that these effects may be attributed to its main two active ingredients.

Keywords: caffeine; dyslipidemia; GC/MS; lead toxicity; neurobehavior

1. Introduction

Lead is a chemical toxicant, which can cause severe damage to the blood [1] and many body organs like the liver, kidney, brain, spleen, and lungs [2]. It is one of the most important toxic heavy elements in the environment [3] which can penetrate the blood–brain barrier (BBB) and hence results in lead poisoning that can cause non-traumatic brain injury (NTBI) [4]. Not only does it break through the BBB, but it also increases its permeability, thus leaving the brain vulnerable to other toxic substances [5]. In humans, lead poisoning can lead to persistent cognitive deficit, which is more disabling than the physical complications [6]. In such cases, comprehensive neuropsychological testing and cognitive rehabilitation are essential in the treatment of lead poisoning [7]. In mice models, lead toxicity has several pathological consequences as liver injury [8] and brain damage [9].

The search for new safe natural therapeutic agents is now attracting many research groups globally [10]. Many natural compounds have revealed positive effects against different pathological conditions [10,11]. Green tea (*Camellia sinensis*) is a popular and commonly consumed drink, and its extract is found in many herbal and dietary supplements [12]. The protective effects of green tea and its main constituents against natural and chemical toxins are now being extensively investigated [13]. Green tea improved some neuro-functional and biochemical signs of arsenic toxicity in rats [14]. The protective effect of green tea on lead-induced oxidative and DNA damage in rat blood and brain tissue homogenates has also been demonstrated [15,16]. It can also protect against lead toxicity in rat kidneys [17], livers [18], and brains [19]. Human studies have provided evidence of an association between lead concentrations in blood and green tea consumption [20]. This may explain the reason why natural products, micronutrients, and nutraceuticals are now being used to treat depression [21,22]. The current study aimed to investigate the positive effects of green tea on the physiological and neurobehavioral changes induced by lead intoxication in mice.

2. Results

2.1. Ameliorating Effect of Green Tea Intake on Body Weight in Lead-Intoxicated Mice

As illustrated in Table 1, the group of mice that have been supplemented with green tea has shown no significant change in the body weight (33.29 ± 2.42) as compared to the control group (34.30 ± 1.72). On the other hand, a significant increase ($p < 0.05$) in body weight has been observed in the lead-intoxicated group (37.9 ± 1.07) in comparison to the control group (34.30 ± 1.72). Surprisingly, the lead-intoxicated group that has been supplemented with green tea, has shown restoration in the bodyweight increase (35.19 ± 1.14) as compared to the lead-intoxicated group (37.9 ± 1.07). Water consumption was decreased in both of lead intoxicated group (10.02 ± 0.17) and the green tea supplemented-lead intoxicated group (12.10 ± 0.16) as compared to the control (13.75 ± 0.15). Meanwhile, no significant change could be detected between the green tea supplemented group (14.49 ± 0.34) and the control group (13.75 ± 0.15).

Table 1. Effect of green tea supplementation on the body weight and water consumption in the lead-intoxicated mice.

Groups	Body wt. (g)	Water Consumption (mL)
Control	34.30 ± 1.72	13.75 ± 0.15
Green tea group (GTE)	33.29 ± 2.42	14.49 ± 0.34
Lead group (Pb)	37.9 ± 1.07 *	10.02 ± 0.17 *
Lead + GT group (GTE + Pb)	35.19 ± 1.14	12.10 ± 0.16

Mean ± SEM., n = 10. * $p < 0.05$ for lead intoxicated group vs. control.

2.2. Green Tea Supplementation Has Improved Hematological Parameters in the Lead-Intoxicated Mice

The effect of green tea supplementation on the cell blood count (CBC) was investigated in the experimental groups. As illustrated in Table 2, the green tea supplemented group has shown a CBC resembling that of the control with slight but no significant change in the red blood corpuscles (RBCs) count, white blood corpuscles (WBCs) count, hemoglobin (Hb) content, hematocrit (HCT), mean corpuscular volume (MCV), mean corpuscular hemoglobin (MCH), mean corpuscular hemoglobin concentration (MCHC), and platelets (PLT) count. Conversely, the lead intoxicated group of mice has shown a disturbed CBC characterized by a significant increase ($p < 0.05$) in the number of WBCs (9.34 ± 1.55), HCT (49.60 ± 1.33), and PLT (793 ± 78.37). In the lead-intoxicated group that has been supplemented with green tea (Lead + GT group), the green tea supplementation ameliorated the lead-associated hematological abnormalities to some extent. Here, the number of RBCs (8.16 ± 0.33), WBCs (8.03 ± 1.28), HCT (43.86 ± 1.62), MCV (53.80 ± 0.66), and

PLT (685.6 ± 23.8) has been reached to near control values when compared to the lead intoxicated group.

Table 2. Effect of green tea supplementation on the cell blood count (CBC) of mice.

Parameters	Control	Green Tea Group	Lead Group	Lead + GT Group
RBCs	7.73 ± 0.13	8.10 ± 0.15	8.45 ± 0.24	8.16 ± 0.33
WBCs	7.01 ± 0.22	6.36 ± 1.20	9.34 ± 1.55 *	8.03 ± 1.28
Hb	14.02 ± 0.51	15.08 ± 0.21	12.58 ± 0.35	13.74 ± 0.60
HCT	43.92 ± 1.65	47.66 ± 0.49 #	49.60 ± 1.33 *	43.86 ± 1.62
MCV	56.08 ± 2.11	59.00 ± 0.95	58.80 ± 0.86	53.80 ± 0.66
MCH	18.16 ± 0.62	18.62 ± 0.31	18.44 ± 0.39	16.84 ± 0.21
MCHC	31.96 ± 0.28	31.66 ± 0.14	31.42 ± 0.26	31.32 ± 0.22
PLT	657.8 ± 17.9	663.6 ± 19.2	793 ± 78.37 *	685.6 ± 23.8 &

Mean ± SEM., n = 10. * $p < 0.05$ for lead intoxicated group vs. control; # $p < 0.05$ for green tea group vs. control; & $p < 0.05$ for lead +GT group vs. control.

2.3. Hypoglycemic Effect of Green Tea Supplementation against Elevated Blood Glucose Levels in Mice

As illustrated in Figure 1, the supplementation with green tea has a hypoglycemic effect represented by a decreased blood glucose level (119.20 ± 11.00) in the group of mice receiving green tea only as compared to the control group (132.94 ± 6.72). Lead intoxication has increased blood glucose levels (144.12 ± 9.82) in comparison to the control. However, the oral supplementation with green tea in the lead-intoxicated group has exerted a significant blood-glucose-lowering effect ($p < 0.05$) compared to the lead intoxicated group. Moreover, this group (Lead + GT) also presented glucose values comparable to the normal control group. Overall, the glucose levels followed the pattern of Lead > Lead + GT > Control > Green tea among different mice groups.

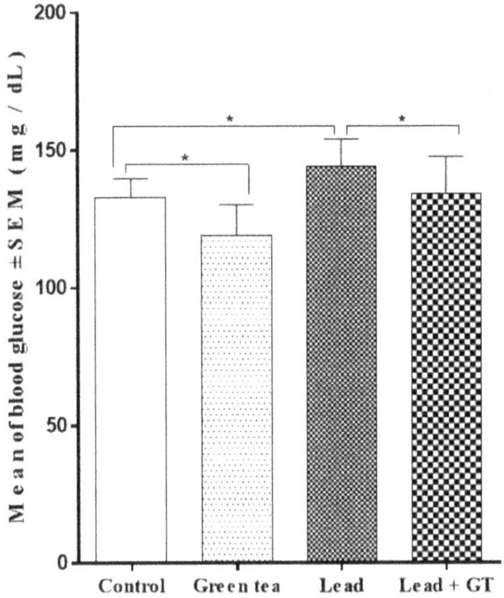

Figure 1. Effect of green tea supplementation on the blood glucose (mg/dL) level. Values are represented as Mean ± SEM., n = 10. * = $p < 0.05$.

2.4. Ameliorating Effect of Green Tea on the Lead Intoxication-Associated Dyslipidemia

The effect of green tea on the blood levels of cholesterol, triglycerides, and LDL was also investigated. The green tea supplemented group of animals has lower levels of cholesterol (89.20 ± 2.27) (Figure 2a), triglycerides (83.52 ± 5.99) (Figure 2b), and LDL (46.32 ± 4.32) (Figure 2c) compared to the control group. The lead-intoxicated group has shown a dyslipidemic profile represented as increased blood levels of cholesterol (114.00 ± 9.12), triglycerides (145.20 ± 10.18), and LDL (79.30 ± 9.09) compared to the control group. However, the lead-intoxicated group supplemented with green tea has shown an amelioration in the lipid profile as represented by the restoration of near-normal values of blood levels of cholesterol (108.60 ± 7.86), triglycerides (104.96 ± 6.12), and LDL (64.54 ± 7.40) compared to the control group. Overall, Lead + GT group showed a more significant decrease ($p < 0.01$) in the triglyceride and LDL content than the cholesterol level ($p < 0.01$), which is initially increased by lead poisoning.

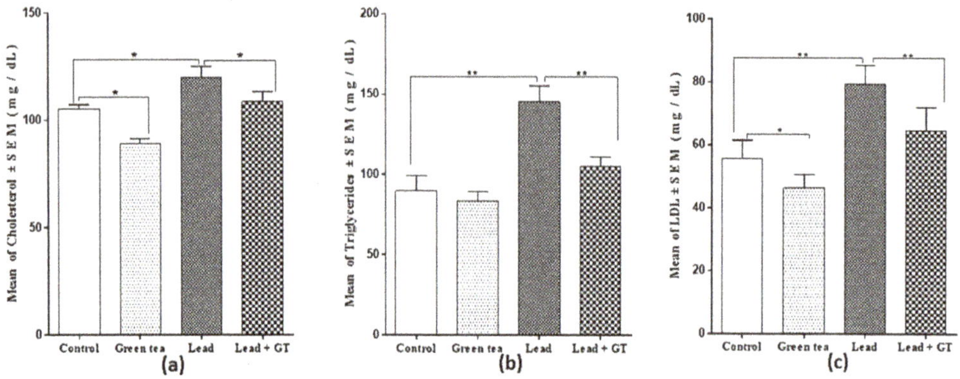

Figure 2. (**a**) Effect of green tea supplementation on the cholesterol level. (**b**) Effect of green tea supplementation on the triglycerides level. (**c**) Effect of green tea supplementation on the low-density lipoprotein (LDL) level. Mean ± SEM., n = 10. * = $p < 0.05$, ** = $p < 0.01$.

2.5. Decreased Accumulation of Lead in Blood and Brain of Green Tea-Supplemented Mice

The concentration of lead in both blood and brain was estimated. As illustrated in Figure 3, no significant change in the blood level of lead could be detected between both control (1.20 ± 0.026) and green tea supplemented (1.36 ± 0.028) groups. Conversely, a highly elevated ($p < 0.01$) level of lead (7.09 ± 0.266) was found in the blood samples from the lead-intoxicated group of mice. The supplementation with green tea has significantly decreased ($p < 0.05$) this elevated level of lead in blood (6.32 ± 0.128). However, this level is still significantly higher ($p < 0.01$) compared to either the control group (1.20 ± 0.026) or the green tea-supplemented group (1.36 ± 0.028).

On the other hand, the cerebral concentration of lead (Figure 4) in the lead-intoxicated group (0.40 ± 0.172) has been significantly elevated ($p < 0.01$) in comparison to both control (0.09 ± 0.026) and the green tea (0.10 ± 0.027) groups of animals. Similar to the situation in blood, green tea supplementation significantly decreased ($p < 0.01$) the cerebral lead concentration. Additionally, the lead-intoxicated group of animals that have received the green tea had a significantly elevated level ($p < 0.01$) of lead (0.35 ± 0.128) in comparison to the control (0.09 ± 0.026).

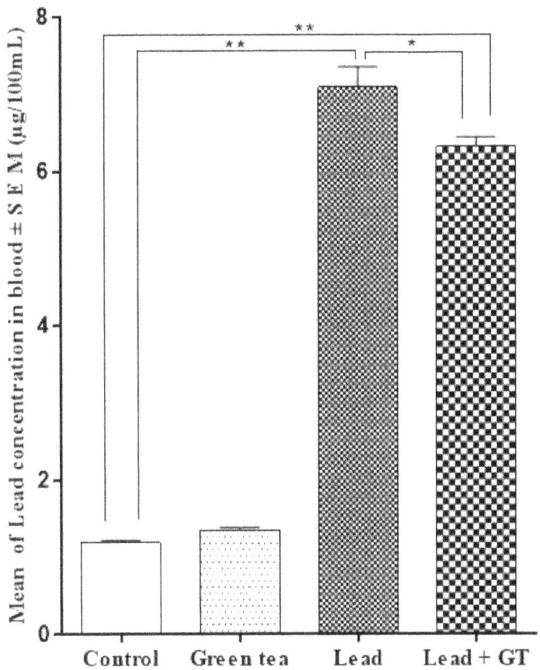

Figure 3. Effect of green tea supplementation on the lead concentration in blood. Mean ± SEM., n = 10. * = $p < 0.05$, ** = $p < 0.01$.

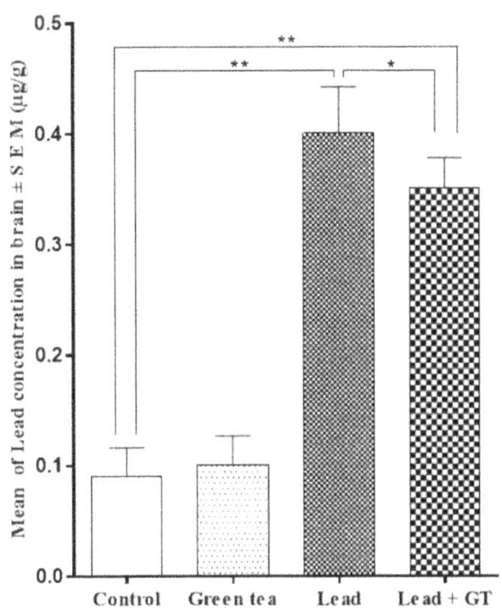

Figure 4. Effect of green tea supplementation on the lead concentration in the brain. Mean ± SEM., n = 10. * = $p < 0.05$, ** = $p < 0.01$.

2.6. Ameliorating Effect of Green Tea on the Locomotory Behavior of Lead-Intoxicated Mice

When investigating the locomotory behavior of the green tea supplemented group in the open area, all the number of squares-crossed, the number of rears, the number of wall-rears, and the locomotion duration have been increased compared to the control group (Table 3). The immobility duration has decreased while the number of washings was not affected. Conversely, in the lead-intoxicated group, the locomotory behavior was negatively affected. All of the number of squares-crossed, the number of rears, the number of wall-rears, and the locomotion duration have been decreased in comparison to the control group. The immobility duration and the number of washings have been increased. Green tea supplementation has positively affected the lead-associated disturbed behavioral responses. The recorded values of the number of squares-crossed, the number of rears, the number of wall-rears, and the locomotion duration have indicated ameliorating effect which sometimes may reach to near the control group values.

Table 3. Locomotory behavior testing in the experimental groups in the open area. * $p < 0.05$ for lead group vs. control.

Test	Control	Green Tea Group	Lead Group	Lead + GT Group
No. of squares-crossed	385	393	297 *	358 &
No. of rears	19	21	12 *	18
No. of wall-rears	38	40	24 *	33 &+
No. of washings	7	6	10 *	8
Locomotion duration (s)	190.6	197.7	104.8 *	139.4 &+
Immobility duration	110.5	103.9	198 *	121.3 &+

Values are Mean ± SEM., $n = 10$ * $p < 0.05$ for lead group vs. green tea group. & $p < 0.05$ for lead +GT group vs. control group. + $p < 0.05$ for lead +GT group vs. green tea group.

2.7. *Antianxiety Effect of Green Tea Supplementation on the Lead-Intoxicated Mice in the Plus-Maze*

The anxiety behavior of all experimental groups of mice was analyzed with the aid of a plus-maze. In the green tea supplemented group, the number of entries into the open arm has been increased (9 ± 0.92) while that of entries into closed-arm has decreased (4.5 ± 1.09) when compared with the control group. Similarly, the time spent in the open arm has increased (173 ± 0.19) while the time spent in the closed arm (83 ± 0.24) and the maze center (38 ± 0.34) has decreased. Oppositely, the lead-intoxicated group have shown a greater anxiety as represented by the decrease in the number of entries into the open arm (6.8 ± 0.64), the increase in the number of entries into the closed arm (6.5 ± 0.66), and the similar disturbance in the time spent in the open arm (60 ± 0.2), closed arm (150 ± 0.23), and in the maze center (57 ± 0.29) compared to the control (Table 4). Green tea supplementation has a clear positive effect on this behavior as represented by the number of entries into the open arm (7.1 ± 0.81), that of entries into the closed arm (5.6 ± 1.78), and the time spent in the open arm (129 ± 0.19), in the closed arm (110 ± 0.27) and the maze center (48 ± 0.16).

Table 4. Anxiety in plus-maze.

Parameter	Control	Green Tea Group	Lead Group	Lead + GT Group Pb
No. of entries into Open arm	7.3 ± 1.71	9 ± 0.92	6.8 ± 0.64	7.1 ± 0.81
Time spent in the open arm (s)	170 ± 0.28	173 ± 0.19	60 ± 0.2 *#	129 ± 0.19 &+
No. of entries into closed arm	4.9 ± 0.87	4.5 ± 1.09	6.5 ± 0.66 *#	5.6 ± 1.78
Time spent in the closed arm (s)	85.8 ± 0.56	83 ± 0.24	150 ± 0.23 *#	110 ± 0.27 &+
Time spent on the maze center (s)	40 ± 0.47	38 ± 0.34	57 ± 0.29 *#	48 ± 0.16 &+

Values are Mean ± SEM., $n = 10$. * $p < 0.05$ for lead group vs. control. # $p < 0.05$ for lead group vs. green tea group. & $p < 0.05$ for lead +GT group vs. control group. + $p < 0.05$ for lead +GT group vs. green tea group.

2.8. Neuroprotective Effects of Green Tea Supplementation as Indicated from Learning and Memory Testing in Automatic Reflex Conditioner

Learning and memory responses were tested in all groups using the automatic reflex conditioner (shuttle box). In the green tea supplemented group, the total latency time (La) and the number of reinforced crossings (Re) during the shock have decreased while the number of intertrial crossing (Ic), the number of the crossing during light and sound stimulus (St) and the number of no crossing during the shock (Tr) have increased compared to the control group (Table 5). The lead-intoxicated group has shown an opposite response, whereas La and the Re have increased while Ic, St, and Tr have decreased compared to the control group. Supplementation with green tea has improved the lead-associated disturbance in learning and memory behaviors. Both La and Re have increased compared to the control, while Ic, St, and Tr have decreased compared to the control group.

Table 5. Learning and memory test in automatic reflex conditioner (shuttle box).

Test	Control	Green Tea Group	Lead Group	Lead + GT Group
La (s)	120 ± 1.22	100 ± 0.99	150 ± 1.45 *#	125 ± 1.69 +
Ic	60 ± 1.6	63 ± 1.76	35 ± 0.37 *#	49 ± 1.32 &+
St	2.7 ± 0.4	2.9 ± 0.45	0.3 ± 0.33 *#	2.3 ± 0.26
Re	21.8 ± 2.41	21 ± 0.47	20.4 ± 1.76	20.7± 0.96
Tr	5.5 ± 2.45	6.1 ± 0.37	9.3 ± 1.9 *#	7.1 ± 0.99 &

Values are mean ± SEM., $n = 10$. * $p < 0.05$ for lead group vs. control. # $p < 0.05$ for lead group vs. green tea group. & $p < 0.05$ for lead +GT group vs. control group. + $p < 0.05$ for lead +GT group vs. green tea group.

2.9. GC/MS Analysis

GC/MS analysis for both aqueous and ethanolic extracts of green tea has revealed two important active ingredients as illustrated in the chromatograms (Figure 5a,b). The first one was the caffeine that appeared at retention time (t_R 36.06 min) and the second one was 1,2-benzenedicarboxylic acid, mono(2-ethylhexyl) ester that appeared at t_R45.57 min. Besides these two compounds, other compounds have appeared at different retention times with relatively small peaks. Surprisingly, both the aqueous and the ethanolic extracts have completely matched their chromatograms.

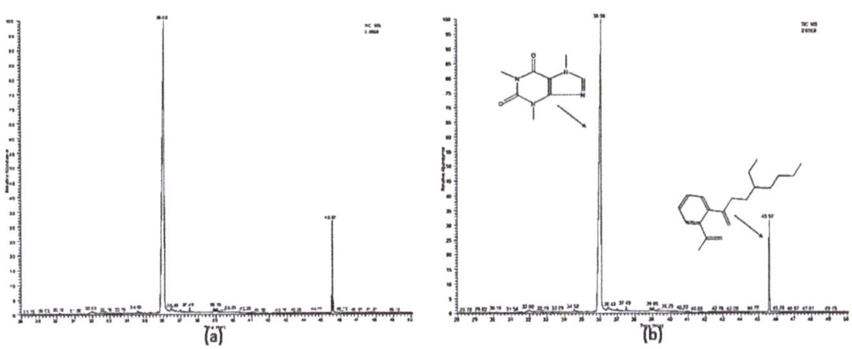

Figure 5. Chromatograms for the (**a**) aqueous extracts and (**b**) the ethanolic extract of green tea. (x-axis = Retention time; y-axis = % intensity/% abundance).

The cyclic structure of both compounds is illustrated below in Figure 6.

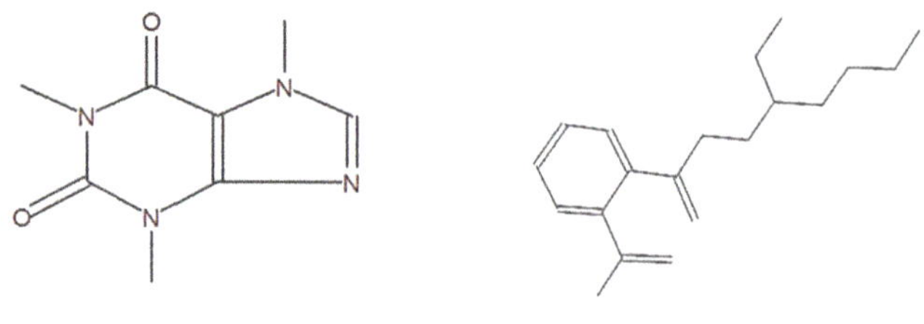

Caffeine 1,2-Benzenedicarboxylic acid, mono(2-ethylhexyl) ester

Figure 6. Chemical structure of caffeine and 1,2-benzenedicarboxylic acid, mono(2-ethylhexyl) ester.

3. Discussion

Green tea is a popular natural beverage used around the world. The current study investigated the positive effects of green tea extracts on the body systems in general and the nervous system particularly. For this, we orally administered lead, green tea, and a combination of lead and green tea to the experimental mice for 6 weeks. We found that the use of green tea restored the body weight gain initially induced by lead intoxication. Previous reports have also shown that lead exposure results in dose-specific weight gain in adult Westar rats [23] and that green tea has positive effects on weight maintenance [24]. Moreover, we noticed that supplementation with green tea has a positive effect on the hematological parameters in lead-intoxicated animals. Previous studies also reported the hematological changes induced by lead intoxication in albino mice [1,25]. It was found that RBCs have a high affinity for lead and approximately 99% of the lead present in the blood is bound to erythrocytes, which makes them more vulnerable to oxidative damage than many other cells [26].

Further, we observed the blood-glucose-lowering effects of green tea on the lead intoxicated group of mice. Previous data also suggest that green tea has an antidiabetic effect both in animal models and even in humans [12]. Interestingly, polysaccharide conjugates from the cell wall of green tea leave exert hypoglycemic effects by augmented insulin secretion. These polysaccharides are also reported to fine-tune the impaired glucose tolerance in vivo [27]. Green tea not only affects glucose metabolism, but also regulates lipid metabolism. Recently, the effects of various crude extracts of green tea on hepatic lipid metabolism and lipid profile have been investigated in mice and rats, respectively [28,29]. Our results further support these observations as we reported ameliorating effect of green tea on lead-associated dyslipidemia.

Our data display higher levels of lead in the blood and brain of lead intoxicated mice groups, which corresponds to many previous studies reporting similar results [30,31]. However, here, green tea supplementation exerted an obvious positive effect on the blood and brain levels of lead. Besides, in an earlier investigation, a mild lead exposure produced severe neurotoxic effects in rats [9]. Generally, lead mimics the calcium dependent pathways because of their similar charge and size. Lead binds to the calcium channels, thus substituting all calcium-dependent processes leading to disruption of brain and behavior [32]. Nevertheless, the protective effects of green tea extract on lead-induced oxidative and DNA damage in rat brains have been well documented [15,18]. In our previous study, we reported the genoprotective and anti-apoptotic effects of green tea [33]. Our current study elucidates that green tea has neurobehavioral positive effects as represented by the locomotory, anxiety, memory, and learning behaviors in green tea supplemented mice. Here, the lead-induced deficits were markedly reversed by the green tea administration. Likewise, another study on old-aged Winstar rats revealed that green tea supplementation can reverse the age-related deficits in learning and memory processes by inhibiting acetylcholinesterase activity [34].

The observed neuroprotective effects of green tea may be attributed to its chemical composition. It contains many polyphenolic compounds, generally known as flavanols or catechins [35] which are effective scavengers of superoxide, hydroxyl, and peroxyl radicals. In addition, catechins are known to chelate metal ions, another contributory mechanism in their protective action against the destructive effect of free radicals on the components of the blood cells [36]. Many studies reported the presence of (-)-epigallocatechin-3-gallate (EGCG), a gallate form of phenolics, in green tea extracts as the most potent and abundant compound in green tea. It possesses anti-proliferative, anti-mutagenic, antioxidant, antibacterial, antiviral, and chemopreventive effects [37,38]. Moreover, EGCG from green tea provides neuroprotection against β-amyloid-induced neuronal death. This protective mechanism often involves the inhibition of overproduction of reactive oxygen species (ROS) by EGCG in a dose-dependent manner [39,40]. Similarly, ECGC protected the Alzheimer-like rat model from neurodegeneration and oxidative damage in their hippocampus [41].

In the light of these findings, we analyzed the aqueous and ethanolic extracts of green tea to know the active ingredients that are responsible for its therapeutic effects. Our findings have revealed two important components which are caffeine and 1,2-benzenedicarboxylic acid, mono(2-ethylhexyl) ester besides other compounds with variable levels. The primary documented effects of caffeine are improvement of memory, concentration, and physical performance [42]. A recent report suggests that moderate and regular consumption of caffeine, a hydrophobic agent, protects against seizure and spatial memory loss [43]. Caffeine can easily cross the BBB, thus providing better neuroprotection against cognitive dysfunction due to excellent bioavailability in the brain [40]. This can rationalize the positive neurological effects observed in our current study. On the other hand, 1,2-benzenedicarboxylic acid mono(2-ethylhexyl) ester has antioxidant and anti-inflammatory properties [44]. Additionally, it exhibited in vitro anticancer potential against liver (HepG2) cancer cells [45]. We did not find ECGC in both crude extracts, which may account for selective solubility of polyphenols in different solvent systems. Taken together, our current data clearly illustrates that green tea has an ameliorating effect on the lead toxicity in mice and hence could be considered as a candidate for human application, especially in lead-polluted areas.

4. Materials and Methods

4.1. Animals

Forty (40) adult male albino mice (*Mus musculus*) with an average weight of 20.2 ± 2.24 g (8–9 weeks' age) were obtained from the animal house at the College of Pharmacy—King Saud University and maintained and monitored in a specific pathogen-free environment. All animal procedures were performed as described before [46]. All animals were allowed to acclimatize in plastic cages in a well-ventilated room for one week before the experiment. The animals were maintained under the standard laboratory conditions described earlier [46], fed a diet of standard commercial pellets, and given water ad libitum.

4.2. Green Tea Extract Preparation

Green tea leaves were purchased from the "Two Leaves" company (Naya Bazar, Pahari Dhira Delhi-110006, India). The crude extract was obtained through the method of cold extraction by taking 120 g of powder in 400 mL of 98% ethanol into a clean dried glass vessel for two weeks at room temperature with shaking. After maceration, the mixture was filtered utilizing Whatman filter paper no.1 with a pore size of 11 μm to separate the extract from the plant debris. The extract was concentrated initially by rotary evaporator at reduced pressure and finally by open air. Animals were daily dosed with the leaves extract of green tea for six weeks.

4.3. Lead Preparation and Dosing Schedule

Lead acetate ($C_4H_6O_4PbH_2O$) with a molecular weight of 379.33 as pure crystals was obtained from Riedel-De HaenAGSeelze-Hannover Germany. Each mouse received 0.5 gm/kg of the extract. This dosage for animals' treatment was chosen based on previous LD_{50} calculations in our lab. Animals were divided into 4 groups (10 mice/group) as follows: Group (I) control group (distilled water); Group (II) green tea group (0.5 mg/kg);

Group (III) lead group (0.2 mg/kg); Group (IV) green tea + lead group (green tea and lead with the same concentrations as taken by groups II and III). Lead acetate and green tea extracts were delivered to the animals through oral intubation.

4.4. Monitoring of Water Consumption and Body Weight Changes

For six weeks, daily water consumption was monitored for all animals in the four groups. The animals were weighed before the commencement of administration and in subsequent weeks during the experiment period.

4.5. Behavioral Studies

Ten animals from each group were used in the current study. For testing, the animals were brought into a room (25 °C) of dim red light reserved for that purpose. All experiments were carried in the early morning for 6 weeks. In all of the behavioral tests, we followed the methods of Ajarem et al. [23].

Locomotory Behavior: The testing was done as described previously [23]. Briefly, the mice are placed in the center of the wooden cage and allowed to freely move for 300 s whereas the number of squares crossed, number of rears, number of wall rears, number of cleanings, duration of locomotion, and duration of immobility are monitored.

Fear and Anxiety Testing in Plus Maze: The testing was done as described previously [23]. Briefly, the mice are placed in the center of the maze and allowed to freely move for 300 s, whereas the number of entries into a closed arm, the number of entries into the open arm, the number of entries into the center, and the time spent for each one is monitored.

Learning and Memory Testing in Shuttle Box: Using the UGO BASILE shuttle box (Italy), whereas the mice are allowed to make thirty trials and the total latency time (La), the number of the reinforced crossing during the shock (Re), the number of intertrial crossing (Ic), the number of no crossing during the shock (Tr), and the number of the crossing during light/sound stimulus (St) are monitored.

4.6. Biochemical Assays

After six weeks, the mice were subjected to anesthesia by putting them in an anesthesia drop-chamber. Inside the chamber, a cotton gauze that was previously soaked with 2 mL of 20% isoflurane in propylene glycol was present. After that, cervical dislocation was done, and blood samples were collected from all animals, and plasma was separated. For cell blood count (CBC), heparinized blood samples were analyzed using Vet abc™ Animal Blood Counter (Horiba ABX, Montpellier, France). For glucose, cholesterol, triglycerides, and low-density lipoprotein (LDL) estimation; Reflotron®® plus was utilized using 30-µL plasma samples being put in specific device strips.

4.7. Lead Estimation Assay

After completion of the behavioral experiment and blood collection, the mice were euthanized and their brain tissues were surgically removed, dissected, and preserved in normal saline. For the analytical determination of Pb, a 10% homogenate of the brain was prepared in Tris-HCl (50 mM). The homogenates were re-centrifuged at 17,000 rpm at 4 °C for 5 min and supernatants were filtered using 0.45 µm pore filters and analyzed by ICP–MS (Inductively Coupled Plasma Mass Spectrometer): ELAN 9000 (Perkin Elmer SciexInstrumento, Concord, Ontario, Canada). Nitric acid (69% *v/v*), super purity grade was supplied from Romil, Cambridge, UK. Hydrochloric acid (37% *v/v*) and hydrofluoric acid (40% *v/v*) were supplied by Merck (Darmstadt, Germany). High purity water obtained from the Millipore Milli-Q water purification system was used throughout the work. The ICP–MS calibration was carried out by external calibration with the blank solution and three working standard solutions (20, 40, and 60 ppm), starting from 1000 mg/L single standard solutions for ICP–MS (A ristar grade, BDH Laboratory Supplies, Poole, UK, for Cd).

4.8. GC-MS Instrumentation

Thermo Trace GC Ultra gas chromatograph coupled with TSQ Quantum mass spectrometer (triple quadrupole) (Thermo Scientific). The experimental conditions were optimized to set the main parameters of chromatographic separation, identification. GC conditions Temperature program: 50 °C (1 min), 10°/min, 250 °C (10 min), Split/splitless injector: 250 °C, Splitless mode, split flow: 50 mL/min, split time: 1 min, Carrier gas: helium, constant flow-rate: 1 mL/min, Transfer line temperature: 250 °C, Column Thermo Trace TR-5MS, Length: 30 m, i.d.: 0.25 mm, film thickness: 0.25 µm, Software: XCalibr software version 2.1, Mass Spectrometry (MS) conditions: full scan, mass range: 40–400 Da, scan time: 0.132 s, ionization mode: electron impact at 70 eV, emission current: 50 µA, compound identification was based on comparison of their mass spectra to those of reference standards obtained from spectral libraries NIST Mass Spectral Libraries v2.1. (National Institute of Standards and Technology) and Wiley.

4.9. Statistical Analysis

Before further statistical analysis, the data were tested for normality using the Anderson–Darling test, as well as for homogeneity variances. The data were normally distributed and is expressed as the mean ± standard error of the mean (SEM). Significant differences among the groups were analyzed by one- or two-way ANOVA followed by Tukey's post-test using SPSS software, version 17. Differences were considered statistically significant at $p < 0.05$.

5. Conclusions

This study adds further evidence of the protective effects of green tea supplementation in lead-induced mice. Oral administration of green tea extracts improved the lead-associated pathological changes in the biochemical and neurobehavioral responses of treated mice in a significant manner. However, further studies are needed to better understand the underlying molecular mechanism for the neuroprotective effects of green tea.

Author Contributions: Conceptualization, A.A.-Q., J.A. and M.A.A.-M.; Data curation, S.N.M.; Formal analysis, W.H.A.-Q. and N.A.; Funding acquisition, M.K.O.; Investigation, A.R.A.-H.; Methodology, A.A.-Q. and J.A.; Project administration, M.K.O.; Resources, and M.A.A.-M.; Software, H.A.E.; Validation, S.A.A.; Visualization, A.A.-Q.; Writing—original draft, A.A.-Q. and J.A.; Writing—review & editing, S.R. and M.A.A.-M. All authors have read and agreed to the published version of the manuscript.

Funding: This research was funded by Researchers Supporting Project number (RSP-2021/374) King Saud University, Riyadh, Saud Arabia.

Institutional Review Board Statement: The study was conducted according to the guidelines of the Declaration of Helsinki and approved by the Institutional Review Board (IRB-HAPO-01-R-011), Al-Imam Mohammad Ibn Saud Islamic University, Saudi Arabia (179-2021).

Informed Consent Statement: Not applicable.

Data Availability Statement: The datasets used and/or analyzed during the current study are available from the corresponding author on reasonable request.

Acknowledgments: The authors extend their appreciation to the Researchers Supporting Project number (RSP-2021/374) King Saud University, Riyadh, Saudi Arabia.

Conflicts of Interest: The authors declare no conflict of interest.

Sample Availability: Samples of the plant extracts are available from the authors.

References

1. Alwaleedi, S.A. Hematobiochemical Changes Induced by Lead Intoxication in Male and Female Albino Mice. *Natl. J. Physiol. Pharm. Pharmacol.* **2016**, *6*, 46–51. [CrossRef]
2. Dart, R.C.; Hurlbut, K.M.; Boyer-Hassen, L.V. Lead. In *Medical Toxicology*, 3rd ed.; Dart, R.C., Ed.; Lippincot Williams and Wilkins: Philadelphia, PA, USA, 2004.

3. Wani, A.; Ara, A.; Usmani, J.A. Lead Toxicity: A Review. *Interdiscip. Toxicol.* **2015**, *8*, 55–64. [CrossRef] [PubMed]
4. Sanders, T.; Liu, Y.; Buchner, V.; Tchounwou, P. Neurotoxic Effects and Biomarkers of Lead Exposure: A Review. *Rev. Environ. Health* **2009**, *24*, 15–45. [CrossRef] [PubMed]
5. Flora, G.; Gupta, D.; Tiwari, A. Toxicity of Lead: A Review with Recent Updates. *Interdiscip. Toxicol.* **2012**, *5*, 47–58. [CrossRef] [PubMed]
6. McKay, C.A. Role of Chelation in the Treatment of Lead Poisoning: Discussion of the Treatment of Lead-Exposed Children Trial (TLC). *J. Med. Toxicol.* **2013**, *9*, 339–343. [CrossRef]
7. Miracle, V.A. Lead Poisoning in Children and Adults. *Dimens. Crit. Care Nurs.* **2017**, *36*, 71–73. [CrossRef]
8. Laamech, J.; El-Hilaly, J.; Fetoui, H.; Chtourou, Y.; Gouitaa, H.; Tahraoui, A.; Lyoussi, B. *Berberis vulgaris* L. Effects on Oxidative Stress and Liver Injury in Lead-Intoxicated Mice. *J. Complement. Integr. Med.* **2017**, *14*, 1–14. [CrossRef]
9. Zhi-wei, Z.; Ru-Lai, Y.; Gui-juan, D.; Zheng-yan, Z. Study on the Neurotoxic Effects of Low-level Lead Exposure in Rats. *J. Zhejiang Univ. Sci. B* **2005**, *6*, 686–692.
10. Necyk, C.; Zubach-Cassano, L. Natural Health Products and Diabetes: A Practical Review. *Can. J. Diabetes* **2017**, *41*, 642–647. [CrossRef]
11. Yarla, N.S.; Bishayee, A.; Sethi, G.; Reddanna, P.; Kalle, A.M.; Dhananjaya, B.L.; Dowluru, K.S.; Chintala, R.; Duddukuri, G.R. Targeting Arachidonic Acid Pathway by Natural Products for Cancer Prevention and Therapy. *Semin. Cancer Biol.* **2016**, *40*, 48–51. [CrossRef]
12. Tsuneki, H.; Ishizuka, M.; Terasawa, M.; Wu, J.B.; Toshiyasu, S.; Ikuko, K. Effect of Green Tea on Blood Glucose Levels and Serum Proteomic Patterns in Diabetic (db/db) Mice and on Glucose Metabolism in Healthy Humans. *BMC Pharmacol.* **2004**, *4*, 18. [CrossRef]
13. Rameshrad, M.; Razavi, B.M.; Hosseinzadeh, H. Protective Effects of Green Tea and its Main Constituents against Natural and Chemical Toxins: A Comprehensive Review. *Food Chem. Toxicol.* **2017**, *100*, 115–137. [CrossRef]
14. Sárközi, K.; Papp, A.; Horváth, E.; Máté, Z.; Ferencz, Á.; Hermesz, E.; Krisch, J.; Paulik, E.; Szabó, A. Green Tea and Vitamin C Ameliorate some Neuro-functional and Biochemical Signs of Arsenic Toxicity in Rats. *Nutr. Neurosci.* **2016**, *19*, 102–109. [CrossRef] [PubMed]
15. Hamed, E.; Meki, A.M.; Abd El-Mottaleb, N.A. Protective Effect of Green Tea on Lead-induced Oxidative Damage in Rat's Blood and Brain Tissue Homogenates. *J. Physiol. Biochem.* **2010**, *66*, 143–151. [CrossRef] [PubMed]
16. Khalaf, A.A.; Moselhy, W.A.; Abdel-Hamed, M.I. The Protective Effect of Green Tea Extract on Lead Induced Oxidative and DNA Damage on Rat Brain. *Neurotoxicology* **2012**, *33*, 280–289. [CrossRef] [PubMed]
17. Abdel-Moneim, A.H.; Meki, A.; Gabr, A.M.; Mobasher, A.; Lutfi, M.F. The Protective Effect of Green Tea Extract against Lead Toxicity in Rats Kidneys. *Asian J. Biomed. Pharm. Sci.* **2014**, *4*, 30–34. [CrossRef]
18. Mehana, E.E.; Meki, A.R.; Fazili, K.M. Ameliorated Effects of Green Tea Extract on Lead Induced Liver Toxicity in Rats. *Exp. Toxicol. Pathol.* **2012**, *64*, 291–295. [CrossRef]
19. Meki, A.; Alghasham, A.; EL-Deeb, E. Effect of Green Tea Extract on Lead Toxicity in Different Organs of Rats. *Int. J. Health Sci.* **2011**, *5*, 12–15.
20. Colapinto, C.; Arbuckle, T.; Dubois, L.; Fraser, W. Is There a Relationship between Tea Intake and Maternal Whole Blood Heavy Metal Concentrations? *J. Expo. Sci. Environ. Epidemiol.* **2016**, *26*, 503–509. [CrossRef] [PubMed]
21. Deacon, G.; Kettle, C.; Hayes, D.; Dennis, C.; Tucci, J. Omega 3 Polyunsaturated Fatty Acids and the Treatment of Depression. *Crit. Rev. Food Sci. Nutr.* **2017**, *57*, 212–223. [CrossRef]
22. Nabavi, S.M.; Daglia, M.; Braidy, N.; Nabavi, S.F. Natural Products, Micronutrients, and Nutraceuticals for the Treatment of Depression: A Short Review. *Nutr. Neurosci.* **2017**, *20*, 180–194. [CrossRef] [PubMed]
23. Ajarem, J.; Naif, G.A.; Ahmed, A.A.; Saleh, N.M.; Mostafa, A.A.; Billy, K.C.C. Oral Administration of Potassium Bromate Induces Neurobehavioral Changes, Alters Cerebral Neurotransmitters Level and Impairs Brain Tissue of Swiss Mice. *Behav. Brain Funct.* **2016**, *12*, 14. [CrossRef]
24. Sun, H.; Ningjian, W.; Xiaomin, N.; Li, Z.; Qin, L.; Zhen, C.; Chi, C.; Meng, L.; Jing, C.; Hualing, Z. Lead Exposure Induces Weight Gain in Adult Rats, Accompanied by DNA Hypermethylation. *PLoS ONE* **2017**, *12*, e0169958. [CrossRef]
25. Hursel, R.; Viechtbauer, W.; Westerterp-Plantenga, M.S. The Effects of Green Tea on Weight Loss and Weight Maintenance: A Meta-analysis. *Int. J. Obes.* **2009**, *33*, 956–961. [CrossRef] [PubMed]
26. LaBreche, H.G.; Sarah, K.M.; Joseph, R.N.; John, P.C. Peripheral Blood Signatures of Lead Exposure. *PLoS ONE* **2011**, *6*, e23043. [CrossRef] [PubMed]
27. Wang, H.; Shi, S.; Bao, B.; Li, X.; Wang, S. Structure Characterization of an Arabinogalactan from Green Tea and its Anti-diabetic Effect. *Carbohydr. Polym.* **2015**, *124*, 98–108. [CrossRef]
28. Guo, Q.; Zhao, B.; Li, M.; Shen, S.; Xin, W. Studies on Protective Mechanisms of Four Components of Green Tea Polyphenols against Lipid Peroxidation in Synaptosomes. *Biochim. Biophys. Acta* **1996**, *1304*, 210–222. [CrossRef]
29. Nam, M.; Choi, M.S.; Choi, J.Y.; Kim, N.; Kim, M.S.; Jung, S.; Kim, J.; Ryu, D.H.; Hwang, G.S. Effect of Green Tea on Hepatic Lipid Metabolism in Mice Fed a High-fat Diet. *J. Nutr. Biochem.* **2018**, *51*, 1–7. [CrossRef]
30. Amanolahi, F.; Rakhshande, H. Effects of Ethanolic Extract of Green Tea on Decreasing the Level of Lipid Profile in Rat. *Avicenna J. Phytomed.* **2013**, *3*, 98–105.

31. Oskarsson, A.; Olson, L.; Michael, R.P.; Birger, L.; Hakan, B.J.; Rkilund, A.; Barry, H. Increased Lead Concentration in Brain and Potentiation of Lead-Induced Neuronal Depression in Rats after Combined Treatment with Lead and Disulfiram. *Environ. Res.* **1986**, *41*, 623–632. [CrossRef]
32. Rocha, A.; Trujillo, K.A. Neurotoxicity of Low-level Lead Exposure: History, Mechanisms of Action, and Behavioral Effects in Humans and Preclinical Models. *Neurotoxicology* **2019**, *73*, 58–80. [CrossRef] [PubMed]
33. Allam, A.A.; Gabr, S.A.; Ajarem, J.; Alghadir, A.H.; Sekar, R.; Chow, B.K. Geno Protective and Anti-Apoptotic Effect of Green Tea against Perinatal Lipopolysaccharide-Exposure Induced Liver Toxicity in Rat Newborns. *Afr. J. Tradit. Complement. Altern. Med.* **2017**, *14*, 166–176. [CrossRef]
34. Kaur, T.; Pathak, C.M.; Pandhi, P.; Khanduja, K.L. Effects of Green Tea Extract on Learning, Memory, Behavior and Acetylcholinesterase Activity in Young and Old Male Rats. *Brain Cogn.* **2008**, *67*, 25–30. [CrossRef] [PubMed]
35. Gurer, H.; Neal, R.; Yang, P.; Oztezcan, S.; Erçal, N. Captopril as An Antioxidant in Lead-exposed Fischer 344 rats. *Hum. Exp. Toxicol.* **1999**, *18*, 27–32. [CrossRef] [PubMed]
36. Campbell, E.L.; Chebib, M.; Johnston, G.A.R. The Dietary Flavonoids Apigenin and (−)-epigallocatechin Gallate Enhance the Positive Modulation by Diazepam of the Activation by GABA of Recombinant GABAA Receptors. *Biochem. Pharmacol.* **2004**, *68*, 1631–1638. [CrossRef]
37. Schramm, L. Going Green: The Role of the Green Tea Component EGCG in Chemoprevention. *J. Carcinog. Mutagen.* **2013**, *4*, 1000142. [CrossRef] [PubMed]
38. Rafieian-Kopaei, M.; Movahedi, A. Breast Cancer Chemopreventive and Chemotherapeutic Effects of *Camellia sinensis* (Green Tea): An Updated Review. *Electron. Physician* **2017**, *9*, 3838–3844. [CrossRef]
39. Bastianetto, S.; Yao, Z.X.; Papadopoulos, V.; Quirion, R. Neuroprotective Effects of Green and Black Teas and Their Catechin Gallate Esters against β-amyloid-induced Toxicity. *Eur. J. Neurosci.* **2006**, *23*, 55–64. [CrossRef]
40. Chen, S.Q.; Wang, Z.S.; Ma, Y.X.; Zhang, W.; Lu, J.L.; Liang, Y.R.; Zheng, X.Q. Neuroprotective Effects, and Mechanisms of Tea Bioactive Components in Neurodegenerative Diseases. *Molecules* **2018**, *23*, 512. [CrossRef]
41. Schimidt, H.L.; Garcia, A.; Martins, A.; Mello-Carpes, P.B.; Carpes, F.P. Green Tea Supplementation Produces Better Neuroprotective Effects than Red and Black Tea in Alzheimer-like Rat Model. *Food Res. Int.* **2017**, *100*, 442–448. [CrossRef]
42. Simone, C.; Piacentino, D.; Gabriele, S.; Mariarosaria, A. Caffeine: Cognitive and Physical Performance Enhancer or Psychoactive Drug? *Curr. Neuropharmacol.* **2015**, *13*, 71–88.
43. Zhou, X.; Zhang, L. The Neuroprotective Effects of Moderate and Regular Caffeine Consumption in Alzheimer's Disease. *Oxid. Med. Cell. Longev.* **2021**, *2021*, 5568011. [CrossRef] [PubMed]
44. Bagavathi, P.E.; Ramasamy, N. GC-MS Analysis of Phytocomponents in the Ethanol Extract of *Polygonum chinense* L. *Pharmacogn. Res.* **2012**, *4*, 11–14.
45. Selvakumar, J.N.; Chandrasekaran, S.D.; Doss, G.P.C.; Kumar, T.D. Inhibition of the ATPase Domain of Human Topoisomerase IIa on HepG2 Cells by 1, 2-benzenedicarboxylic Acid, Mono (2-ethylhexyl) Ester: Molecular Docking and Dynamics Simulations. *Curr. Cancer Drug Targets* **2019**, *19*, 495–503. [CrossRef] [PubMed]
46. Altoom, N.G.; Ajarem, J.S.; Allam, A.A.; Maodaa, S.N.; Abdel-Maksoud, M.A. Deleterious effects of Potassium Bromate Administration on Renal and Hepatic Tissues of Swiss Mice. *Saudi J. Biol. Sci.* **2018**, *25*, 278–284. [CrossRef]

molecules

MDPI

Article

In Vitro Study of the Bioavailability and Bioaccessibility of the Main Compounds Present in Ayahuasca Beverages

Joana Gonçalves [1,2], Miguel Castilho [1], Tiago Rosado [1,2], Ângelo Luís [1,2,*], José Restolho [1], Nicolás Fernández [3], Eugenia Gallardo [1,2,*] and Ana Paula Duarte [1,2,*]

1 Centro de Investigação em Ciências da Saúde (CICS-UBI), Universidade da Beira Interior, Av. Infante D. Henrique, 6200-506 Covilhã, Portugal; joanadgoncalves13@gmail.com (J.G.); miguel.castilho@ubi.pt (M.C.); tiagorosadofful@hotmail.com (T.R.); jrestolho@gmail.com (J.R.)
2 Laboratório de Fármaco-Toxicologia, UBIMedical, Universidade da Beira Interior, Estrada Municipal 506, 6200-284 Covilhã, Portugal
3 Cátedra de Toxicología y Química Legal, Laboratorio de Asesoramiento Toxicológico Analítico (CENATOXA), Facultad de Farmacia y Bioquímica, Universidad de Buenos Aires, Junín 956, Ciudad Autónoma de Buenos Aires (CABA), Buenos Aires C1113AAD, Argentina; nfernandez@ffyb.uba.ar
* Correspondence: angelo.luis@ubi.pt (Â.L.); egallardo@fcsaude.ubi.pt (E.G.); apcd@ubi.pt (A.P.D.); Tel.: +351-275-329-003 (Â.L.& E.G.); +351-275-329-002 (A.P.D.)

Abstract: Ayahuasca is a psychoactive beverage that contains the psychoactive compound *N*,*N*-dimethyltryptamine and *β*-carboline alkaloids. This study aims at determining in vitro the bioavailability and bioaccessibility of the main compounds present in decoctions of four individual plants, in a commercial mixture and in four mixtures of two individual plants used in the preparation of Ayahuasca. The samples were subjected to an in vitro digestion process, and the Caco-2 cell line was used as an absorption model. The integrity and permeability of the cell monolayer were evaluated, as well as the cytotoxicity of the extracts. After digestion and cell incubation, the compounds absorbed by the cell monolayer were quantified by high-performance liquid chromatography coupled to a diode array detector. The results showed that compounds such as *N*,*N*-dimethyltryptamine, Harmine, Harmaline, Harmol, Harmalol and Tetrahydroharmine were released from the matrix during the in vitro digestion process, becoming bioaccessible. Similarly, some of these compounds, after being incubated with the cell monolayer, were absorbed, becoming bioavailable. The extracts did not show cytotoxicity after cell incubation, and the integrity and permeability of the cell monolayer were not compromised.

Keywords: ayahuasca; bioavailability; bioaccessibility; PAMPA; HPLC-DAD

Citation: Gonçalves, J.; Castilho, M.; Rosado, T.; Luís, Â.; Restolho, J.; Fernández, N.; Gallardo, E.; Duarte, A.P. In Vitro Study of the Bioavailability and Bioaccessibility of the Main Compounds Present in Ayahuasca Beverages. *Molecules* **2021**, *26*, 5555. https://doi.org/10.3390/molecules26185555

Academic Editors: Salvatore Genovese and Celestino Santos-Buelga

Received: 22 August 2021
Accepted: 10 September 2021
Published: 13 September 2021

1. Introduction

Ayahuasca is a psychoactive beverage traditionally consumed in the Amazon Basin of South America [1]. This word of Quechua origin, means "vine of the soul" or "vine of the dead" and is composed of the terms "*aya*" and "*wasca*", which means "spirit" and "vine", respectively [2,3]. This psychoactive beverage consists of a thick, oily and brownish decoction, which is prepared from the leaves of *Psychotria viridis* (*P. viridis*) and scraps from the stem of *Banisteriopsis caapi* (*B. caapi*) [4,5]. However, over the years, the preparation of Ayahuasca has undergone variations [1,2,6]. Thus, some species of natural origin have been used in the preparation of the beverage, namely *Brunfelsia* spp., *Daturaolens*, *Malouetia tamarquina*, *Psychotria carthagenesis*, *Brugmansia suaveolens*, *Tabernaemontana* spp., or *Nicotiana tabacum*, replacing *P. viridis*, or *Peganum harmala*, Harmine freebase/HCl, Moclobemide and Tetrahydroharmine freebase/HCl, replacing *B. caapi* [6,7].

This psychoactive beverage was traditionally used by native healers for divine cults and in the cure of psychological disorders, stimulation of visual creativity and creative thinking [1,6]. Its hallucinogenic effects are due to the presence of N,N-dimethyltryptamine

(DMT) (Figure 1) from *P. viridis*, which behaves as an agonist of serotonin receptors (5-HT1A/2A/2C) [3]. When ingested alone, this compound is inactive, as it is rapidly metabolized by peripheral monoamine oxidase A (MAO-A) [8]. However, in the presence of β-carbolinic alkaloids, such as Harmaline, Harmine and Tetrahydroharmine (THH), from *B. caapi*, DMT can access the central nervous system, since these are temporary inhibitors of hepatic and intestinal MAO-A [2]. THH also acts as a serotonin reuptake inhibitor, enhancing the effects of DMT [2,3]. The knowledge about this synergy between compounds present in the two plants has been known by indigenous peoples for about 3000 years [9].

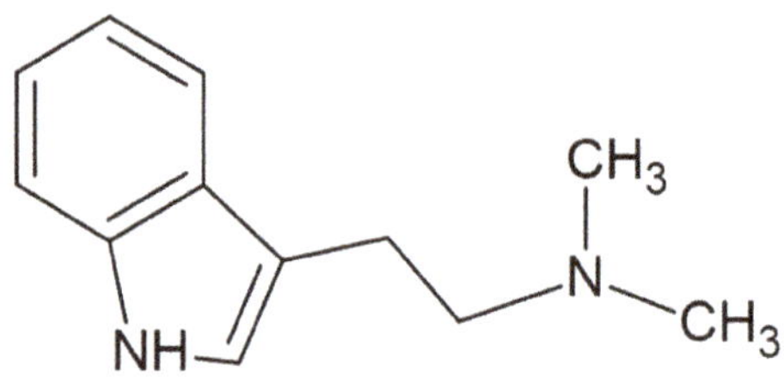

Figure 1. Molecular structure of DMT.

Although Ayahuasca has been consumed for centuries, in the last 25 years, its use has been increasing in different parts of the world [10,11]. The consumption of Ayahuasca in controlled environments such as religious rituals and experimental procedures is not associated with psychotic episodes [10]. However, the expansion of Ayahuasca has raised some concerns about the possible adverse effects associated with consumption, but also an interest in its potential therapeutic effects [6,10].

Bioavailability and bioaccessibility are important concepts that make it possible to understand the behaviour of some compounds in the body. Bioaccessibility consists of the amount of a compound that is released from a matrix, being available to be absorbed, after ingestion and consequent digestion [12]. On the other hand, bioavailability is defined as the fraction that reaches the bloodstream and that, after metabolization and distribution, produces an effect [13]. In vitro digestion is a procedure that has been used to determine the fraction of compounds that are released from the matrix and become bioaccessible [14]. In vitro digestion models aim to mimic the digestive process along the digestive tract (mouth, stomach and intestine), simulating physiological conditions such as pH, salt concentration, digestion time, among others [15,16]. Regarding the assessment of bioaccessibility, cell lines are frequently used, namely the line derived from a human colon carcinoma Caco-2, due to its similar morphology with the cells of the small intestine [17]. In addition, these cells have narrow intracellular junctions and express enzymes similar to those that are present in the small intestine, allowing to mimic the transport mechanisms that occur therein [18–20].

There are no studies concerning the fate of the active ingredients of Ayahuasca formulations after ingestion, namely concerning their absorption to the general circulation for distribution. Therefore, with this study we aimed at evaluating the bioavailability and bioaccessibility of the active compounds present in four individual plants, a commercial mixture and four plant mixtures, used to prepare the Ayhauasca decoction. For that, an in vitro digestion process was used, as well as the parallel artificial membrane permeability assay (PAMPA) using Caco-2 cells, in order to know better their path in the human body and part of the mechanisms regulating their passage to the blood stream.

2. Results and Discussion

Considering the several potential bioactive properties of Ayahuasca, four individual decoctions of each plant used in the preparation of Ayahuasca were prepared in this work, as well as four decoctions of a mixture of plants (with two different plant materials, one source of DMT and the other of β-carboline alkaloids). A decoction of a commercial mixture was also prepared. The bioavailability and bioaccessibility of the main compounds present in Ayahuasca was evaluated in the nine samples.

2.1. Characterization of Main Compounds in Initial Samples and after Digestive Process

DMT is present in some plants used in the preparation of Ayahuasca beverages. Given the use of plant samples containing this compound in religious rituals, it has received some attention over the years due to its psychoactive effects [21]. Besides that, the access of this compound to the bloodstream is dependent on the β-carboline alkaloids [1,8,22,23]. Thus, an analytical method by high performance liquid chromatography coupled diode array detector (HPLC-DAD) was developed, which allowed the quantification of the main active compounds present in the samples of Ayahuasca beverages (Table 1). This analytical method was developed and validated in accordance with the standards of the Food and Drug Administration [24]. Thus, it was linear between 0.16 and 10.00 μg/mL for Harmol, THH, Harmaline and Harmine, between 0.31 and 10.00 μg/mL for Harmalol and between 0.031 and 1.00 μg/mL for DMT, with coefficients of determination (R^2) higher than 0.997. The intra- and inter-day precision revealed coefficients of variation below 15% and the accuracy was within the range of ±15%. The LOD and LLOQ obtained were 0.31 μg/mL for all compounds, except for DMT (0.031 μg/mL).

All samples from a mixture of two plants showed substantial concentrations of DMT, with the mixture of *M. hostilis* and *B. caapi* having the highest concentration, and the mixture of *P. viridis* and *B. caapi* having the lowest concentration. Regarding the individual samples, both the *P. viridis* and *M. hostilis* decoctions and the commercial mixture showed substantial amounts of DMT. In contrast, in the decoctions of *B. caapi* and *P. harmala*, this compound was not detected. Moreover, all mixtures presented considerable concentrations of β-carboline alkaloids, with the mixture of *M. hostilis* and *P. harmala* presenting the highest amount. Regarding the individual samples, these compounds were not detected in the decoctions of *P. viridis* and *M. hostilis*. On the other hand, in the decoctions of *B. caapi* and *P. harmala*, all β-carboline alkaloids were detected, with THH and Harmol being in greater quantity in *B. caapi* and Harmine, Harmalol and Harmaline in greater quantity in *P. harmala*. Regarding the commercial mixture, it was possible to detect all β-carbolines, except for Harmalol. Bensalem et al. [25] carried out the quantification of Harmine, Harmaline, Harmol and Harmalol in samples of *P. harmala*, having verified that, similarly to what was observed in the present study, the compound with the highest concentration was Harmaline, followed by Harmine, Harmalol and, the least concentrated, Harmol. In addition, Avula et al. [26] carried out the quantification of Harmol, Harmine, Harmaline, among other compounds, using ultra-performance liquid chromatography-triple quadrupole mass spectroscopy with ultraviolet detection (UPLC-UV-MS) and high-performance thin-layer chromatography (HPTLC). It was found that, similarly to the results now obtained, Harmine was found in a higher quantity than Harmol, being Harmaline not detected [26]. Several studies were also carried out, with the aim of determining the concentration of DMT and β-carbolines in Ayahuasca samples. Pires et al. [27] used gas chromatography equipment with nitrogen/phosphorous detector to quantify DMT, Harmine, Harmaline and THH in eight samples of Ayahuasca. Similar to what was observed in the present work, the four compounds were detected in all samples [27]. Moreover, Souza et al. [28] analysed 38 Ayahuasca samples using liquid chromatography coupled to tandem-mass spectrometry (LC-MS/MS) verified the presence of THH, DMT, Harmine and Harmaline. Recently, Chambers et al. [29] quantified the DMT present in 6 samples of Ayahuasca, obtaining values between 45.7 and 230.5 mg/L. It is important to point out that the concentrations of each compound in the Ayahuasca samples can be very variable. This fact can be due to a number of factors, namely the variability of the proportion used by each user, as well as the different preparation methods [27,28]. Additionally, the concentration of the compounds in each plant can also be very variable [27]. According to Kaasik et al. [30], the average variations of concentrations of DMT, THH, Harmine and Harmaline, can be, respectively, 26.2%, 29.8%, 41.5% and 2.5%. The samples used in this study were acquired online, making it difficult to know their degree of purity.

Table 1. Concentration of the main compounds of ayahuasca in different vegetal samples. The values are expressed as mean (μg/mg extract) ± SD.

Samples	Compounds	Initial Concentrations
P. viridis	DMT	6.50 ± 0.01
B. caapi	THH	5.00 ± 0.10
	Harmol	0.14 ± 0.00
	Harmine	10.00 ± 0.28
	Harmalol	0.05 ± 0.00
	Harmaline	4.68 ± 0.14
P. harmala	THH	3.05 ± 0.04
	Harmol	0.02 ± 0.00
	Harmine	12.00 ± 0.00
	Harmalol	0.66 ± 0.01
	Harmaline	17.00 ± 0.00
M. hostilis	DMT	10.50 ± 0.02
Commercial mixture	DMT	10.40 ± 0.01
	THH	2.09 ± 0.07
	Harmol	0.01 ± 0.00
	Harmine	0.02 ± 0.00
	Harmalol	ND
	Harmaline	0.37 ± 0.02
P. viridis + *B. caapi*	DMT	4.50 ± 0.01
	THH	2.50 ± 0.07
	Harmol	0.01 ± 0.00
	Harmine	0.48 ± 0.00
	Harmalol	ND
	Harmaline	0.07 ± 0.00
P. viridis + *P. harmala*	DMT	6.50 ± 0.09
	THH	0.63 ± 0.05
	Harmol	0.02 ± 0.00
	Harmine	0.30 ± 0.01
	Harmalol	0.08 ± 0.00
	Harmaline	0.48 ± 0.01
M. hostilis + *B. caapi*	DMT	8.00 ± 0.02
	THH	1.90 ± 0.06
	Harmol	0.03 ± 0.00
	Harmine	0.82 ± 0.02
	Harmalol	0.04 ± 0.00
	Harmaline	0.12 ± 0.00
M. hostilis + *P. harmala*	DMT	8.50 ± 0.01
	THH	3.44 ± 0.05
	Harmol	0.06 ± 0.00
	Harmine	9.00 ± 0.00
	Harmalol	0.36 ± 0.00
	Harmaline	13.5 ± 0.06

ND-not detected.

After quantifying the main compounds present in samples of Ayahuasca beverages, the same compounds were quantified over the three stages of the in vitro digestion process (salivary, gastric and duodenal). By observing the aliquots collected in each step, it is possible to verify that there were colour variations throughout the process. Likewise, the concentrations of DMT and β-carboline alkaloids also varied between samples and, within the same sample, between digestion steps (Table 2).

Table 2. Concentration of the main compounds of ayahuasca in different digestion steps. The values are expressed as mean (μg/mL) ± SD.

Samples	Compounds	Salivary	Gastric	Duodenal
P. viridis	DMT	0.84 ± 0.60	7.77 ± 0.08	7.49 ± 0.19
B. caapi	THH	0.83 ± 0.00	0.78 ± 0.13	1.05 ± 0.09
	Harmol	ND	ND	ND
	Harmine	1.56 ± 0.00	4.13 ± 0.03	1.98 ± 0.03
	Harmalol	ND	ND	ND
	Harmaline	0.19 ± 0.00	0.33 ± 0.00	0.21 ± 0.00
P. harmala	THH	1.52 ± 0.09	1.66 ± 0.12	1.32 ± 0.01
	Harmol	ND	ND	ND
	Harmine	18.38 ± 0.18	19.52 ± 0.05	10.02 ± 0.01
	Harmalol	1.47 ± 0.04	1.54 ± 0.02	1.03 ± 0.07
	Harmaline	29.66 ± 0.10	26.18 ± 0.14	22.88 ± 0.26
M. hostilis	DMT	9.55 ± 0.03	8.96 ± 0.17	8.33 ± 0.00
Commercial mixture	DMT	4.28 ± 0.05	4.09 ± 0.02	3.38 ± 0.08
	THH	0.95 ± 0.05	1.42 ± 0.00	0.50 ± 0.03
	Harmol	ND	ND	ND
	Harmine	ND	ND	ND
	Harmalol	1.21 ± 0.00	1.02 ± 0.01	0.81 ± 0.01
	Harmaline	1.29 ± 0.03	1.12 ± 0.01	0.85 ± 0.00
P. viridis + *B. caapi*	DMT	2.37 ± 0.01	1.61 ± 0.05	2.00 ± 0.02
	THH	3.33 ± 0.05	4.05 ± 0.11	2.82 ± 0.02
	Harmol	0.29 ± 0.00	0.34 ± 0.00	0.26 ± 0.01
	Harmine	1.23 ± 0.03	3.09 ± 0.22	1.80 ± 0.05
	Harmalol	0.26 ± 0.00	0.27 ± 0.02	0.25 ± 0.00
	Harmaline	0.19 ± 0.01	0.40 ± 0.00	0.26 ± 0.01
P. viridis + *P. harmala*	DMT	4.30 ± 0.08	3.94 ± 1.33	4.56 ± 0.15
	THH	ND	ND	ND
	Harmol	ND	ND	ND
	Harmine	1.64 ± 0.01	6.70 ± 0.12	3.05 ± 0.10
	Harmalol	0.37 ± 0.01	0.38 ± 0.01	0.34 ± 0.04
	Harmaline	4.62 ± 0.02	8.93 ± 0.04	4.05 ± 0.05
M. hostilis + *B. caapi*	DMT	9.07 ± 0.04	5.52 ± 0.09	7.36 ± 0.05
	THH	2.89 ± 0.21	2.37 ± 0.11	2.45 ± 0.04
	Harmol	0.28 ± 0.02	0.24 ± 0.01	0.22 ± 0.00
	Harmine	4.02 ± 0.04	10.34 ± 0.07	7.69 ± 0.27
	Harmalol	ND	ND	ND
	Harmaline	0.28 ± 0.01	0.42 ± 0.01	0.29 ± 0.01
M. hostilis + *P. harmala*	DMT	9.65 ± 0.12	4.76 ± 0.07	6.68 ± 0.16
	THH	0.89 ± 0.01	ND	ND
	Harmol	ND	ND	ND
	Harmine	4.41 ± 0.07	10.92 ± 0.23	9.38 ± 0.05
	Harmalol	0.60 ± 0.03	0.87 ± 0.01	0.63 ± 0.04
	Harmaline	11.45 ± 0.20	16.96 ± 0.11	12.08 ± 0.02

ND-not detected.

Analysing the results, it was possible to verify that the amount of DMT varies throughout the in vitro digestion process. In general, the concentration of DMT at the end of the entire process decreased in samples of *M. hostilis*, in the commercial mixture and in the mixtures of *M. hostilis* and *B. caapi* and *M. hostilis* and *P. harmala*. Conversely, there was an increase in DMT in the sample of *P. viridis*, while in the mixtures of *P. viridis* and *B. caapi* and *P. viridis* and *P. harmala* there were no noticeable changes. With respect to β-carbolines, there was a variation from compound to compound. The concentration of Harmol remained constant throughout the digestion process of the sample of *M. hostilis* and *B. caapi*, increased in the mixture of *P. viridis* and *B. caapi* and was not detected in the other samples. It was also not possible to detect Harmalol during the digestion of the samples of *B. caapi* and in the mixture of *M. hostilis* and *B. caapi*. In the other samples where this compound was initially detected, its concentration decreased slightly. Regarding THH, it was verified that its concentration increased in the samples of *B. caapi* and decreased in the commercial mixture and in mixtures of *P. viridis* and *B. caapi* and *M. hostilis* and *P. harmala*. A slight decrease of this compound was also observed in the sample of *P. harmala* and in the mixture of *M. hostilis* and *B. caapi*. In the mixture of *P. viridis* and *P. harmala* this compound was not detected. It was verified that the concentration of Harmine increased, except in the commercial mixture (not detected) and in *P. harmala* (decreased). The concentration of Harmaline remained constant in the mixture of *M. hostilis* and *B. caapi* and decreased in the sample of *P. harmala*, in the commercial mixture and in the mixture of *P. viridis* and *P. harmala*. In the samples of *B. caapi*, and in the mixtures of *P.viridis* and *B. caapi* and *M. hostilis* and *P. harmala*, there was a slight increase in the concentration of Harmaline. These variations in the concentrations of β-carboline alkaloids may be due to the fact that these compounds degrade and easily give rise to another β-carboline (Figure 2) [31].

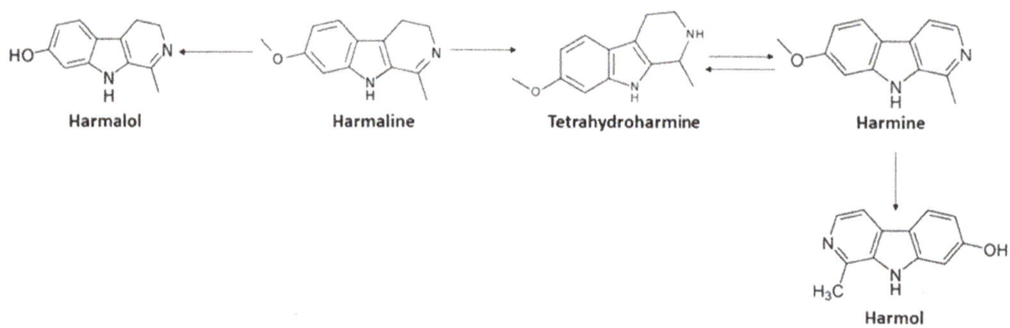

Figure 2. Reactions between β-carboline alkaloids.

So far, no bioaccessibility studies have been carried out on Ayahuasca or plants used in its preparation. Digestion studies including this type of samples have not been carried out so far, so it is not possible to make a comparison.

2.2. Cell Culture

2.2.1. Evaluation of Cell Viability

The cytotoxicity of each sample was assessed using the MTT assay. In analysing the results, it was verified that there was a slight decrease in cell viability in the samples of the digested commercial mixture and in the crude extract of *B. caapi*. The other samples showed no decrease in cell viability (Table 3). These results are in agreement with those obtained by Katchborian-Neto et al. [32], which evaluated the cytotoxicity of Ayahuasca samples in SH-SY5Y cells. Additionally, three of the samples intensely increased cell viability within the first 48 h [32]. In addition, Samoylenko et al. [33] evaluated the cytotoxicity of *B. caapi* extracts in six cell lines, verifying that the extracts did not show cytotoxicity.

Table 3. Cell viability after exposure to extracts. The values are expressed as mean ± SD.

Samples	Cell Viability (%)
P. viridis Crude	156.01 ± 27.31
P. viridis Digested	128.85 ± 9.03
B. caapi Crude	95.92 ± 1.83
B. caapi Digested	113.66 ± 11.59
P. harmala Crude	171.46 ± 28.88
P. harmala Digested	117.62 ± 3.59
M. hostilis Crude	148.28 ± 14.18
M. hostilis Digested	96.04 ± 12.23
Commercial mixture Crude	101.50 ± 13.25
Commercial mixture Digested	79.52 ± 0.93
P. viridis + *B. caapi* Crude	148.07 ± 26.83
P. viridis + *B. caapi* Digested	103.74 ± 3.43
P. viridis + *P. harmala* Crude	162.55 ± 15.63
P. viridis + *P. harmala* Digested	127.75 ± 9.97
M. hostilis + *B. caapi* Crude	126.61 ± 16.39
M. hostilis + *B. caapi* Digested	118.50 ± 1.59
M. hostilis + *P. harmala* Crude	138.41 ± 17.63
M. hostilis + *P. harmala* Digested	120.70 ± 3.12

2.2.2. Evaluation of the Electrical Resistance of the Cell Transendothelial Membrane

The integrity of the cell monolayer was evaluated by the TEER assay, before and after cell incubation with the extracts (Table 4). The TEER assay allows monitoring the integrity of cell layers in in vitro assays, as well as possible changes in intercellular junctions, by evaluating transendothelial electrical resistance [34]. Analysing the results of the TEER measurements before incubation with the extracts, it was observed that the monolayer was intact, since the values were above the 150–200 Ω cm^2 range, minimum acceptable limit [35]. After incubation with the extracts, a new TEER measurement was performed, with no significant differences between the values of the first and second measurements. Additionally, the values of the second measurement were also above the minimum acceptable limit. Therefore, the integrity of the cell monolayer was confirmed [35]. So far, there are no studies with Ayahuasca samples where the TEER assay has been performed.

Table 4. TEER values obtained before and after incubation with the extracts. The values are expressed as mean ± SD. Statistically significant values were considered if $p < 0.05$ (*).

Samples	TEER (Ω cm^2)		
	Before	After	*p*-Value
Control	990 ± 31.11	1034 ± 31.11	0.293
P. viridis Crude	1298 ± 155.56	1628 ± 207.94	0.239
P. viridis Digested	1518 ± 93.34	2046 ± 155.56	0.054
B. caapi Crude	1166 ± 155.56	1408 ± 110.73	0.146
B. caapi Digested	1232 ± 134.42	1276 ± 116.41	0.317
P. harmala Crude	1386 ± 217.79	1408 ± 124.45	0.913
P. harmala Digested	1188 ± 186.68	1496 ± 110.73	0.107
M. hostilis Crude	1254 ± 155.56	1298 ± 31.11	0.733
M. hostilis Digested	1584 ± 177.82	1496 ± 116.41	0.112
Commercial mixture Crude	1694 ± 155.56	1716 ± 232.83	0.754
Commercial mixture Digested	1232 ± 44.00	1232 ± 25.40	0.643
P. viridis + *B. caapi* Crude	1364 ± 186.68	1496 ± 0.00	0.423
P. viridis + *B. caapi* Digested	1166 ± 93.34	1386 ± 31.11	0.087
P. viridis + *P. harmala* Crude	1415 ± 93.34	1408 ± 91.59	0.936
P. viridis + *P. harmala* Digested	1100 ± 141.44	1144 ± 116.41	0.795
M. hostilis + *B. caapi* Crude	1232 ± 76.21	1276 ± 25.40	0.189
M. hostilis + *B. caapi* Digested	1188 ± 248.90	1408 ± 127.02	0.619
M. hostilis + *P. harmala* Crude	1254 ± 93.34	1430 ± 31.11	0.127
M. hostilis + *P. harmala* Digested	1232 ± 177.82	1452 ± 0.00	0.246

2.2.3. Evaluation of Cell Monolayer Permeability

Cell monolayer permeability was assessed by Lucifer Yellow permeability assay (Table 5). The Lucifer Yellow permeability assay allows evaluating the permeability characteristics of a cell monolayer, by measuring the passive diffusion of different molecules through it [36]. This assay was performed after exposing the cells to extracts. Analysing the results, it was shown that there were no significant changes in cell permeability, when compared to the control. These results are in agreement with those obtained in the TEER assay, suggesting that there were no changes in intracellular spaces, nor in cell barrier function and in membrane permeability [37,38]. Previous studies also suggest that both TEER measurement and permeability are related, and that a significant increase in the permeability is accompanied by a decrease in TEER values [37,38]. Similarly to what was observed in the TEER assay, no studies were found where the Lucifer Yellow permeability assay was performed with Ayahuasca samples.

Table 5. Percentage of permeability of Caco-2 cells after incubation with the extracts. The values are expressed as mean ± SD. Statistically significant values were considered if $p < 0.05$ (*).

Samples	Permeability (%)	*p*-Value
Control	16.94 ± 2.35	-
P. viridis Crude	19.59 ± 3.00	0.281
P. viridis Digested	13.49 ± 1.03	0.165
B. caapi Crude	16.79 ± 0.14	0.879
B. caapi Digested	16.11 ± 0.49	0.823
P. harmala Crude	15.01 ± 0.46	0.462
P. harmala Digested	17.97 ± 1.37	0.523
M. hostilis Crude	14.38 ± 0.72	0.322
M. hostilis Digested	14.10 ± 0.41	0.267
Commercial mixture Crude	19.88 ± 2.84	0.383
Commercial mixture Digested	16.13 ± 1.83	0.865
P. viridis + *B. caapi* Crude	16.42 ± 0.40	0.959
P. viridis + *B. caapi* Digested	16.03 ± 1.50	0.463
P. viridis + *P. harmala* Crude	13.81 ± 0.49	0.225
P. viridis + *P. harmala* Digested	13.47 ± 1.85	0.283
M. hostilis + *B. caapi* Crude	13.07 ± 1.89	0.139
M. hostilis + *B. caapi* Digested	13.18 ± 0.16	0.074
M. hostilis + *P. harmala* Crude	12.40 ± 1.64	0.069
M. hostilis + *P. harmala* Digested	11.65 ± 1.79	0.058

2.2.4. Characterization of the Main Compounds after Cell Incubation

The amount of compounds present in the collected aliquots after cell incubation of crude and digested extracts were also quantified by HPLC-DAD (Tables 6 and 7). It was verified that DMT, Harmine and Harmaline are the compounds, from those present in the digested extract, which cross the cell monolayer the most. Similarly, in the crude extract the same results were observed. The concentration of these three compounds in all samples and for both extracts, increased gradually in the basolateral compartment throughout the incubation period, except in the digested extract in the mixture of *P. viridis* and *B. caapi*, where DMT increases after 2 h of incubation, remaining approximately constant until 4 h of cell incubation. In general, in the digested extract, all the compounds gradually increased during cell incubation, except for Harmol and Harmalol, which were not detected during the entire process. Similarly, in the crude extract, Harmalol was not detected, but Harmol was detected after 2 h of incubation in the *P. harmala* sample. Moreover, as in what was observed in the digested extract, in the crude extract all compounds gradually increased during cell incubation, except for the THH present in the mixture of *M. hostilis* and *B. caapi*, which decreases slightly after 2 h of incubation, increasing again after 4 h.

Table 6. Concentration of the main compounds of Ayahuasca in the aliquots collected after the different incubation times with the crude extract (Mean µg/mL Extract) ± SD.

Samples	Compounds	Time		
		1 h	2 h	4 h
P. viridis	DMT	0.50 ± 0.01	0.89 ± 0.21	1.22 ± 0.05
B. caapi	THH	ND	0.66 ± 0.02	0.59 ± 0.06
	Harmol	ND	ND	ND
	Harmine	0.33 ± 0.03	1.00 ± 0.01	1.35 ± 0.03
	Harmalol	ND	ND	ND
	Harmaline	ND	ND	ND
P. harmala	THH	ND	ND	0.72 ± 0.12
	Harmol	ND	0.19 ± 0.02	ND
	Harmine	2.09 ± 0.02	4.41 ± 0.23	5.90 ± 0.04
	Harmalol	ND	ND	ND
	Harmaline	3.65 ± 0.06	5.71 ± 0.60	8.55 ± 0.17
M. hostilis	DMT	ND	1.18 ± 0.10	1.42 ± 0.06
Commercial mixture	DMT	ND	0.55 ± 0.04	0.73 ± 0.04
	THH	ND	ND	ND
	Harmol	ND	ND	ND
	Harmine	ND	ND	ND
	Harmalol	ND	ND	ND
	Harmaline	ND	ND	ND
P. viridis + *B. caapi*	DMT	0.16 ± 0.01	0.47 ± 0.06	0.63 ± 0.07
	THH	0.65 ± 0.02	0.78 ± 0.06	0.93 ± 0.04
	Harmol	ND	ND	ND
	Harmine	ND	0.42 ± 0.06	0.76 ± 0.04
	Harmalol	ND	ND	ND
	Harmaline	ND	ND	ND
P. viridis + *P. harmala*	DMT	0.33 ± 0.03	1.01 ± 0.16	1.10 ± 0.07
	THH	ND	ND	ND
	Harmol	ND	ND	ND
	Harmine	0.35 ± 0.02	0.39 ± 0.01	0.78 ± 0.04
	Harmalol	ND	ND	ND
	Harmaline	0.31 ± 0.00	0.67 ± 0.09	0.76 ± 0.03
M. hostilis + *B. caapi*	DMT	0.69 ± 0.07	1.67 ± 0.11	1.87 ± 0.04
	THH	0.74 ± 0.08	0.53 ± 0.03	0.65 ± 0.07
	Harmol	ND	ND	ND
	Harmine	0.98 ± 0.11	1.64 ± 0.11	2.42 ± 0.08
	Harmalol	ND	ND	ND
	Harmaline	ND	ND	ND
M. hostilis + *P. harmala*	DMT	0.29 ± 0.01	0.72 ± 0.04	1.24 ± 0.18
	THH	ND	ND	ND
	Harmol	ND	ND	ND
	Harmine	0.47 ± 0.02	1.07 ± 0.10	1.63 ± 0.27
	Harmalol	ND	ND	ND
	Harmaline	0.55 ± 0.08	1.07 ± 0.14	1.42 ± 0.22

ND-not detected.

Table 7. Concentration of the main compounds of Ayahuasca in the aliquots collected after the different incubation times with the digested extract (Mean μg/mL) ± SD.

Samples	Compounds	Time		
		1 h	2 h	4 h
P. viridis	DMT	0.73 ± 0.00	1.48 ± 0.02	1.99 ± 0.03
B. caapi	THH	ND	ND	ND
	Harmol	ND	ND	ND
	Harmine	ND	ND	1.14 ± 0.03
	Harmalol	ND	ND	ND
	Harmaline	ND	ND	ND
P. harmala	THH	ND	ND	ND
	Harmol	ND	ND	ND
	Harmine	1.83 ± 0.01	3.08 ± 0.09	4.19 ± 0.03
	Harmalol	ND	ND	ND
	Harmaline	3.81 ± 0.13	4.75 ± 0.12	5.63 ± 0.08
M. hostilis	DMT	ND	1.61 ± 0.07	1.90 ± 0.02
Commercial mixture	DMT	ND	0.61 ± 0.02	0.67 ± 0.02
	THH	ND	ND	ND
	Harmol	ND	ND	ND
	Harmine	ND	ND	ND
	Harmalol	ND	ND	ND
	Harmaline	ND	ND	ND
P. viridis + B. caapi	DMT	ND	0.50 ± 0.07	0.48 ± 0.01
	THH	ND	ND	ND
	Harmol	ND	ND	ND
	Harmine	ND	ND	0.70 ± 0.02
	Harmalol	ND	ND	ND
	Harmaline	ND	ND	ND
P. viridis + P. harmala	DMT	ND	0.65 ± 0.00	0.86 ± 0.03
	THH	ND	ND	ND
	Harmol	ND	ND	ND
	Harmine	ND	1.02 ± 0.01	1.26 ± 0.03
	Harmalol	ND	ND	ND
	Harmaline	0.74 ± 0.00	0.99 ± 0.01	1.26 ± 0.01
M. hostilis + B. caapi	DMT	0.78 ± 0.01	1.74 ± 0.01	2.11 ± 0.02
	THH	ND	ND	ND
	Harmol	ND	ND	ND
	Harmine	0.77 ± 0.00	1.96 ± 0.01	2.54 ± 0.64
	Harmalol	ND	ND	ND
	Harmaline	ND	ND	ND
M. hostilis + P. harmala	DMT	0.60 ± 0.03	1.57 ± 0.06	1.86 ± 0.02
	THH	ND	ND	ND
	Harmol	ND	ND	ND
	Harmine	0.85 ± 0.03	2.38 ± 0.09	3.34 ± 0.04
	Harmalol	ND	ND	ND
	Harmaline	1.10 ± 0.02	2.21 ± 0.01	3.02 ± 0.03

ND-not detected.

In general, it was possible to observe that all the analysed compounds managed to cross the cell monolayer, except Harmalol and Harmol. In the digested samples the bioavailability percentages ranged between 8.30–28.9% for DMT, 0–29.63% for Harmaline and 33.03–57.58% for Harmine. So far, no studies have been carried out on the bioavailability of Ayahuasca, so it is not possible to make comparisons with the present study. However, differences in β-carboline concentrations can be explained by the rapid mutual conversion of these compounds (Figure 2) [31]. Additionally, DMT easily crosses the barriers of the body, since it is a small and hydrophobic molecule with a low molecular weight [21]. It was also observed that the amount of the compounds decreased after crossing the cell monolayer, when compared to the values obtained after in vitro digestion. This fact has already been verified in bioavailability studies with other compounds of natural origin [39,40]. In a study that evaluated the bioavailability and bioaccessibility of *Prunus avium* L., carried out by our research group, this same decrease in the amount of compounds after crossing the Caco-2 cell monolayer was also verified [14].

3. Materials and Methods

3.1. Chemicals and Materials

The analytical standards DMT, Harmine, Harmaline, THH, Harmol and Harmalol were kindly provided by Nal von Minden, GmbH (Regensburg, Germany). Lucifer Yellow, 3-[4,5-dimethylthiazol-2-yl]-2,5-diphenyltetrazolium bromide (MTT) and Roswell Park Memorial Institute (RPMI) medium were obtained from Sigma-Aldrich (Sintra, Portugal). Methanol (HPLC grade) was obtained from Fischer Chemical (Loughborough, UK). Formic acid and dimethyl sulfoxide (99.9% of purity) were purchased from Sigma-Aldrich (Sintra, Portugal). Deionized water was obtained from a Milli-Q System (Millipore, Billerica, MA, USA).

3.2. Sample and Working Solutions Preparation

All vegetal samples were acquired online from Shayana Shop (https://www.shayanashop.com, Amsterdam, The Netherlands) (accessed on 25 May 2019). The decoctions of Ayahuasca were prepared according to a traditional recipe kindly provided by Dr. Nicolás Fernández. Thus, 0.210 g of each of the five vegetal samples were weighed (*P. viridis* (leaves), *P. harmala* (seeds), *B. caapi* (scraps from the stem), *M. hostilis* (root bark) and commercial mixture) and were then macerated in a mortar with a few drops of water. After that, 250 mL of ultra-pure water was added, and the mixture was transferred to a Schott flask. This preparation was boiled at 100 °C for 4 h. Similarly, four decoctions were prepared where two of the above vegetal samples were mixed (*P. viridis* and *P. harmala*; *P. viridis* and *B. caapi*; *M. hostilis* and *P. harmala*; *M. hostilis* and *B. caapi*). After boiling, the samples were cooled, filtered, frozen at −80 °C and freeze-dried.

Individual stock solutions of DMT, Hamine, Harmaline, Harmol and Harmalol were prepared at 1 mg/mL in methanol. Working solutions were prepared by serial dilutions in methanol.

3.3. In Vitro Simulation of Human Digestion Process

The in vitro digestion assay was carried out as described in a previous work [14]. Initially, salivary fluid (potassium chloride, monosodium phosphate, sodium sulphate, sodium chloride, sodium bicarbonate, urea, α-amylase, mucin and uric acid), gastric fluid (sodium chloride, monosodium phosphate, potassium chloride, calcium chloride, ammonium chloride, hydrochloric acid, glucose, urea, pepsin, mucin and bovine serum albumin), duodenal fluid (sodium chloride, sodium bicarbonate, potassium dihydrogen phosphate, potassium chloride, magnesium chloride, hydrochloric acid, urea, calcium chloride dihydrate, bovine serum albumin, pancreatin and lipase) and bile fluid (sodium chloride, sodium bicarbonate, potassium chloride, hydrochloric acid, urea, calcium chloride dihydrate, bile and bovine serum albumin) were prepared. For the assay, each freeze-dried decoction was dissolved in 100 mL of deionized water. To each of the nine samples, 6 mL of simulated salivary fluid (pH 6.8) was added, being this mixture was incubated at 37 °C for 5 min with orbital shaking at 90 rpm. Then, 12 mL of simulated gastric fluid (pH 1.3) was added, followed by incubation in the same conditions for 2 h. After this time, 6 mL of simulated bile fluid (pH 8.2), 12 mL of simulated duodenal fluid (pH 8.1) and 2 mL of sodium bicarbonate solution (1M) was added. The solution was incubated again at 37 °C with orbital shaking at 90 rpm for 2 h.

Aliquots were collected at the end of each stage of the in vitro digestion process, which were immediately cooled, then frozen at −80 °C for 30 min and later sonicated at 4 °C in an ice bath for 30 min. Subsequently, the samples were filtered through a 0.22 μm cellulose acetate pore filter and subsequently analysed by high-performance liquid chromatography (HPLC). Furthermore, an additional aliquot of the last stage of the in vitro digestion of each sample was collected, which was also placed on ice, frozen at −80 °C and sonicated. Then, their pH was measured and corrected, when necessary, to physiological pH. Subsequently, these aliquots were used in the PAMPA assay.

3.4. *Cell Culture*

The Caco-2 cell line (Database name: American Type Culture Collection (ATCC) Accession numbers: HTB-37)[41] was cultured in RPMI medium supplemented with 1% antibiotic mixture and 10% foetal bovine serum, at passages between 33 and 37. Subsequently, the cells were incubated at 37 °C in a humidified atmosphere containing 5% CO_2.

For the MTT assay, the cells were seeded in 96 multi-well plates (cat. number 734-2802 avantor, VWR, Amadora Portugal) at a cell density of 0.5×10^4. For the PAMPA assay, the cells were seeded in culture inserts, placed in 12 multi-well plates (cat. number 734-2731 avantor, Laborspirit, Santo Antão de Tojal, Portugal) at a cell density of 6×10^4, remaining for a period of 21 days in order to form a confluent monolayer. After that time, 500 μL of each of the nine samples (digested and undigested) was added to the apical chamber, to be in contact with the cell monolayer (*P. viridis*—0.278 mg/mL; *B. caapi*—0.062 mg/mL; *P. harmala*—0.226 mg/mL; *M. hostilis*—0.382 mg/mL; commercial mixture—0.156 mg/mL; *P. viridis* + *B. caapi*—0.203 mg/mL; *P. viridis* + *P. harmala*—0.344 mg/mL; *M. hostilis* + *B. caapi*—0.555 mg/mL e *M. hostilis* + *P. harmala*—0.4 mg/mL). After 1, 2 and 4 h, 250 μL was collected in the basolateral chamber. The collected aliquots were analysed by HPLC. All tests were performed in triplicate.

3.4.1. MTT Cell Viability Assay

The cytotoxicity of the samples was assessed by the MTT assay. For that, after the cells became confluent, they were exposed to the samples (digested and undigested) 1, 2 and 4 h. RPMI medium was used as a negative control. After incubation, the medium was removed and an MTT solution was added. Then, the cells were incubated for 3 h. After that time, the MTT solution was removed, and the formazan crystals formed were dissolved in dimethyl sulfoxide (DMSO), being the absorbance measured using a microplate reader at 570 nm.

3.4.2. Transepithelial Electrical Resistance Assay

The integrity of the cell monolayer was evaluated by measuring the transepithelial electrical resistance (TEER). Before the incubation of the cells' monolayer with the extracts (digested and undigested), the TEER was measured. Initially, the electrode of the transepithelial resistance meter (EVOM2, World Precision Instrument, Sarasora, FL, USA) was equilibrated with RPMI medium and then was placed in each well to form an angle of 90°. The procedure was performed in triplicate and the TEER was determined according to the following equation:

$$\text{TEER value} = (\text{mean of the resistances of each well} - \text{mean of the resistance of blank}) \times \text{insert area} \quad (1)$$

3.4.3. Lucifer Yellow Permeability Assay

The Lucifer Yellow Permeability Assay allows evaluating changes in the permeability characteristics of the cell monolayer after passive passage of compounds. This test was performed as described in a previous work [14]. Briefly, the RPMI medium of the chambers delimited by the insert (apical and basolateral) was removed and replaced by 500 μL of the Lucifer Yellow solution in the apical chamber and 1.5 mL of Hank's balanced salt solution (HBSS) in the basolateral chamber. After that, the multi-well was incubated for 1 h, and then 200 μL of each basolateral chamber was pipetted to another culture plate, being the fluorescence measured at 485 nm (excitation) and 535 nm (emission) using a spectrofluorimeter. HBSS was used as a blank and a Lucifer Yellow solution (0.1 mg/mL) was used as a positive control. The permeability percentage was calculated as follows:

$$\% \text{ permeability} = (\text{mean of fluorescence of each well} - \text{fluorescence of blank})/(\text{fluorescence of positive control} - \text{fluorescence of blank}) \times 100 \quad (2)$$

3.5. Instrumental and Chromatographic Conditions

The quantification of main compounds present in Ayahuasca beverages was performed on an HPLC system coupled to a diode array detector (DAD) (Agilent technologies Soquímica, Lisbon, Portugal). The mobile phase was composed of 0.1% formic acid in methanol (A) and 0.1% formic acid in water (B). The elution was carried out in gradient mode and included 5% A (0–2 min), 50% A (2–32 min) and again, 5% A (32–40 min). The flow rate was 1.5 mL/min, and the injection volume was 50 μL. The stationary phase consisted of an YMC-Triart PFP (5 μm, 4.6 i.d. × 150 mm) analytical column coupled to a Guard-c holder (4 × 10 mm) and a Triart PFP (5 μm, 3 ×10 mm) pre-column, all from YMC Europe GMBH (Solítica, Lisbon, Portugal), being maintained at 25 °C. Harmine and Harmol were detected at 246 nm, DMT and THH at 278 nm and Harmaline and Harmalol at 360 nm. The temperature of the sampler was set at 4 °C.

3.6. Statistical Analysis

The results are expressed as mean values with standard deviations (SD). The Student's *t*-test was employed and statistically significant values were considered when $p < 0.05$ (*).

4. Conclusions

During the in vitro digestion, the compounds were released from the matrix, becoming bioaccessible. The concentration of β-carboline alkaloids shows an appreciable transformation, while the variation of DMT is smaller. After the in vitro digestion, the detected compounds could be absorbed by the cell monolayer, becoming bioavailable but in lower concentrations. Likewise, the compounds present in the extracts that did not undergo in vitro digestion also became bioavailable. So, it can be inferred that digestion is not essential to occur absorption at the intestine level.

After cell incubation with the extracts, it was verified that they were not cytotoxic, and the integrity and the permeability of the cell monolayer remained unchanged, suggesting that the compounds did not interfere with intercellular junctions.

Further studies in which a more realistic approximation of the intestinal matrix is used should be carried out in order to overcome possible flaws of the used model.

Author Contributions: Conceptualization, Â.L., E.G. and A.P.D.; plant materials, J.R.; sample preparation, N.F.; methodology, J.G. and T.R.; chromatographic analysis and data, M.C. and T.R.; formal analysis, J.G.; investigation, J.G.; writing—original draft preparation, J.G.; writing—review and editing, J.G., Â.L., E.G. and A.P.D.; supervision, Â.L., E.G. and A.P.D.; funding acquisition, Â.L., J.R., E.G. and A.P.D. All authors have read and agreed to the published version of the manuscript.

Funding: This work was partially supported by CICS-UBI, which is financed by National Funds from Fundação para a Ciência e a Tecnologia (FCT) and by Fundo Europeu de Desenvolvimento Re-gional (FEDER) under the scope of PORTUGAL 2020 and Programa Operacional do Centro (CENTRO 2020), with the project reference UIDB/00709/2020. Joana Gonçalves acknowledges the PhD fellowship from FCT (Reference: SFRH/BD/149360/2019). Ângelo Luís acknowledges the contract of Scientific Employment in the scientific area of Microbiology financed by FCT.

Institutional Review Board Statement: Not applicable.

Informed Consent Statement: Not applicable.

Data Availability Statement: The data presented in this study are available on request from the corresponding authors.

Conflicts of Interest: The authors declare no conflict of interest.

Sample Availability: Not available.

References

1. Estrella-Parra, E.A.; Almanza-Pérez, J.C.; Alarcón-Aguilar, F.J. Ayahuasca: Uses, Phytochemical and Biological Activities. *Nat. Prod. Bioprospect.* **2019**, *9*, 251–265. [CrossRef]
2. Hamill, J.; Hallak, J.; Dursun, S.M.; Baker, G. Ayahuasca: Psychological and Physiologic Effects, Pharmacology and Potential Uses in Addiction and Mental Illness. *Curr. Neuropharmacol.* **2018**, *17*, 108–128. [CrossRef] [PubMed]
3. Gonçalves, J.; Luís, Â.; Gallardo, E.; Duarte, A.P. Psychoactive substances of natural origin: Toxicological aspects, therapeutic properties and analysis in biological samples. *Molecules* **2021**, *26*, 1397. [CrossRef] [PubMed]
4. Andrade, T.S.; de Oliveira, R.; da Silva, M.L.; Von Zuben, M.V.; Grisolia, C.K.; Domingues, I.; Caldas, E.D.; Pic-Taylor, A. Exposure to ayahuasca induces developmental and behavioral alterations on early life stages of zebrafish. *Chem. Biol. Interact.* **2018**, *293*, 133–140. [CrossRef] [PubMed]
5. Gable, R.S. Risk assessment of ritual use of oral dimethyltryptamine (DMT) and harmala alkaloids. *Addiction* **2007**, *102*, 24–34. [CrossRef]
6. Gonçalves, J.; Luís, Â.; Gradillas, A.; García, A.; Restolho, J.; Fernández, N.; Domingues, F.; Gallardo, E.; Duarte, A.P. Ayahuasca beverages: Phytochemical analysis and biological properties. *Antibiotics* **2020**, *9*, 731. [CrossRef] [PubMed]
7. Simão, A.Y.; Gonçalves, J.; Duarte, A.P.; Barroso, M.; Cristóvão, A.C.; Gallardo, E. Toxicological Aspects and Determination of the Main Components of Ayahuasca: A Critical Review. *Medicines* **2019**, *6*, 106. [CrossRef]
8. Riba, J.; Valle, M.; Urbano, G.; Yritia, M.; Morte, A.; Barbanoj, M.J. Human pharmacology of ayahuasca: Subjective and cardiovascular effects, monoamine metabolite excretion, and pharmacokinetics. *J. Pharmacol. Exp. Ther.* **2003**, *306*, 73–83. [CrossRef]
9. Callaway, J.C.; McKenna, D.J.; Grob, C.S.; Brito, G.S.; Raymon, L.P.; Poland, R.E.; Andrade, E.N.; Andrade, E.O.; Mash, D.C. Pharmacokinetics of Hoasca alkaloids in healthy humans. *J. Ethnopharmacol.* **1999**, *65*, 243–256. [CrossRef]
10. dos Santos, R.G.; Bouso, J.C.; Hallak, J.E.C. Ayahuasca, dimethyltryptamine, and psychosis: A systematic review of human studies. *Ther. Adv. Psychopharmacol.* **2017**, *7*, 141–157. [CrossRef]
11. Gaujac, A.; Dempster, N.; Navickiene, S.; Brandt, S.D.; De Andrade, J.B. Determination of N,N-dimethyltryptaminein beverages consumed in religious practices by headspace solid-phase microextraction followed by gas chromatography ion trap mass spectrometry. *Talanta* **2013**, *106*, 394–398. [CrossRef]
12. Li, Y.; Padoan, E.; Ajmone-Marsan, F. Soil particle size fraction and potentially toxic elements bioaccessibility: A review. *Ecotoxicol. Environ. Saf.* **2021**, *209*, 111806. [CrossRef] [PubMed]
13. Carbonell-Capella, J.M.; Buniowska, M.; Barba, F.J.; Esteve, M.J.; Frígola, A. Analytical methods for determining bioavailability and bioaccessibility of bioactive compounds from fruits and vegetables: A review. *Compr. Rev. Food Sci. Food Saf.* **2014**, *13*, 155–171. [CrossRef] [PubMed]
14. Gonçalves, J.; Ramos, R.; Luís, Â.; Rocha, S.; Rosado, T.; Gallardo, E.; Duarte, A.P. Assessment of the Bioaccessibility and Bioavailability of the Phenolic Compounds of Prunus avium L. by in Vitro Digestion and Cell Model. *ACS Omega* **2019**, *4*, 7605–7613. [CrossRef]
15. Versantvoort, C.H.M.; Oomen, A.G.; Van De Kamp, E.; Rompelberg, C.J.M.; Sips, A.J.A.M. Applicability of an in vitro digestion model in assessing the bioaccessibility of mycotoxins from food. *Food Chem. Toxicol.* **2005**, *43*, 31–40. [CrossRef]
16. Minekus, M.; Alminger, M.; Alvito, P.; Ballance, S.; Bohn, T.; Bourlieu, C.; Carrière, F.; Boutrou, R.; Corredig, M.; Dupont, D.; et al. A standardised static in vitro digestion method suitable for food-an international consensus. *Food Funct.* **2014**, *5*, 1113–1124. [CrossRef] [PubMed]
17. Sambuy, Y.; De Angelis, I.; Ranaldi, G.; Scarino, M.L.; Stammati, A.; Zucco, F. The Caco-2 cell line as a model of the intestinal barrier: Influence of cell and culture-related factors on Caco-2 cell functional characteristics. *Cell Biol. Toxicol.* **2005**, *21*, 1–26. [CrossRef]
18. Barthe, L.; Woodley, J.; Houin, G. Gastrointestinal absorption of drugs: Methods and studies. *Fundam. Clin. Pharmacol.* **1999**, *13*, 154–168. [CrossRef] [PubMed]
19. Kosińska-Cagnazzo, A.; Diering, S.; Prim, D.; Andlauer, W. Identification of bioaccessible and uptaken phenolic compounds from strawberry fruits in in vitro digestion/Caco-2 absorption model. *Food Chem.* **2015**, *170*, 288–294. [CrossRef]
20. Devkar, S.; Kandhare, A.; Sloley, B.; Jagtap, S.; Lin, J.; Tam, Y.; Katyare, S.; Bodhankar, S.; Hegde, M. Evaluation of the bioavailability of major withanolides of Withania somnifera using an in vitro absorption model system. *J. Adv. Pharm. Technol. Res.* **2015**, *6*, 159–164.
21. Cameron, L.P.; Olson, D.E. Dark Classics in Chemical Neuroscience: N, N-Dimethyltryptamine (DMT). *ACS Chem. Neurosci.* **2018**, *9*, 2344–2357. [CrossRef]
22. Riba, J.; Romero, S.; Grasa, E.; Mena, E.; Carrió, I.; Barbanoj, M.J. Increased frontal and paralimbic activation following ayahuasca, the pan-amazonian inebriant. *Psychopharmacology* **2006**, *186*, 93–98. [CrossRef] [PubMed]
23. De Araujo, D.B.; Ribeiro, S.; Cecchi, G.A.; Carvalho, F.M.; Sanchez, T.A.; Pinto, J.P.; de Martinis, B.S.; Crippa, J.A.; Hallak, J.E.C.; Santos, A.C. Seeing with the eyes shut: Neural basis of enhanced imagery following ayahuasca ingestion. *Hum. Brain Mapp.* **2012**, *33*, 2550–2560. [CrossRef] [PubMed]

24. Food and Drug Administration Bioanalytical Method Validation Guidance for Industry. Available online: https://www.fda.gov/drugs/guidance-compliance-regulatory-information/guidances-drugs (accessed on 9 June 2021).
25. Bensalem, S.; Soubhye, J.; Aldib, I.; Bournine, L.; Nguyen, A.T.; Vanhaeverbeek, M.; Rousseau, A.; Boudjeltia, K.Z.; Sarakbi, A.; Kauffmann, J.M.; et al. Inhibition of myeloperoxidase activity by the alkaloids of *Peganum harmala* L. (Zygophyllaceae). *J. Ethnopharmacol.* **2014**, *154*, 361–369. [CrossRef] [PubMed]
26. Avula, B.; Wang, Y.H.; Rumalla, C.S.; Smillie, T.J.; Khan, I.A. Simultaneous determination of alkaloids and flavonoids from aerial parts of Passiflora species and dietary supplements using UPLC-UV-MS and HPTLC. *Nat. Prod. Commun.* **2012**, *7*, 1177–1180. [CrossRef]
27. Pires, A.P.S.; De Oliveira, C.D.R.; Moura, S.; Dörr, F.A.; Silva, W.A.E.; Yonamine, M. Gas chromatographic analysis of dimethyltryptamine and β-carboline alkaloids in Ayahuasca, an amazonian psychoactive plant beverage. *Phytochem. Anal.* **2009**, *20*, 149–153. [CrossRef]
28. Souza, R.C.Z.; Zandonadi, F.S.; Freitas, D.P.; Tófoli, L.F.F.; Sussulini, A. Validation of an analytical method for the determination of the main ayahuasca active compounds and application to real ayahuasca samples from Brazil. *J. Chromatogr. B Anal. Technol. Biomed. Life Sci.* **2019**, *1124*, 197–203. [CrossRef]
29. Chambers, M.I.; Appley, M.G.; Longo, C.M.; Musah, R.A. Detection and Quantification of Psychoactive N, N-Dimethyltryptamine in Ayahuasca Brews by Ambient Ionization High-Resolution Mass Spectrometry. *ACS Omega* **2020**, *5*, 28547–28554. [CrossRef]
30. Kaasik, H.; Souza, R.C.Z.; Zandonadi, F.S.; Tófoli, L.F.; Sussulini, A. Chemical Composition of Traditional and Analog Ayahuasca. *J. Psychoact. Drugs* **2021**, *53*, 65–75. [CrossRef]
31. Taylor, W.I. Simple Derivatives Of Tryptophan. In *Indole Alkaloids*; Taylor, W.I., Ed.; Pergamon Press, Elsevier: Amsterdam, The Netherlands, 1966; Volume 1, pp. 29–51. ISBN 9781483196718.
32. Katchborian-Neto, A.; Santos, W.T.; Nicácio, K.J.; Corrêa, J.O.A.; Murgu, M.; Martins, T.M.M.; Gomes, D.A.; Goes, A.M.; Soares, M.G.; Dias, D.F.; et al. Neuroprotective potential of Ayahuasca and untargeted metabolomics analyses: Applicability to Parkinson's disease. *J. Ethnopharmacol.* **2020**, *255*, 112743. [CrossRef]
33. Samoylenko, V.; Rahman, M.M.; Tekwani, B.L.; Tripathi, L.M.; Wang, Y.H.; Khan, S.I.; Khan, I.A.; Miller, L.S.; Joshi, V.C.; Muhammad, I. Banisteriopsis caapi, a unique combination of MAO inhibitory and antioxidative constituents for the activities relevant to neurodegenerative disorders and Parkinson's disease. *J. Ethnopharmacol.* **2010**, *127*, 357–367. [CrossRef] [PubMed]
34. Elbrecht, D.H.; Long, C.J.; Hickman, J.J. Transepithelial/endothelial Electrical Resistance (TEER) theory and ap-plications for microfluidic body-on-a-chip devices Keywords TEER Body-on-a-chip Barrier tissue Blood-brain barrier Organ Endothelial cells Epithelial cells Human-on-a-chip. *J. Rare Dis. Res. Treat.* **2016**, *1*, 46–52.
35. Hellinger, É.; Veszelka, S.; Tóth, A.E.; Walter, F.; Kittel, Á.; Bakk, M.L.; Tihanyi, K.; Háda, V.; Nakagawa, S.; Dinh Ha Duy, T.; et al. Comparison of brain capillary endothelial cell-based and epithelial (MDCK-MDR1, Caco-2, and VB-Caco-2) cell-based surrogate blood-brain barrier penetration models. *Eur. J. Pharm. Biopharm.* **2012**, *82*, 340–351. [CrossRef] [PubMed]
36. Colombini, A.; Perego, S.; Ardoino, I.; Marasco, E.; Lombardi, G.; Fiorilli, A.; Biganzoli, E.; Tettamanti, G.; Ferraretto, A. Evaluation of a possible direct effect by casein phosphopeptides on paracellular and vitamin D controlled transcellular calcium transport mechanisms in intestinal human HT-29 and Caco2 cell lines. *Food Funct.* **2013**, *4*, 1195–1203. [CrossRef] [PubMed]
37. Satsu, H.; Yokoyama, T.; Ogawa, N.; Fujiwara-Hatano, Y.; Shimizu, M. The changes in the neuronal PC12 and the intestinal epithelial Caco-2 cells during the coculture. The functional analysis using an in vitro coculture system. *Cytotechnology* **2001**, *35*, 73–79. [CrossRef] [PubMed]
38. Piccolino, M.; Neyton, J.; Gerschenfeld, H.M. Decrease of gap junction permeability induced by dopamine and cyclic adenosine 3′:5′ -monophosphate in horizontal cells of turtle retina. *J. Neurosci.* **1984**, *4*, 2477–2488. [CrossRef]
39. Fazzari, M.; Fukumoto, L.; Mazza, G.; Livrea, M.A.; Tesoriere, L.; Di Marco, L. In vitro bioavailability of phenolic compounds from five cultivars of frozen sweet cherries (Prunus avium L.). *J. Agric. Food Chem.* **2008**, *56*, 3561–3568. [CrossRef]
40. Toydemir, G.; Boyacioglu, D.; Capanoglu, E.; Van Der Meer, I.M.; Tomassen, M.M.M.; Hall, R.D.; Mes, J.J.; Beekwilder, J. Investigating the transport dynamics of anthocyanins from unprocessed fruit and processed fruit juice from sour cherry (Prunus cerasus L.) across intestinal epithelial cells. *J. Agric. Food Chem.* **2013**, *61*, 11434–11441. [CrossRef]
41. ATCC Caco-2 [Caco2] | ATCC. Available online: https://www.atcc.org/products/htb-37 (accessed on 2 September 2021).

molecules

MDPI

Article

Chemical Characterization and Bioactivity of Commercial Essential Oils and Hydrolates Obtained from Portuguese Forest Logging and Thinning

Ana Ruas [1], Angelica Graça [2], Joana Marto [2], Lídia Gonçalves [2], Ana Oliveira [3], Alexandra Nogueira da Silva [3], Madalena Pimentel [2,3], Artur Mendes Moura [4], Ana Teresa Serra [5,6], Ana Cristina Figueiredo [7,*] and Helena M. Ribeiro [2]

1 Departamento de Biologia Vegetal, Faculdade de Ciências da Universidade de Lisboa, C2, Campo Grande, 1749-016 Lisbon, Portugal; anaruas_@hotmail.com
2 Research Institute for Medicines (iMed.UL), Faculty of Pharmacy, Universidade de Lisboa, 1649-003 Lisbon, Portugal; angelicagraca@campus.ul.pt (A.G.); jmmarto@ff.ulisboa.pt (J.M.); lgoncalves@ff.ulisboa.pt (L.G.); mpimentel@ff.ulisboa.pt (M.P.); hribeiro@campus.ul.pt (H.M.R.)
3 Laboratory of Microbiological Control, ADEIM—Faculty of Pharmacy, University of Lisbon, Av. Forças Armadas, 1649-019 Lisbon, Portugal; anaoliveira@adeim.pt (A.O.); ansilva@ff.ulisboa.pt (A.N.d.S.)
4 Faculdade de Farmácia, Universidade de Lisboa, 1749-016 Lisbon, Portugal; arturmoura@ff.ulisboa.pt
5 IBET—Instituto de Biologia Experimental e Tecnológica, Apartado 12, 2780-901 Oeiras, Portugal; tserra@ibet.pt
6 Instituto de Tecnologia Química e Biológica, Universidade Nova de Lisboa, Av. da República, 2780-157 Oeiras, Portugal
7 Centro de Estudos do Ambiente e do Mar (CESAM Lisboa), Faculdade de Ciências da Universidade de Lisboa, BV, DBV, C2, Campo Grande, 1749-016 Lisbon, Portugal
* Correspondence: acsf@fc.ul.pt

Citation: Ruas, A.; Graça, A.; Marto, J.; Gonçalves, L.; Oliveira, A.; da Silva, A.N.; Pimentel, M.; Moura, A.M.; Serra, A.T.; Figueiredo, A.C.; et al. Chemical Characterization and Bioactivity of Commercial Essential Oils and Hydrolates Obtained from Portuguese Forest Logging and Thinning. *Molecules* **2022**, *27*, 3572. https://doi.org/10.3390/molecules27113572

Academic Editor: Daniela Rigano

Received: 13 May 2022
Accepted: 30 May 2022
Published: 2 June 2022

Abstract: Essential oils (EOs) and hydrolates (Hds) are natural sources of biologically active ingredients with broad applications in the cosmetic industry. In this study, nationally produced (mainland Portugal and Azores archipelago) EOs (11) and Hds (7) obtained from forest logging and thinning of *Eucalyptus globulus*, *Pinus pinaster*, *Pinus pinea* and *Cryptomeria japonica*, were chemically evaluated, and their bioactivity and sensorial properties were assessed. EOs and Hd volatiles (HdVs) were analyzed by GC-FID and GC-MS. 1,8-Cineole was dominant in *E. globulus* EOs and HdVs, and α- and β-pinene in *P. pinaster* EOs. Limonene and α-pinene led in *P. pinea* and *C. japonica* EOs, respectively. *P. pinaster* and *C. japonica* HVs were dominated by α-terpineol and terpinen-4-ol, respectively. The antioxidant activity was determined by DPPH, ORAC and ROS. *C. japonica* EO showed the highest antioxidant activity, whereas one of the *E. globulus* EOs showed the lowest. Antimicrobial activity results revealed different levels of efficacy for *Eucalyptus* and *Pinus* EOs while *C. japonica* EO showed no antimicrobial activity against the selected strains. The perception and applicability of emulsions with 0.5% of EOs were evaluated through an in vivo sensory study. *C. japonica* emulsion, which has a fresh and earthy odour, was chosen as the most pleasant fragrance (60%), followed by *P. pinea* emulsion (53%). In summary, some of the studied EOs and Hds showed antioxidant and antimicrobial activities and they are possible candidates to address the consumers demand for more sustainable and responsibly sourced ingredients.

Keywords: essential oils; hydrolates; chemical composition; antioxidant activity; antimicrobial activity; sensory evaluation

1. Introduction

Nowadays, consumers have a growing interest in substances of natural origin, such as essential oils (EOs), which have been widely used for various purposes. There has been a growing interest from different industries such as pharmaceuticals, cosmetics, and food, in using EOs, mainly due to their biological properties, such as antifungal, antibacterial and antioxidant activities [1]. In the European Union (EU), EOs have been mainly used as flavoring agents in the food industry, in perfumes and aftershaves, in the cosmetics industry, and as functional ingredients in the pharmaceutical industry [2]. In the cosmetics industry, EOs and their isolated constituents are widely used, mainly due to their pleasant scents, as well as their preservative and antioxidant properties [3–5]. Moreover, essential oils are used in topical formulations owing to other recognized properties, such as anti-inflammatory, antimicrobial, antioxidant, healing, anti-mutagenic, and anti-aging effects, protection against damage caused by UV-B radiation, and potential use as emollients, dyes, humectants, etc. [5].

In the context of a circular economy, there is an emergent concern in obtaining added value from biomass resulting from forest maintenance, namely from forest logging and thinning. The Mediterranean Forest can provide natural resources that can be exploited and constitute an additional sustainable income to local producers [6,7].

A hydrolate (Hd) is an EO isolation procedure co-product that has a very similar, although less intense, odour compared to its corresponding EO. Unlike EOs, they are water-soluble extracts and can be added to formulas with a high-water content. The characteristics of Hds, especially their biological properties, make them widely used in various industries, such as cosmetics and food. These compounds are promising natural raw materials in many different products [8,9], and several types are already commercially used, mainly as cosmetic and food ingredients. Analysis of their chemical composition show that Hds usually contain less than 1 g/L (i.e., 0.10%) of EO water-soluble compounds. Oxygen-containing compounds are usually dominant in the Hds volatiles, and they may reveal some similarity with EO compositions, although several studies reveal differences, particularly in Hds from hydrocarbon rich EOs [8–10].

In this work, EOs and Hds from *Eucalyptus globulus* Labill., *Pinus pinaster* Aiton, *Pinus pinea* L. and *Cryptomeria japonica* D. Don., obtained from forest logging and thinning, were evaluated to gain further insight into their potential use, bioactivity, and likely acceptability in the cosmetic industry. The EOs and their Hds were selected based on their economic and forestry importance in Portugal. In addition, *E. globulus* [11–13], *P. pinaster* [14–16], *P. pinea* [17–20] and *C. japonica* [21–24] EOs and some Hds [8] have shown important biological activities.

The EOs and Hds were obtained from local producers, from mainland Portugal and the Azores archipelago. This study aimed at (1) characterizing the chemical composition of the EOs, and hydrolate volatiles (HdVs), (2) determining their antioxidant and antimicrobial capacity, and (3) assessing only the EOs sensory properties and their acceptability in EOs-topical formulations.

2. Results and Discussion

2.1. Essential Oil Composition

All EOs were fully chemically characterized (detailed relative amounts of all the identified components are listed in Supplementary Tables S1–S4), although Table 1 reports only their main components (≥5%).

Eucalyptus globulus. In total, 44 to 49 compounds were identified in *E. globulus* EOs, accounting for 98–99% of the total composition. Oxygen-containing monoterpenes dominated in all *E. globulus* EOs, ranging from 56 to 72% (Table S1). The main component of the *E. globulus* EOs was 1,8-cineole (eucalyptol), ranging from 49 to 65%. α-Pinene (11–22%), limonene (8–18%), and α-terpenyl acetate (traces-5%) were other relevant compounds (Table 1).

The range of the relative amounts of the main components (1,8-cineole and α-pinene) determined in the present study (Table 1) agrees with the values determined in previous studies for *E. globulus* collected in Portugal (63–70% and 13–20%, respectively) [25–29]. Whereas all samples showed an α-pinene content ≥ 10%, only three had a 1,8-cineole percentage ≥ 60% as specified for the quality assessment of *E. globulus* raw EO by ISO 770:2002 [30].

Table 1. Percentage composition of the main components (≥5%) of *Eucalyptus globulus*, *Pinus pinaster*, *Pinus pinea* and *Cryptomeria japonica* essential oils (EOs). For sample codes *vide* Materials and Methods Section.

EOs Main Components (≥5%)	RI	Samples					
Eucalyptus globulus		**Eg_OE_1_G**	**Eg_OE_2_B**	**Eg_OE_3_O**	**Eg_OE_4_E**	**Eg_OE_5_P**	**Eg_OE_6_S**
α-Pinene	930	13.2	13.3	11.0	21.8	14.7	13.8
1,8-Cineole	1005	65.2	63.2	59.5	53.9	58.2	49.4
Limonene	1009	8.2	17.2	13.7	16.6	12.5	18.0
α-Terpenyl acetate	1334	2.2	t	5.4	0.2	0.8	0.9
Pinus pinaster		**Pp_OE_1_G**	**Pp_OE_2_P**	**Pp_OE_3_S**			
α-Pinene	930	27.0	44.6	36.5			
β-Pinene	963	28.0	23.0	18.8			
β-Myrcene	975	11.0	5.0	5.9			
δ-3-Carene	1000	6.6	2.1	1.8			
Limonene	1009	4.5	3.9	3.3			
β-Caryophyllene	1414	4.5	5.0	8.7			
Germacrene-D	1474	6.3	1.7	5.6			
Pinus pinea		**Ppi_OE_1_B**					
α-Pinene	930	7.6					
Limonene	1009	72.8					
Cryptomeria japonica		**Cj_OE_1_M**					
α-Pinene	930	26.1					
Sabinene	958	18.1					
Phyllocladene	2006	13.8					

RI: In-lab calculated retention index of *n*-alkanes on the DB-1 column. t: traces (<0.05%).

Pinus pinaster. Between 63 and 68 components were identified in *P. pinaster*, accounting for 98–99% of the total composition of EOs (Table S2). *P. pinaster* EOs consisted mainly of monoterpene hydrocarbons (70 to 82%). From the three *P. pinaster* EOs, two were dominated by α-pinene (37–45%), and the third showed similar amounts of α-pinene and β-pinene (28% and 29%, respectively) (Table 1). These results agree with previous reports for *P. pinaster* EOs from Portugal, in which α- and β-pinene were the dominant compounds (25–62% and 20–52%, respectively) [31–33].

Pinus pinea. Fifty components were identified in *Pinus pinea* EO accounting for 99% of the total composition (Table S3). *P. pinea* EO was dominated by monoterpenes (95%), namely limonene (73%) (Table 1). *P. pinea* EO is known for its chemical homogeneity and despite some percentual variations, the obtained data is comparable to that previously reported by Rodrigues et al. [32] for younger needles collected in Portugal.

Cryptomeria japonica. Seventy-nine components were identified in *C. japonica* EO, accounting for 97% of the total composition (Table S4). The main components of *C. japonica* EO were monoterpene hydrocarbons (66%) and diterpene hydrocarbons (16%). α-Pinene (26%), sabinene (18%) and phyllocladene (14%) dominated this EO (Table 1), in agreement with previously reported data for *C. japonica* foliage collected in Azores [27,34,35].

2.2. Hydrolates Volatiles Composition

As for all EOs, the hydrolate volatiles (HdVs) were fully chemically characterized, and the detailed relative amounts of all the identified components are listed in Tables S5–S8. Table 2 shows the HdV main components only (≥5%).

Table 2. Percentage composition of the main components (≥5%) of *Eucalyptus globulus*, *Pinus pinaster* and *Cryptomeria japonica* hydrolate volatiles (HdVs). For samples codes *vide* Materials and Methods Section.

HdVs Main Components (≥5%)	RI	Samples			
Eucalyptus globulus		**Eg_Hd_1_G**	**Eg_Hd_2_O**	**Eg_Hd_3_E**	**Eg_Hd_4_P**
1,8-Cineole	1005	80.2	55.5	53.5	4.5
Limonene	1009	7.3	14.1	6.7	1.5
trans-Pinocarveol	1106	4.9	0.4	8.0	36.6
cis-*p*-2-Menthen-1-ol	1114	t	t	t	4.6
Myrtenal	1153		t	t	5.5
α-Terpineol	1159	2.7	17.2	24.7	5.3
Myrtenol	1168		t	t	12.0
cis-Carveol	1202	1.1	0.1	2.8	8.6
Pinus pinaster		**Pp_Hd_1_G**	**Pp_Hd_2_P**		
1,8-Cineole	1005	5.0	t		
cis-*p*-2-Menthen-1-ol	1114	t	14.0		
neo-Isopulegol	1116		14.0		
Terpinen-4-ol	1148	7.5			
p-Cymen-8-ol	1148	7.5	t		
α-Terpineol	1159	43.8	38.1		
Verbenone	1164	17.9	28.7		
Perilla alcohol	1274	6.6	t		
Thymol	1275	6.6	t		
Cryptomeria japonica		**Cj_Hd_1_M**			
1,8-Cineole	1005	6.3			
Terpinen-4-ol	1148	56.2			
α-Terpineol	1159	4.6			
Phyllocladene	2006	4.8			

RI: In-lab calculated retention index of *n*-alkanes on the DB-1 column. t: traces (<0.05%).

Eucalyptus globulus. In total, 46–58 constituents were identified in *E. globulus* HdVs (Table S5). Similarly to *E. globulus* essential oils, three of the HdVs were dominated by 1,8-cineole (54–80%) (Table 2), while the fourth sample was dominated by *trans*-pinocarveol (37%). The second main component varied according to the sample, two samples showing high percentages of α-terpineol (17% and 25%), one limonene (7%) and the fourth sample myrtenol (12%).

As detailed in a recent review [8], previous reports indicated 1,8-cineole (62–93%) and α-terpineol (3–17%) as main components of *E. globulus* HdVs [36,37].

Pinus pinaster. In *P. pinaster* HdVs, 38 or 42 compounds were identified (Table S6). These HdVs were dominated by oxygen-containing monoterpenes (95% in both cases), namely by α-terpineol (38–44%), followed by verbenone (18–29%) (Table 2).

Although there are reports of *P. cembra* and *P. sylvestris* HdVs [8], to the best of our knowledge no previous study addressed *P. pinaster* HdVs. No study was performed with *P. pinea* HdVs as this hydrolate was not available.

Cryptomeria japonica. Forty-four components were identified in *C. japonica* HdVs (Table S7). Oxygen-containing monoterpenes (79%), particularly terpinen-4-ol (56%), dominated *C. japonica* HdVs (Table 2).

These results agree with those reported by Nakagawa et al. [38], which showed that terpinen-4-ol was the main compound of *C. japonica* HdVs, from branches with leaves or just from leaves (32 and 37%, respectively).

2.3. *Antioxidant Activity of Essential Oils and Hydrolates*

2.3.1. DPPH and ORAC Assays

The EOs studied generally showed a weak antioxidant activity determined by the 1,1-diphenyl-2-picrylhydrazyl (DPPH) method. Exceptions were CJ_OE_1_M (23.1 mg/L), which demonstrated a considerable antioxidant activity, followed by Pp_OE_1_G (55.2 mg/L).

The lowest antioxidant activity was observed for EG_OE_2_B (647.3 mg/mL). In contrast, the antioxidant activities determined using the Oxygen-Radical Absorbance Capacity (ORAC) method were higher, namely for Pp_OE_3_S (565450.6 μmol TE/g), Pp_OE_2_P (355575.7 μmol TE/g) and Cj_OE_1_M (224877.9 μmol TE/g). With this method, the lower antioxidant activities were obtained for Eg_OE_2_B (53669.2 μmol TE/g) and Eg_OE_5_P (86174.9 μmol TE/g) (Table 3).

Table 3. Antioxidant capacity of the assessed essential oils and hydrolates. For samples codes *vide* Materials and Methods Section.

Essential Oils	DPPH (IC50, mg/mL)	ORAC (μmol TE/g)	Reduction of ROS * (%)
Eg_OE_1_G	197.6 ± 20.4	113245.9 ± 15003.8	40.0 ± 0.9
Eg_OE_2_B	647.3 ± 5.7	53669.2 ± 8659.3	49.3 ± 0.8
Eg_OE_3_O	151.8 ± 0.0	171891.9 ± 25388.4	−15.7 ± 1.5
Eg_OE_4_E	WA	113884.2 ± 14067.0	27.2 ± 0.8
Eg_OE_5_P	WA	86174.9 ± 9813.9	50.0 ± 0.0
Eg_OE_6_S	246.7 ± 24.5	160532.2 ± 16659.3	6.8 ± 1.2
Pp_OE_1_G	55.2 ± 0.9	161208.7 ± 24896.4	34.3 ± 3.7
Pp_OE_2_P	WA	355575.7 ± 30254.3	29.5 ± 0.5
Pp_OE_3_S	WA	565450.6 ± 70377.8	21.7 ± 1.9
Ppi_OE_1_B	195.7 ± 22.9	165063.9 ± 20907.1	−3.3 ± 1.2
Cj_OE_1_M	23.1 ± 0.2	224877.9 ± 25680.9	83.5 ± 2.8
Hydrolates			
Eg_Hd_1_G	WA	84.1 ± 10.0	81.0 ± 2.3
Eg_Hd_2_O	WA	1129.7 ± 100.6	46.8 ± 5.0
Eg_Hd_3_E	WA	454.6 ± 39.7	-
Eg_Hd_4_P	WA	238.5 ± 24.5	79.2 ± 2.0
Pp_Hd_1_G	WA	212.2 ± 16.9	84.8 ± 1.3
Pp_Hd_2_P	WA	295.1 ± 44.4	80.3 ± 1.9
Cj_Hd_1_M	WA	131.1 ± 10.8	92.8 ± 1.3
Ascorbic Acid	0.04 ± 1.1	-	95.3 ± 0.5

TE: Trolox equivalents. * in vitro ROS reduction generated by 500 μM H_2O_2 in HaCaT cell line. WA: Without Activity. ROS: Reactive Oxygen Species.

C. japonica EO had the highest antioxidant capacity determined by the DPPH assay, and the third best when using the ORAC assay (Table 3). Ho et al. [39] used the DPPH method to evaluate the antioxidant activity of *C. japonica* EO obtained from different plant parts (leaf, heartwood, sapwood, bark, and twigs). The reported IC50 values were higher than those obtained in this study, with the sapwood EO showing the highest radical scavenging capability, and the leaf EO having the lowest antioxidant capacity.

One of the *P. pinaster* EO samples, namely Pp_OE_3_S, showed the highest antioxidant capacity, determined by the ORAC assay (Table 3). Mediavilla et al. [7] evaluated the antioxidant activity of forest species EOs, including from *P. pinaster* and *P. sylvestris*, using the ORAC method. *P. pinaster* EO had a higher antioxidant activity than that of *P. sylvestris* EO, which showed the lowest ORAC value of all the species studied. In contrast, the ORAC results obtained herein showed higher antioxidant activity than that described by Mediavilla et al. [7].

The DPHH method did not reveal any measurable antioxidant activity for the Hds, while the results obtained with the ORAC method showed lower values of antioxidant activity of the Hds compared to the respective EOs (Table 3).

As shown in Table 3, the Eg_Hd_2_O sample of *E. globulus* Hd had the highest antioxidant capacity determined by the ORAC assay (1129.7 μmol TE/g). This high antioxidant capacity may be due to the presence of low percentage compounds such as terpinen-4-ol, which was previously reported to have antioxidant efficacy against AAPH radicals [40]. The two samples of *E. globulus* EO had the lowest antioxidant capacity, as shown by both assays. The major components found in these *E. globulus* EO samples were 1,8-cineole and α-pinene, which are described in the literature as compounds with a weak antioxidant activity against DPPH and

AAPH radicals [41]. In general, *C. japonica* and *P. pinaster* EOs showed higher antioxidant capacity than that determined for *E. globulus* EOs.

As can be observed from the results, the antioxidant efficiencies determined for the EO samples depend on the methods of evaluation. In general, it was observed that the EO's antioxidant capacity was higher when using the ORAC method, compared to the DPPH method. The different results between the methods are possibly related to the distinct mechanisms used to evaluate the antioxidant activity. The ORAC method is included in the hydrogen atom transfer (HAT) group of methods, in which there is a competition reaction between antioxidant substances and a fluorescence probe, by a radical [42]. On the other hand, the DPPH method belongs to the electron transfer (ET) group in which a single electron transfer reaction occurs with DPPH reacting by itself, both as a radical and as a probe [43]. Another possible justification for the results obtained is that the ORAC method is more sensitive, being able to detect the antioxidant activity of an extract even when this contains only a small amount of polyphenols. The main advantage of this method is that it combines both the inhibition time and the degree of inhibition of radical generation as it leads the oxidation reaction to completion and uses the area under the curve to quantify antioxidant activity [44].

2.3.2. Intracellular ROS Measurement

In the intracellular reactive oxygen species (ROS) measurement, for a concentration of 10% (v/v), the EO samples showed a lower capacity to reduce the % ROS, compared to the samples of Hds. In fact, some of the EOs, such as Ppi_OE_1_B and Eg_OE_3_O, even potentiated the formation of ROS (Table 3). The decrease in the % ROS by Hd extracts, for the same concentration, was not significantly different from 1 mg/mL ascorbic acid ($p \geq 0.05$), except for Eg_Hd_4_O, with a significantly lower ($p \leq 0.05$) capacity of ROS reduction (47 $\pm$ 5%) than that of ascorbic acid. The EO and Hd samples at a lower concentration, 1% (v/v), were not able to reduce H_2O_2-induced ROS formation.

C. japonica Hd showed the best antioxidant capacity against peroxyl radicals (Table 3), possibly because of its main chemical compound, terpinen-4-ol, which is an antioxidant, as suggested by Souza et al. [40]. However, this Hd also contains other antioxidant compounds present in lower amounts, such as β-eudesmol [45]. The second-best antioxidant capacity was observed for a *P. pinaster* Hd sample, a result which is possibly due to the presence of terpinen-4-ol and other compounds, such as thymol [46] and perilla alcohol [47].

2.4. *Antimicrobial Activity of Essential Oils and Hydrolates*

The results of antimicrobial activity are presented in Table 4. Only the EO from *P. pinea* was evaluated, since no Hd was obtained.

Table 4. Minimum inhibitory concentrations (MICs) of *Eucalyptus globulus*, *Pinus pinaster*, *Pinus pinea* and *Cryptomeria japonica* essential oils (EOs) against Gram-positive and Gram-negative bacteria, yeast, and mold. For sample codes *vide* Materials and Methods Section.

	Minimum Inhibitory Concentrations (MICs) (μg/mL)					
EOs Samples	*Staphylococcus aureus* **ATCC 6538**	*Bacillus subtilis* **ATCC 6633**	*Pseudomonas aeruginosa* **ATCC 9027**	*Escherichia coli* **ATCC 8739**	*Candida albicans* **ATCC 10231**	*Aspergillus brasiliensis* **ATCC 16404**
Eg_OE_1_G	125	31.25	500	15.62	7.81	>500
Eg_OE_2_B	125	31.25	500	3.90	3.90	>500
Eg_OE_3_O	62.5	15.62	31.25	15.62	31.25	>500
Eg_OE_4_E	125	15.62	500	62.5	31.25	>500
Eg_OE_5_P	62.5	1.95	500	3.90	3.90	>500
Eg_OE_6_S	62.5	15.62	500	15.62	7.81	>500
Pp_OE_1_G	>500	>500	>500	>500	>500	>500
Pp_OE_2_P	31.25	15.62	500	15.62	62.5	>500
Pp_OE_3_S	>500	15.62	>500	125	125	>500
Ppi_OE_1_B	62.5	7.81	>500	125	15.62	>500
Cj_OE_1_M	>500	>500	>500	>500	>500	>500

The antimicrobial activity of the EOs and Hds was tested against Gram-positive and Gram-negative bacteria, a mold, and a yeast. All *E. globulus* EO samples had antimicrobial activity, although to different extents. Concerning Gram-positive bacteria, all *E. globulus* EO samples showed significant antimicrobial activity against *Bacillus subtilis*, with Eg_OE_5_P presenting the lowest MIC (1.95 μg/mL), while for *Staphylococcus aureus*, the MIC observed when this OE was used was 62.5 μg/mL. Concerning Gram-negative bacteria, most EOs in the tested concentration range were not effective against *Pseudomonas aeruginosa*, except Eg_OE_3_O, with a MIC of 31.25 μg/mL. Eg_OE_2_B and Eg_OE_5_P showed the highest activities against *E. coli* and were also the most active against the pathogenic yeast *Candida albicans*. None of the *E. globulus* samples showed activity against *Aspergillus brasiliensis* (Table 4).

The *P. pinaster* EO samples revealed, in general, an antimicrobial activity lower than that of *E. globulus*. The highest activity was observed on *B. subtillis* with Pp_OE_2_P and Pp_OE_3_S, with MIC values of 15.62 μg/mL. In Gram-negative bacteria, and similarly to the *E. globulus* samples, the *P. pinaster* samples were more active against *E. coli*. No activity against *P. aeruginosa* was observed in the tested concentration range, except for Pp_OE_2_P, for which a high MIC value (500 μg/mL) was obtained. The sample Pp_OE_1_G EO did not have any antimicrobial activity against the tested strains. Furthermore, none of the *P. pinaster* EO samples were active against *A. brasiliensis* (Table 4). The results concerning the *P. pinaster* EO samples revealed that only one of the samples, namely, Pp_OE_2_P, had antimicrobial efficacy against all the strains considered. This may be related to a higher percentage of α-pinene, which was previously mentioned as having antimicrobial properties.

P. pinea EO had better activity against *B. subtilis* than against *S. aureus* (Gram-positive) and against *E. coli* than against *P. aeruginosa* (Gram-negative). This sample also had no antifungal activity against *A. brasiliensis*, but was effective against *C. albicans*, with a MIC of 15.62 μg/mL (Table 4).

The *C. japonica* OE did not show any antimicrobial efficacy against any of the tested bacterial and fungal strains. Nevertheless, the variations in the antimicrobial efficacy may be related to other factors that can influence or justify these changes, namely, the culture medium used, the evaluation method, the origin of the botanical species, the plant age, the type of material used (dry or fresh), the amount of EO used in the test, and the isolation technique [48].

The Hds evaluated did not have any detectable antimicrobial activity (data not shown). To the best of our knowledge, the available literature on Hds of the species studied regarding their antimicrobial capacity is scarce. However, the absence of antimicrobial capacity by the Hds may be due to the low concentration of the extracts. Šilha et al. [49] compared Hds of four different plant species, non-concentrated and 50× concentrated and observed that the non-concentrated ones did not have antimicrobial capacity. On the other hand, concentrated hydrolates showed antimicrobial efficacy.

These results agree with what has been described in the literature. For example, Cimanga et al. [50] showed that *S. aureus* ATCC 6538 and *P. aeruginosa* ATCC 9027, were the strains most resistant to the *E. globulus* EO samples. Furthermore, the *B. subtilis* ATCC 6633 strain was considered one of the most susceptible. Hmamouchi et al. [18] showed that, in general, *P. pinaster* EO had higher antimicrobial activity than *P. pinea* EO. Finally, Nakagawa et al. [38] reported that most of the analyzed extracts of *C. japonica* did not show any measurable antimicrobial efficacy.

2.5. Sensorial Evaluation

Questionnaire Results

The sensorial evaluation was only performed for five (5) EOs, selected according to each plant species, except for *P. pinaster*, which presented two samples with very different odours, which were therefore both included in the study. From the six samples of *E. globulus* EOs, the chosen one was that containing the highest percentage of 1,8-cineole (eucalyptol), since this compound confers most of the odour from *Eucalyptus* species. The hydrolates were not selected due to their weak, much less intense odour.

A total of 100 inexperienced participants were questioned, most of which were between 41 and 50 years old (30%), followed by the group of 18 to 30 years olds (22%) (Table 5).

Table 5. Sociodemographic characteristics of the 100 volunteers.

Sociodemographic Characteristics		All Samples (n = 100) n (%)
	Age range (years)	
<18		6 (6%)
18–30		22 (22%)
31–40		13 (13%)
41–50		30 (30%)
51–60		20 (20%)
>60		9 (9%)
	Gender	
Female		67 (67%)
Male		33 (33%)
	Education	
Primary education		13 (13%)
Secondary education		26 (26%)
Higher education		61 (61%)
	Region	
Countryside		70 (70%)
City		30 (30%)

The under 18 class had the lowest number of respondents (6%). In addition, 67% were female and 43% were male. The participants were divided into three categories based on their education level. The most common education level was higher education (66%), while the least common one was primary education (13%). The countryside was the region with the highest predominance of individuals (70). The remaining 30 individuals belonged to a city environment (Table 5).

In this section, questions were asked regarding the odour of the emulsions provided to the participants (Table 5). The first question in this study was: "How do you evaluate the emulsions' odour?". Most participants (46, 38 and 42 individuals) rated *C. japonica* Cj_OE_1_M, *P. pinea* Ppi_OE_1_B and *P. pinaster* Pp_OE_2_P EOs emulsions, respectively, as having a perceptible odour. The *E. globulus* EO emulsion was considered as having the most intense odour among the four available, with 48 responses. The second question was: "In case you identified any odour, how would you classify it?". Most respondents rated the odours as pleasant and fresh for the Cj_OE_1_M and Ppi_OE_1_B EOs emulsions, with 60 and 53 answers, respectively. The *E. globulus* (Eg_OE_1_G) EO emulsion was considered a very unpleasant odour compared to the other emulsions, with just 36 answers. Most respondents considered the Pp_OE_1_G and Pp_OE_2_P emulsion odours unpleasant, with 41% and 44% percentage values, respectively (Table 6).

The third question was: "In your opinion, the odours of the different emulsions belong to the same plant species?". As shown in Table 6, most participants (41%) answered "yes", i.e., they considered that all emulsions belong to the same plant species. The fourth question in this Section was: "Order the emulsions, according to your preference, on a scale of 1–5". Overall, the Cj_OE_1_M and Ppi_OE_1_B EOs emulsions were the participants' favourites, with 28 and 24% of individuals, respectively, considering them to have a pleasant odour, and 19% considering these as their favourite odour (Table 6). The aim of the fifth question was to understand if one or more emulsions caused any feeling of well-being. For this, a single question was asked: "Select which emulsion(s) cause you a feeling of physical or mental well-being." Most of the participants selected the Cj_OE_1_M EO emulsion (20%), while a similar percentage (17%) answered the Eg_OE_1_G EO emulsion or none of the emulsions (Table 7). The sixth and final question in Section 1, related to question five, was "Refer what feelings of well-being the emulsions caused you". Most participants responded "refreshing" and "relaxing" as the sensations of well-being caused by the emulsions mentioned above (23 and 18%, respectively).

Table 6. Participants responses regarding the characterization of emulsions odour.

Section 1. Emulsions' Odour. Odoriferous Characterization of Emulsions			N (%)		
Evaluation of odours	Cj_OE_1_M	Ppi_OE_1_B	Eg_OE_1_G	Pp_OE_1_G	Pp_OE_2_P
1. Without odour	3 (3%)	1 (1%)	0 (0%)	15 (15%)	13 (13%)
2. Slightly perceptible	37 (37%)	8 (8%)	4 (4%)	40 (40%)	28 (28%)
3. Perceptible	46 (46%)	38 (38%)	15 (15%)	28 (28%)	42 (42%)
4. Very perceptible	11 (11%)	34 (34%)	33 (33%)	14 (14%)	11 (11%)
5. Intense odour	3 (3%)	19 (19%)	48 (48%)	3 (3%)	6 (6%)
Classification of odours	Cj_OE_1_M	Ppi_OE_1_B	Eg_OE_1_G	Pp_OE_1_G	Pp_OE_2_P
1. Very unpleasant	0 (0%)	9 (9%)	36 (36%)	4 (4%)	5 (5%)
2. Unpleasant	22 (22%)	25 (25%)	27 (27%)	41 (41%)	44 (44%)
3. Pleasant and hot odour	15 (15%)	12 (12%)	15 (15%)	10 (10%)	13 (13%)
4. Pleasant and fresh odour	60 (60%)	53 (53%)	22 (22%)	30 (30%)	27 (27%)
Ranking in order of preference	Cj_OE_1_M	Ppi_OE_1_B	Eg_OE_1_G	Pp_OE_1_G	Pp_OE_2_P
1. Hateful Odour	3 (3%)	8 (8%)	32 (32%)	9 (9%)	11 (11%)
2. Unpleasant Odour	17 (17%)	18 (18%)	23 (23%)	41 (41%)	43 (43%)
3. Neither pleasant nor unpleasant	33 (33%)	31 (31%)	13 (13%)	30 (30%)	22 (22%)
4. Pleasant Odour	28 (28%)	24 (24%)	20 (20%)	18 (18%)	14 (14%)
5. Favourite Odour	19 (19%)	19 (19%)	12 (12%)	2 (2%)	10 (10%)
Do you think that emulsions belong to the same plant species?					
Positive answers					41 (41%)
Uncertainly answers					34 (34%)
Negative answers					25 (25%)

Table 7. Participants responses about the feelings of well-being caused by the emulsions' odour.

Section 1. Emulsions' Odour. Feelings of Well-Being Caused by the Emulsions' Odour	N (%)
Emulsions that caused feelings of well-being	
Ppi_OE_1_B and Eg_OE_1_G	2 (2%)
Cj_OE_1_M, Pp_OE_1_G and Pp_OE_2_P	3 (3%)
Cj_OE_1_M	20 (20%)
Cj_OE_1_M and Pp_OE_1_G	1 (1%)
Eg_OE_1_G and Pp_OE_2_P	2 (2%)
Cj_OE_1_M and Pp_OE_2_P	2 (2%)
Cj_OE_1_M and Ppi_OE_1_B	5 (5%)
Pp_OE_2_P	6 (6%)
Ppi_OE_1_B	11 (11%)
Cj_OE_1_M, Ppi_OE_1_B and Eg_OE_1_G	5 (5%)
Eg_OE_1_G	17 (17%)
Cj_OE_1_M, Eg_OE_1_G, Pp_OE_2_P and Pp_OE_1_G	1 (1%)
Cj_OE_1_M and Eg_OE_1_G	1 (1%)
Pp_OE_1_G and Pp_OE_2_P	2 (2%)
Ppi_OE_1_B and Pp_OE_2_P	1 (1%)
Cj_OE_1_M, Ppi_OE_1_B and Pp_OE_1_G	2 (2%)
Cj_OE_1_M, Ppi_OE_1_B, Pp_OE_1_G and Pp_OE_2_P	1 (1%)
Pp_OE_1_G	1 (1%)
None	17 (17%)
Feelings of well-being	
Refreshing	23 (23%)
Decongestant	15 (15%)
Decongestant, Stimulating and Refreshing	2 (2%)
Decongestant and Refreshing	4 (4%)
Relaxing, Decongestant and Refreshing	7 (7%)
Relaxing	18 (18%)
Relaxing and Stimulating	1 (1%)
Stimulating and Refreshing	1 (1%)
Relaxing and Refreshing	4 (4%)
Stimulating	3 (3%)
Relaxing, Stimulating and Refreshing	1 (1%)
Relaxing and Decongestant	2 (2%)
Decongestant and Stimulating	2 (2%)
None	17 (17%)

The results from this section are summarized in Table 8. First, participants were asked whether they were likely to purchase a particular product for personal use with the odour of the selected emulsions. The question was: "Rate each of the products below, on a scale of 1–5, considering the probability of buying one with the emulsion's odour". Regarding the Cj_OE_1_M emulsion, participants considered that they would be more likely to purchase an air freshener and massage cream with its odour (34 and 33%, respectively) and 18% of volunteers said they would buy it. Perfume and candy with the Cj_OE_1_M emulsion odour were considered by the majority to be the products they would never buy (43 and

Table 10. Analyzed essential oils (EOs) and hydrolates (Hds) and their codes.

Plant Species	EOs Code *	Hds Code
Eucalyptus globulus	Eg_OE_1_G	Eg_Hd_1_G
	Eg_OE_2_B	-
	Eg_OE_3_O	Eg_Hd_2_O
	Eg_OE_4_E	Eg_Hd_3_E
	Eg_OE_5_P	Eg_Hd_4_P
	Eg_OE_6_S	-
Pinus pinaster	Pp_OE_1_G	Pp_Hd_1_G
	Pp_OE_2_P	Pp_Hd_2_P
	Pp_OE_3_S	-
Pinus pinea	Ppi_OE_1_B	-
Cryptomeria japonica	Cj_OE_1_M	Cj_Hd_1_M

* To ensure data protection each producer was assigned with an arbitrary code letter.

3.3. Essential Oil and Hydrolate Volatiles Composition Analysis

The EOs and the HdVs were analysed by gas chromatography-mass spectrometry (GC-MS) for component identification, and by gas chromatography with flame ionization detector (GC-FID) for components quantification.

3.3.1. Gas Chromatography (GC)-Flame Ionization Detection (FID) Analysis

Gas chromatographic analyses were performed using a Perkin Elmer Clarus 400 gas chromatograph (Perkin Elmer, Shelton, CT, USA) equipped with two flame ionization detectors (FIDs), a data handling system and a vaporizing injector port into which two columns of different polarities were installed: a DB-1 fused-silica column (polydimethylsiloxane, 30 m × 0.25 mm i.d., film thickness 0.25 µm; J & W Scientific, Inc., Rancho Cordova, CA, USA) and a DB-17HT fused-silica column [(50% phenyl)-methylpolysiloxane, 30 m × 0.25 mm i.d., film thickness 0.15 µm; J & W Scientific, Inc., Rancho Cordova, CA, USA]. The oven temperature was programmed at 45–175 °C, at 3 °C/min, subsequently at 15 °C/min up to 300 °C, and then held isothermal for 10 min; injector and detector temperatures were 280 °C and 300 °C, respectively; the carrier gas was hydrogen, adjusted to a linear velocity of 30 cm/s. The samples were injected using split sampling technique at a ratio of 1:50. The percentage composition of the volatiles was computed by the normalization method from the GC peak areas, and calculated as mean values of two injections, from each sample, without using the response factors, in accordance with ISO 7609 [56].

3.3.2. Gas Chromatography-Mass Spectrometry (GC-MS)

The GC-MS unit consisted of a Perkin Elmer Clarus 600 gas chromatograph, equipped with DB-1 fused-silica column (30 m × 0.25 mm i.d., film thickness 0.25 µm; J & W Scientific, Inc., Rancho Cordova, CA, USA), and interfaced with a Perkin-Elmer 600 T mass spectrometer (software version 5.4.2.1617, Perkin Elmer, Shelton, CT, USA). Injector and oven temperatures were as above; the transfer line temperature was 280 °C; the ion source temperature was 220 °C; the carrier gas was helium, adjusted to a linear velocity of 30 cm/s; the split ratio was 1:40; the ionization energy was 70 eV; the scan range was 40–300 u; the scan time was 1 s. The identity of the components was assigned by comparison of their retention indices, calculated in accordance with ISO 7609 [56], with C_8–C_{27} *n*-alkane indices and with a GC-MS spectra from a lab-made library, created with reference essential oils, laboratory-synthesized components, laboratory isolated compounds and commercially available standards.

3.4. Determination of Antioxidant Activity

The antioxidant activity was determined by evaluating the 1,1-diphenyl-2-picrylhydrazyl (DPPH) radical scavenging activity and oxygen radical absorbance capacity (ORAC) according to Ribeiro et al. [57] and Freitas et al. [58], respectively.

3.4.1. DPPH Radical Scavenging Activity

To assess the antioxidant activity, an ethanolic solution containing DPPH radicals with a concentration of 1.60×10^{-3} mol/L was prepared. Several dilutions in absolute ethanol were prepared from each analyzed sample's stock solution of EOs (10, 20, 30, 40, 50, 60, 70, 80 μL/μL). The same dilutions were repeated for the hydrolates but in phosphate-buffered saline (PBS).

Several dilutions of the EOs were prepared in ethanol, while the same dilutions of Hds were prepared in PBS. The negative control was an ethanolic solution of DPPH, 8×10^{-5} M. The positive control was 1×10^{-3} M aqueous solution of ascorbic acid.

The absorbance was measured at 517 nm using a fluorescence microplate reader (FLUOstar BMGLabtech, Ortenberg, Germany). The antioxidant activity was calculated as percentage inhibition of DPPH, using Equation (1)

$$\% \text{ Inhibition} = \left(\frac{A_{DPPH} - A_S}{A_{DPPH}} \right) \times 100 \quad (1)$$

where A_{DPPH} is the absorbance of the DPPH solution and A_S is the absorbance of the solution when the EO samples were added. Each experiment was performed in triplicate, and results are expressed as mean ± standard deviation (SD). The half-maximal effective concentration (EC_{50}), i.e., the required sample concentration to scavenge 50% of the DPPH radicals present in the solution, was calculated using the GraphPad Prism 5.0 software.

3.4.2. Oxygen Radical Absorbance Capacity (ORAC)

The ORAC method followed the protocol described by Freitas et al. [58]. Briefly, a mixture consisting of 5.18 M from 2,2'azobis (2-methylpropionamidine) dihydrochloride (AAPH) and 4×10^{-3} mM of disodium fluorescein (DF) (both prepared in 75 mM PBS pH: 7.4) was added to each well of a 96-well black plate. A calibration curve was performed with 6-hydroxy-2,5,7,8-tetramethylchroman-2-carboxylic acid (Trolox), in a range of concentrations from 0 μM to 40 μM. The final dilutions of the EOs and Hds samples used to make the readings were obtained after several attempts, until a concentration within the range of values of the Trolox curve was found. The fluorescence of the samples was read in a microplate reader (FLx800 from Biotek® Instruments, Inc., Winnooski, VT, USA), after a 10 min incubation at 37 °C. The fluorescence (excitation 485 nm, emission 527 nm) was determined every minute, for 40 min at 37 °C.

The fluorimeter control software used was Gen5 (version 3.10, Biotek® Instruments, Inc., Winnooski, VT, USA. 2006). The net areas under the curve (AUC) of the standard (Trolox) and samples were calculated. The standard curve was obtained by plotting Trolox concentrations against the average net AUC of each concentration measurements. The results were expressed in micromoles of Trolox equivalent antioxidant capacity per gram of EO/Hd (μmol TEAC/g EO/Hd). All data were presented as mean ± SD of two replicates.

3.4.3. In Vitro Antioxidant Activity

The ability of EOs and Hds samples to reduce the ROS production was determined using a well-characterized probe, 20,70-dichlorofluorescein diacetate (H_2-DCFDA; Life Technologies, Glasgow, UK), as described by Marques et al. [59] and Carriço et al. [60], with some adaptations. Briefly, the human keratinocytes HaCaT cell line (Cell line Service GmbH, Germany) was seeded at 2×10^4 cells per well in 96-well plates with 100 μL of the cell culture medium per well and incubated at 37 °C for 24 h. Cells were pre-incubated for 30 min with 20 μM of H_2-DCFDA, in the dark, at 37 °C. Then the probe solution was removed, and a fresh medium was added containing the different samples to be tested. Then, 1 mg/mL of ascorbic acid was used as positive control and the culture medium was the negative control. Cells were incubated with the different samples at 10% (v/v) for 1 h at 37 °C prior to the addition of 500 μM H_2O_2.

The DCF levels were determined by fluorescence (excitation 485 nm, emission 520 nm) in a microplate reader (FLUOstar BMGLabtech, Ortenberg, Germany). Data from six replicates were reported as the relative mean of % ROS reduction determined by relative fluorescence units (RFU) of culture medium with H_2O_2 as 100% and the % ROS reduction as in Equation (2):

$$\left[100 - \left(\frac{fluorescence\ of\ sample\ exposed\ cells}{fluorescence\ of\ unexposed\ control\ from\ the\ same\ experiement}\right)\right] \times 100 \quad (2)$$

The data were expressed as mean ± SD of experiments ($n = 8$). Statistical evaluation of data was performed using a one way analysis of variance (ANOVA). Tukey–Kramer multiple comparison test (GraphPad PRISM software, version 5.01, La Jolla, CA, USA) was used to compare the difference between the groups, and differences were considered significant for $p < 0.05$.

3.5. Evaluation of Antimicrobial Activity

The minimum inhibition concentration (MIC) was determined to evaluate the antimicrobial activity of the EOs and Hds against Gram-positive and Gram-negative bacteria, yeast, and mold.

3.5.1. Microbial Strains

The microbial strains selected for the study were: *Staphylococcus aureus* ATCC 6538, *Bacillus subtilis* ATCC 6633, *Pseudomonas aeruginosa* ATCC 9027, *Escherichia coli* ATCC 8739, *Candida albicans* ATCC 10,231 and *Aspergillus brasiliensis* ATCC 16404, available at ADEIM/Faculty of Pharmacy, University of Lisbon.

3.5.2. Determination of Minimum Inhibitory Concentration by the Microdilution Method

The MIC determinations for each EO and Hd were performed by the broth microdilution method. Microbial suspensions of each strain were prepared in PBS to a final concentration of 1.5×10^8 CFU/mL for bacterial strains and 1.5×10^6 CFU/mL for the yeast and mold. Then, 100 µL Mueller Hinton (MH) for bacteria and Sabourad dextrose broth (SBD) for the yeast and mold were added to each well of a 96-well microplate. Afterwards, 100 µL of the EO and Hd solutions (1 mg/mL) prepared in the appropriate culture media were added to the first well followed by a twofold serial dilution to obtain final concentrations ranging from 500 mg/mL to 0.48 µg/mL. Finally, 10 µL of the bacterial/fungi suspension diluted in the appropriate culture medium were added to each well to obtain a final concentration of 10^5 CFU/mL. The microplates were incubated at 35–37 °C for 24 h for bacteria or 48 h for yeast, and at 22 °C for 5 days for *Aspergillus brasiliensis*. Bacterial growth and culture medium were used as controls. The minimal inhibitory concentration (MIC) was defined as the lowest concentration where no visible growth was observed, and growth was monitored by measuring OD_{600nm} in a microplate reader (Varioskan™ multimode microplate reader, Thermo Scientific, Massachusetts, USA). All experiments were performed in triplicate.

3.6. Sensory Evaluation

A sensory double-blind evaluation of *E. globulus*, *P. pinaster*, *P. pinea* and *C. japonica* EOs in skin care emulsions was performed by a group of inexperienced volunteers ($n = 100$), females and males aged 18 to 60 years old, with a signed informed consent. The academic objective of the research was reported to the participants, maintaining the privacy of the volunteers and the confidentiality of the information collected. The exclusion criteria included persons who had or have an infection with SARS-CoV-2, respiratory health problems or olfactory diseases in which the olfactory part was affected, which could compromise the sensory questionnaire results.

The emulsions were perfumed using five EOs selected from each species under study, except for *P. pinaster* EO, in which two samples from different producers were chosen. To prepare the emulsions, the oily phase (decyl oleate, cetyl alcohol, ceteareth-11, ceteareth-20 and paraffinum liquidum) and aqueous phase (purified water and glycerin) were heated separately to 75 °C. Then the oily phase was added to the water phase and the system was mixed (130 rpm) with constant agitation in a VMI bench mixer until a 30 °C temperature was reached. Finally, the EOs were added and manually mixed. The emulsions were formulated with 0.5% of EO from the selected samples according to the protocol published by Neves et al. [3]. The excipients used in formulations and the percentage composition of the emulsions prepared with the EOs are described in Table 11.

Table 11. Qualitative and quantitative (%, *w*/*w*) composition of the emulsions prepared with *Eucalyptus globulus*, *Pinus pinaster*, *Pinus pinea* and *Cryptomeria japonica* essential oils.

INCI *	Trade Name	Function	(%, *w*/*w*)
Phase A			
Ceteareth-11	Eumulgin B1®	Non-ionic O/W emulsifier	1.5
Ceteareth-20	Eumulgin B2	Non-ionic O/W emulsifier	1.5
Cetyl alcohol	Cetyl alcohol	Thickener	2.0
Paraffinum liquidum	Mineral oil	Emollient	2.5
Decyl oleate	Tegosoft DO®		4.5
Phase B			
Glycerin	Glycerin	Humectant	5.0
Aqua	Purified water	Solvent	82.0
Phase C			
Parfum	Essential Oil	Fragrance	0.5

* Ingredients' names according to the International Nomenclature of Cosmetics Ingredients (INCI).

The formulations of the different samples were coded with different colours: *E. globulus*, Eg_OE_1_G, orange; *P. pinaster*, Pp_OE_1_G, blue; *P. pinaster*, Pp_OE_2_P, purple; *P. pinea*, Ppi_OE_1_B, green; and *C. japonica*, Cj_OE_1_M, pink.

To evaluate the acceptability of all formulations, a questionnaire (Table S1) was answered by each volunteer. The questionnaire was divided into two sections: regarding the Emulsions' Odour, which aimed to classify emulsions according to the olfactory preferences of the selected volunteers, and concerning the Emulsions' Applicability with the main objective to evaluate the acceptability of the emulsions considering the probability of purchasing different personal care products (perfume; air freshener; massage cream; toothpaste; shampoo; candy) with the odour and colour of the respective emulsion.

The data collected using the sensory questionnaire during the experimental study were subjected to statistical treatment, using the statistical program IBM® SPSS® software (Statistical Package for the Social Sciences version 27 for Windows 10).

4. Conclusions

This study focused on the chemical analysis of essential oils and hydrolates from *Eucalyptus globulus* L., *Pinus pinaster* A., *Pinus pinea* L. and *Cryptomeria japonica* D. Don, as well as on assessing their antioxidant and antimicrobial potential and sensorial properties.

Some of the EOs and Hds showed relevant antioxidant activity and antimicrobial activity. Furthermore, the sensory evaluation revealed the odours favoured by the participants and which products could use such odours. Thus, it can be concluded that essential oils could be used as natural antioxidant substances or cosmetic preservatives, for example. Moreover, such products address the demand for sustainable and responsibly sourced odours accepted by consumers, but further testing will be needed to ensure consumer safety and satisfaction.

Supplementary Materials: The following supporting information can be downloaded at: https://www.mdpi.com/article/10.3390/molecules27113572/s1, Table S1. Percentage composition of *Eucalyptus globulus* essential oils. For samples codes *vide* Table 10 in Materials and Methods section. Table S2. Percentage composition of *Pinus pinaster* essential oils. For samples codes *vide* Table 10 in Materials and Methods section. Table S3. Percentage composition of *Pinus pinea* essential oil. For samples codes *vide* Table 10 in Materials and Methods section. Table S4. Percentage composition of

Cryptomeria japonica essential oil. For samples codes *vide* Table 10 in Materials and Methods section. Table S5. Percentage composition of *Eucalyptus globulus* hydrolates volatiles. For samples codes *vide* Table 10 in Materials and Methods section. Table S6. Percentage composition of *Pinus pinaster* hydrolates volatiles. For samples codes *vide* Table 10 in Materials and Methods section. Table S7. Percentage composition of *Cryptomeria japonica* hydrolates volatiles. For samples codes *vide* Table 10 in Materials and Methods section. Sensory Questionnaire (English Version).

Author Contributions: Formal analysis, methodology, and investigation, A.R., A.G., J.M., L.G., A.O., A.T.S., A.N.d.S., M.P., A.M.M., A.C.F. and H.M.R.; conceptualization, resources, and supervision, A.C.F. and H.M.R. All authors were involved in writing—review, editing. All authors have read and agreed to the published version of the manuscript.

Funding: This research was funded by the Fundação para a Ciência e Tecnologia, Portugal to CESAM UIDP/50017/2020+UIDB/50017/2020+LA/P/0094/2020, UIDB/04138/2020 and UIDP/04138/2020 to iMed.ULisboa, CEECINST/00145/2018 to J.M. and principal investigator grants CEECIND/03143/2017 to L.G.L, and iNOVA4Health—UIDB/04462/2020 and UIDP/04462/2020, INTERFACE Program, through the Innovation, Technology and Circular Economy Fund (FITEC), is gratefully acknowledged. ATS acknowledges Grant CEECIND/04801/2017.

Institutional Review Board Statement: The study was conducted in accordance with the Declaration of Helsinki and approved by the Institutional Review Board (or Ethics Committee) of Faculdade de Farmácia da Universidade de Lisboa (protocol code no. 1_2022 May 2022).

Informed Consent Statement: Informed consent was obtained from all subjects involved in the study.

Data Availability Statement: Not applicable.

Acknowledgments: The authors acknowledge the producers that provided the essential oils and hydrolates for analysis. The authors thank Ana M. Martins, from iMed. ULisboa for language editing and proofreading. The authors acknowledge the Fundação para a Ciência e Tecnologia, Portugal for support to CESAM UIDP/50017/2020+UIDB/50017/2020+LA/P/0094/2020.

Conflicts of Interest: The authors declare no conflict of interest.

Sample Availability: Samples of EOs and Hds are available from the authors.

References

1. Nehme, R.; Andrés, S.; Pereira, R.B.; Jenmaa, M.B.; Bouhallab, S.; Ceciliani, F.; López, S.; Rahali, F.Z.; Ksouri, R.; Pereira, D.M.; et al. Essential Oils in Livestock: From Health to Food Quality. *Antioxidants* **2021**, *10*, 330. [CrossRef] [PubMed]
2. Surburg, H.; Panten, J. *Common Fragrance and Flavor Materials: Preparation, Properties and Uses*, 6th ed.; Wiley-VCH Verlag GmbH & Co. KGaA: Weinheim, Germany, 2016; pp. 267–271.
3. Neves, A.; Marto, J.; Duarte, A.; Gonçalves, L.M.; Pinto, P.; Figueiredo, A.C.; Ribeiro, H.M. Characterization of Portuguese *Thymbra capitata*, *Thymus caespititius* and *Myrtus communis* essential oils in topical formulations. *Flavour Fragr. J.* **2017**, *32*, 392–402. [CrossRef]
4. Sarkic, A.; Stappen, I. Essential oils and their single compounds in cosmetics—A critical review. *Cosmetics* **2018**, *5*, 11. [CrossRef]
5. Cunha, C.; Ribeiro, H.M.; Rodrigues, M.; Aranjo, A.R.T.S. Essential oils used in dermocosmetics: Review about its biological activities. *J. Cosmet. Dermatol.* **2021**, *21*, 513–529. [CrossRef] [PubMed]
6. Tavares, C.S.; Martins, A.; Faleiro, M.L.; Miguel, M.G.; Duarte, L.C.; Gameiro, J.A.; Roseiro, L.B.; Figueiredo, A.C. Bioproducts from Forest Biomass: Essential oils and hydrolates from wastes of *Cupressus lusitanica* Mill. and *Cistus ladanifer* L. *Ind. Crops Prod.* **2020**, *144*, 112034. [CrossRef]
7. Mediavilla, I.; Guillamón, E.; Ruiz, A.; Esteban, L.S. Essential oils from residual foliage of forest tree and shrub species: Yield and antioxidant capacity. *Molecules* **2021**, *26*, 3257. [CrossRef] [PubMed]
8. Tavares, C.S.; Gameiro, J.A.; Roseiro, L.B.; Figueiredo, A.C. Hydrolates: A review on their volatiles composition, biological properties and potential uses. *Phytochem. Rev.* **2022**. [CrossRef]
9. Wajs-Bonikowska, A.; Sienkiewicz, M.; Stobiecka, A.; Maciag, A.; Szoka, Ł.; Karna, E. Chemical Composition and Biological Activity of *Abies alba* and *A. koreana* Seed and Cone Essential Oils and Characterization of Their Seed Hydrolates. *Chem. Biodivers.* **2015**, *15*, 407–418.
10. Baydar, H.; Sangun, M.K.; Erbas, S.; Kara, N. Comparison of Aroma Compounds in Distilled and Extracted Products of Sage (*Salvia officinalis* L.). *J. Essent. Oil Bear. Plants* **2013**, *16*, 39–44. [CrossRef]
11. Luís, Â.; Duarte, A.; Gominho, J.; Domingues, F.; Duarte, A.P. Chemical composition, antioxidant, antibacterial and anti-quorum sensing activities of *Eucalyptus globulus* and *Eucalyptus radiata* essential oils. *Ind. Crops Prod.* **2015**, *79*, 274–282. [CrossRef]

12. Hayat, U.; Jilani, M.I.; Rehman, R.; Nadeem, F. A Review on *Eucalyptus globulus*: A New Perspective in Therapeutics. *Int. J. Chem. Biochem. Sci.* **2015**, *8*, 85–91.
13. Bachir, R.G.; Benali, M. Antibacterial activity of the essential oils from the leaves of *Eucalyptus globulus* against *Escherichia coli* and *Staphylococcus aureus*. *Asian Pac. J. Trop Biomed.* **2012**, *2*, 739–742. [CrossRef]
14. Zolfaghari, B.; Iravani, S. Essential oil constituents of the bark of *Pinus pinaster* from iran. *J. Essent. Oil-Bear. Plants* **2012**, *15*, 348–351. [CrossRef]
15. Tümen, İ.; Akkol, E.K.; Taştan, H.; Süntar, I.; Kurtcam, M. Research on the antioxidant, wound healing, and anti-inflammatory activities and the phytochemical composition of maritime pine (*Pinus pinaster* Ait). *J. Ethnopharmacol.* **2018**, *211*, 235–246. [CrossRef]
16. Lugovic, L.; Mirna, S.; Ozanic-Bulic, S.; Sjerobabski-Masnec, I. Phototoxic and Photoallergic Skin Reactions. *Coll. Antropol.* **2011**, *31*, 63–67.
17. Amri, I.; Gargouri, S.; Hamrouni, L.; Hanana, M.; Fezzani, T.; Jamoussi, B. Chemical composition, phytotoxic and antifungal activities of *Pinus pinea* essential oil. *J. Pest. Sci.* **2012**, *85*, 199–207. [CrossRef]
18. Hmamouchi, M.; Hamamouchi, J.; Zouhdi, M. Chemical and Antimicrobial Properties of Essential Oils of Five Moroccan Pinaceae. *J. Essent. Oil Res.* **2001**, *13*, 298–302. [CrossRef]
19. Ulukanli, Z.; Karabörklü, S.; Bozok, F.; Ates, B.; Erdogan, S.; Cenet, M.; Karaaslan, M.G. Chemical composition, antimicrobial, insecticidal, phytotoxic and antioxidant activities of Mediterranean *Pinus brutia* and *Pinus pinea* resin essential oils. *Chin. J. Nat. Med.* **2014**, *12*, 901–910. [CrossRef]
20. Süntar, I.; Tumen, I.; Ustün, O.; Keleş, H.; Küpeli, A.E. Appraisal on the wound healing and anti-inflammatory activities of the essential oils obtained from the cones and needles of *Pinus* species by in vivo and in vitro experimental models. *J. Ethnopharmacol.* **2012**, *139*, 533–540. [CrossRef]
21. Cheng, S.S.; Lin, H.Y.; Chang, S.T. Chemical composition and antifungal activity of essential oils from different tissues of Japanese cedar (*Cryptomeria japonica*). *J. Agric. Food Chem.* **2005**, *53*, 614–619. [CrossRef]
22. Matsunaga, T.; Hasegawa, C.; Toru, K.; Suzuki, H.; Saito, H.; Sagioka, T.; Takahashi, R.; Tsukamoto, H.; Morikawa, T.; Akiyama, T. Isolation of the Antiulcer Compound in Essential Oil from the Leaves of *Cryptomeria japonica*. *Chem. Pharm. Bull.* **2000**, *23*, 595–598. [CrossRef]
23. Wang, S.Y.; Lai, W.C.; Chu, F.H.; Lin, C.T.; Shen, S.Y.; Chang, S.T. Essential oil from the leaves of *Cryptomeria japonica* acts as a silverfish (*Lepisma saccharina*) repellent and insecticide. *J. Wood Sci.* **2006**, *52*, 522–526. [CrossRef]
24. Cha, J.D.; Jeong, M.R.; Jeong, S.; Moon, S.; Kil, B.; Yun, S.; Lee, K.; Song, Y. Chemical composition and antimicrobial activity of the essential oil of *Cryptomeria japonica*. *Phyther. Res.* **2007**, *21*, 295–299. [CrossRef] [PubMed]
25. Silvestre, A.J.D.; Cavaleiro, J.A.S.; Delmond, B.; Filliatre, C.; Bourgeois, G. The Essential Oil of *Eucalyptus globulus* Labill. from Portugal. *Flavour Fragr. J.* **1994**, *9*, 51–53. [CrossRef]
26. Faria, J.M.S.; Lima, A.S.; Mendes, M.D.; Leiria, R.; Geraldes, D.A.; Figueiredo, A.C.; Trindade, H.; Pedro, L.G.; Barroso, J.G.; Sanches, J. *Eucalyptus* from Mata Experimental do Escaroupim (Portugal): Evaluation of the essential oil composition from sixteen species. *Acta Hortic.* **2011**, *925*, 61–66. [CrossRef]
27. Faria, J.M.S.; Barbosa, P.; Bennett, R.N.; Mota, M.; Figueiredo, A.C. Bioactivity against *Bursaphelenchus xylophilus*: Nematotoxics from essential oils, essential oils fractions and decoction waters. *Phytochemistry* **2013**, *94*, 220–228. [CrossRef] [PubMed]
28. Vieira, M.; Bessam, L.J.; Martins, M.R.; Arantes, S.; Teixeira, A.P.S.; Mendes, Â.; Costa, P.M.; Belo, A.D.F. Chemical Composition, Antibacterial, Antibiofilm and Synergistic Properties of Essential Oils from *Eucalyptus globulus* Labill. and Seven Mediterranean Aromatic Plants. *Chem. Biodivers.* **2017**, *14*, e1700006. [CrossRef] [PubMed]
29. Miguel, M.; Gago, C.; Antunes, M.; Lagoas, S.; Faleiro, M.L.; Megías, C.; Cortés-Giraldo, I.; Vioque, J.; Figueiredo, A.C. Antibacterial, Antioxidant, and Antiproliferative Activities of *Corymbia citriodora* and the Essential Oils of Eight *Eucalyptus* Species. *Medicines* **2018**, *5*, 61. [CrossRef]
30. *ISO 770:2002*; Crude or rectified oils of Eucalyptus globulus (Eucalyptus globulus Labill.); International Organization for Standardization: Geneva, Switzerland, 2002.
31. Carmo, M.M.; Frazão, S. The essencial oil of portuguese Pine needles. First Results. In *Progress in Essential Oil Research*; Brunke, E.-J., Ed.; Walter de Gruyter: New York, NY, USA, 1986; pp. 169–174.
32. Rodrigues, A.M.; Mendes, M.D.; Lima, A.S.; Barbosa, P.M.; Ascensão, L.; Barroso, J.G.; Pedro, L.G.; Mota, M.M.; Figueiredo, A.C. *Pinus halepensis, Pinus pinaster, Pinus pinea* and *Pinus sylvestris* Essential Oils Chemotypes and Monoterpene Hydrocarbon Enantiomers, before and after Inoculation with the Pinewood Nematode Bursaphelenchus xylophilus. *Chem. Biodivers.* **2016**, *14*, e1600153. [CrossRef]
33. Miguel, M.G.; da Silva, C.I.; Farah, L.; Braga, F.C.; Figueiredo, A.C. Effect of essential oils on the release of TNF-α and CCL2 by LPS-stimulated THP-1 cells. *Plants* **2021**, *10*, 50. [CrossRef]
34. Moiteiro, C.; Esteves, T.; Ramalho, L.; Rojas, R.; Alvarez, S.; Zacchino, S.; Bragança, H. Essential Oil Characterization of Two Azorean *Cryptomeria japonica* Populations and Their Biological Evaluations. *Nat. Prod. Commun.* **2013**, *8*, 1785–1790. [CrossRef]
35. Figueiredo, A.C.; Moiteiro, C.; Rodrigues, M.C.S.M.; Almeida, A.J.R.M. Essential Oil Composition From *Cryptomeria japonica* D.Don Grown in Azores: Biomass Valorization From Forest Management. *Nat. Prod. Commun.* **2021**, *16*, 1–10. [CrossRef]
36. Paolini, J.; Leandri, C.; Desjobert, J.M.; Barboni, T.; Costa, J. Comparison of liquid-liquid extraction with headspace methods for the characterization of volatile fractions of commercial hydrolats from typically Mediterranean species. *J. Chromatogr. A* **2008**, *1193*, 37–49. [CrossRef]

37. Inouye, S.; Takahashi, M.; Abe, S. A comparative study on the composition of forty-four hydrosols and their essential oils. *Int. J. Essent. Oil Ther.* **2008**, *2*, 89–104.
38. Nakagawa, T.; Zhu, Q.; Ishikawa, H.; Ohnuki, K.; Kakino, K.; Horiuchi, N.; Shinotsuka, H.; Naito, T.; Matsumoto, T.; Minamisawa, N.; et al. Multiple uses of Essential Oil and by-products from various parts of Yakushima native cedar (*Cryptomeria japonica*).
J. Wood Chem. Technol. **2016**, *36*, 42–55. [CrossRef]
39. Ho, C.; Wang, E.-C.; Yu, H.-T.; Yu, H.-M.; Su, C.-Y. Compositions and antioxidant activities of essential oils of different tissues from *Cryptomeria japonica* D. Don. *For. Res. Q.* **2010**, *32*, 63–76.
40. Souza, C.F.; Baldissera, M.D.; Silva, L.L.; Geihs, M.A.; Baldisserotto, B. Is monoterpene terpinen-4-ol the compound responsible for the anesthetic and antioxidant activity of *Melaleuca alternifolia* essential oil (tea tree oil) in silver catfish? *Aquaculture* **2018**, *486*, 217–223. [CrossRef]
41. Cutillas, A.; Carrasco, A.; Martinez-Gutierrez, R.; Tomas, V.; Tudela, J. Rosmarinus officinalis L. essential oils from Spain: Composition, antioxidant capacity, lipoxygenase and acetylcholinesterase inhibitory capacities, and antimicrobial activities. *Plant Biosyst. Int. J. Deal. All Asp. Plant Biol.* **2018**, *152*, 1282–1292. [CrossRef]
42. Cao, G.; Alessio, H.M.; Cutler, R.G. Oxygen-radical absorbance capacity assay for antioxidants. *Free. Radic. Biol. Med.* **1993**, *14*, 303–311. [CrossRef]
43. Mohammadi, A.; Sani, T.A.; Ameri, A.A.; Imani, M.; Golmakani, E.; Kamali, H. Seasonal variation in the chemical composition, antioxidant activity, and total phenolic content of Artemisia absinthium essential oils. *Phcog. Res.* **2015**, *7*, 329–334.
44. Prior, R.L. Fruits and vegetables in the prevention of cellular oxidative damage. *Am. J. Clin. Nutr.* **2003**, *78*, 570S–578S. [CrossRef]
45. Kim, K.Y. Anti-inflammatory and ECM gene expression modulations of β-eudesmol via NF-κB signaling pathway in normal human dermal fibroblasts. *Biomed. Dermatol.* **2018**, *2*, 3. [CrossRef]
46. Lee, B.H.; Nam, T.G.; Park, W.J.; Kang, H.; Heo, H.J.; Chung, D.K.; Kim, G.H.; Kim, D. Antioxidative and Neuroprotective Effects of Volatile Components in Essential Oils from Chrysanthemum indicum Linné Flowers. *Food Sci. Biotechnol.* **2015**, *24*, 717–723. [CrossRef]
47. Ambrosio, C.M.S.; Diaz-Arenas, G.L.; Agudelo, L.P.A.; Stashenko, E.; Contreras-Castillo, C.J.; da Glória, E.M. Chemical Composition and Antibacterial and Antioxidant Activity of a Citrus Essential Oil and Its Fractions. *Molecules* **2021**, *26*, 2888. [CrossRef]
48. Janssen, A.M.; Scheffer, J.J.C.; Svendsen, A.B. Antimicrobial activities of essential oils. A 1976–1986 literature review on possible applications. *Pharm. Weekbl. Sci. Ed.* **1987**, *9*, 193–197. [CrossRef] [PubMed]
49. Šilha, D.; Švarcová, K.; Bajer, T.; Královec, K.; Tesařová, E.; Moučková, K.; Pejchalová, M.; Bajerová, P. Chemical Composition of Natural Hydrolates and Their Antimicrobial Activity on *Arcobacter*-Like Cells in Comparison with Other Microorganisms. *Molecules* **2020**, *25*, 5654. [CrossRef] [PubMed]
50. Cimanga, K.; Apers, S.; de Bruyne, T.; Miert, S.V.; Hermans, N.; Totté, J.; Pieters, L.; Vlietinck, A.J. Chemical Composition and Antifungal Activity of Essential Oils of Some Aromatic Medicinal Plants Growing in the Democratic Republic of Congo. *J. Essent. Oil Res.* **2002**, *14*, 382–387. [CrossRef]
51. Sharmeen, J.B.; Mahomoodally, F.M.; Zengin, G.; Maggi, F. Essential Oils as Natural Sources of Fragrance Compounds for Cosmetics and Cosmeceuticals. *Molecules* **2021**, *26*, 666. [CrossRef] [PubMed]
52. El-Zaeddi, H.; Martínez-Tomé, J.; Calín-Sánchez, A.; Burló, F.; Carbonell-Barrachina, A.A. Volatile Composition of Essential Oils from Different Aromatic Herbs Grown in Mediterranean Regions of Spain. *Foods* **2016**, *5*, 41. [CrossRef]
53. Schreiner, L.; Bauer, P.; Buettner, A. Resolving the smell of wood-identification of odour-active compounds in Scots pine (*Pinus sylvestris* L.). *Sci. Rep.* **2018**, *8*, 8294. [CrossRef]
54. Tripathi, A.K.; Mishra, S. Plant Monoterpenoids (Prospective Pesticides). In *Ecofriendly Pest Management for Food Security*, 1st ed.; Omkar, Ed.; Elsevier Inc.: Amsterdam, The Netherlands, 2016; pp. 507–524.
55. Pauwels, M.; Rogiers, V. Human health safety evaluation of cosmetics in the EU: A legally imposed challenge to science. *Toxicol. Appl. Pharmacol.* **2010**, *243*, 260–274. [CrossRef]
56. *ISO 7609:1985*; Essential oils—Analysis by gas chromatography on capillary columns—General method; International Organization for Standardization: Geneva, Switzerland, 1985.
57. Ribeiro, H.M.; Allegro, M.; Marto, J.; Pedras, B.; Oliveira, N.G.; Paiva, A.; Barreiros, S.; Gonçalves, L.M.D.; Simões, P.M.C. Converting Spent Coffee Grounds into Bioactive Extracts with Potential Skin Antiaging and Lightening Effects ACS Sustainable. *Chem. Eng.* **2018**, *6*, 6289–6295.
58. Freitas, R.; Martins, A.; Silva, J.; Alves, C.; Pinteus, S.; Alves, J.; Teodoro, F.; Ribeiro, H.M.; Gonçalves, L.; Petrovski, Ž.; et al. Highlighting the Biological Potential of the Brown Seaweed Fucus spiralis for Skin Applications. *Antioxidants* **2020**, *9*, 611. [CrossRef]
59. Marques, P.; Marto, J.; Gonçalves, L.M.; Pacheco, R.; Fitas, M.; Pinto, P.; Serralheiro, M.L.M.; Ribeiro, H.M. *Cynara scolymus* L.: A promising Mediterranean extract for topical anti-aging prevention. *Ind. Crops Prod.* **2017**, *109*, 699–706.
60. Carriço, C.; Pinto, P.; Graça, A.; Gonçalves, L.M.; Ribeiro, H.M.; Marto, J. Design and Characterization of a New Quercus Suber-Based Pickering Emulsion for Topical Application. *Pharmaceuticals* **2019**, *11*, 131. [CrossRef]

molecules

MDPI

Review

Essential Oils in Respiratory Mycosis: A Review

Mónica Zuzarte [1,2,3,*] and Lígia Salgueiro [4,5]

1 Faculty of Medicine, Coimbra Institute for Clinical and Biomedical Research (iCBR), University of Coimbra, 3000-548 Coimbra, Portugal
2 Center for Innovative Biomedicine and Biotechnology (CIBB), University of Coimbra, 3000-548 Coimbra, Portugal
3 Clinical Academic Centre of Coimbra (CACC), 3000-548 Coimbra, Portugal
4 Faculty of Pharmacy, University of Coimbra, 3000-548 Coimbra, Portugal; ligia@ff.uc.pt
5 Faculty of Sciences and Technology, Department of Chemical Engineering, Chemical Process Engineering and Forest Products Research Centre (CIEPQPF), University of Coimbra, 3030-790 Coimbra, Portugal
* Correspondence: mzuzarte@uc.pt

Abstract: Respiratory mycosis is a major health concern, due to the expanding population of immunosuppressed and immunocompromised patients and the increasing resistance to conventional antifungals and their undesired side-effects, thus justifying the development of new therapeutic strategies. Plant metabolites, namely essential oils, represent promising preventive/therapeutic strategies due to their widely reported antifungal potential. However, regarding fungal infections of the respiratory tract, information is disperse and no updated compilation on current knowledge is available. Therefore, the present review aims to gather and systematize relevant information on the antifungal effects of several essential oils and volatile compounds against the main type of respiratory mycosis that impact health care systems. Particular attention is paid to *Aspergillus fumigatus*, the main pathogen involved in aspergillosis, *Candida auris*, currently emerging as a major pathogen in certain parts of the world, and *Cryptococcus neoformans*, one of the main pathogens involved in pulmonary cryptococcosis. Furthermore, the main mechanisms of action underlying essential oils' antifungal effects and current limitations in clinical translation are presented. Overall, essential oils rich in phenolic compounds seem to be very effective but clinical translation requires more comprehensive in vivo studies and human trials to assess the efficacy and tolerability of these compounds in respiratory mycosis.

Keywords: aspergillosis; candidiasis; cryptococcosis; endemic infections; opportunistic infections; plant volatiles

Citation: Zuzarte, M.; Salgueiro, L. Essential Oils in Respiratory Mycosis: A Review. *Molecules* **2022**, *27*, 4140. https://doi.org/10.3390/molecules27134140

Academic Editors: Ana Paula Duarte, Ângelo Luís and Eugenia Gallardo

Received: 6 June 2022
Accepted: 25 June 2022
Published: 28 June 2022

1. Introduction

Respiratory infections have increased over the past decades, becoming important causes of morbidity among immunocompromised and immunosuppressed patients and accounting for 4.3 million annual deaths worldwide [1]. Among these, fungal infections are underestimated, despite being responsible for mortality rates above 50% [2]. Although exposure to these respiratory pathogens occurs regularly during a lifetime, healthy individuals rarely develop symptomatic infections. However, in individuals with impaired defenses—for example, due to the use of chemotherapeutic and immunosuppressive agents, and antibiotics, or having prosthetic devices, grafts, burns, neutropenia or HIV—systemic life-threatening fungal infections may occur [3].

Overall, two main types of infections can occur: endemic (primary) or opportunistic. The first occur primarily in patients living or traveling in developing countries and, despite being mostly asymptomatic or causing mild infections, they account for huge health-care costs in these regions. Importantly, an exposure to high concentrations of inoculum or in patients with compromised immune defenses, life-threatening infections or the reactivation of latent foci may occur. The most common fungal infections include blastomycosis, coccidioidomycosis, histoplasmosis, paracoccidioidomycosis and sporotrichosis [4]. On the other hand, opportunistic infections generally occur in debilitated patients with impaired defense mechanisms and have a global distribution. These include aspergillosis, candidiasis, cryptococcosis, hyalohyphomycosis, mucormycosis, phaeohyphomycosis, and *Pneumocystis* pneumonia, with *Aspergillus*, *Cryptococcus* and *Pneumocystis* being the major pulmonary fungal pathogens capable of causing life-threatening invasive diseases [5]. In Section 2, a detailed characterization of these infections is presented with information regarding diagnosis, pathological symptoms and treatment.

The diagnosis of respiratory mycosis is quite difficult as it may present nonspecific symptoms and noninvasive diagnostic tests are poorly sensitive. Therefore, a combination of factors is frequently considered for an accurate diagnosis, namely clinical setting, chest imaging, and negative bacterial or viral studies [6]. To treat respiratory mycosis, several antifungals are available and include triazoles, echinocandins and formulations with amphotericin B. The first interfere with cytochrome P450 (CYP450) and lead to the inhibition of lanosterol, thus decreasing ergosterol synthesis and inhibiting cell-membrane development [7]. The most common are fluconazole and itraconazole, which are effective against *Candida* (except *C. glabrata* and *C. krusei*), *Cryptococcus*, *Blastomyces*, *Histoplasma*, and *Coccidioides*. The second-generation triazoles, including voriconazol and posaconazole, are also active against *Aspergillus* and *Mucor*. As these drugs undergo hepatic metabolism, drug–drug interactions may occur through the CYP450 system and side effects such as gastrointestinal disturbances, hepatotoxicity, headache, and rash have been reported [8]. Echinocandins induce osmotic instability and cell death in fungi, by compromising the synthesis of beta-(1,3) glucan synthase, an important component of their cell walls [7]. Examples of these antifungals are caspofungin, micafungin and anidulafungin, primarily used against *Candida*, *Aspergillus* and *Pneumocystis*. Treatment with echinocandins can cause rash, headache, fever, and chills [9]. Amphotericin B (AmB) also causes fungal death by creating channels or pores on the fungal cell membrane that leak cellular components [10]. Side-effects include weight loss, headache, fatigue, and phlebitis and most importantly nephrotoxicity that can be mitigated by using amphotericin B (AmB) formulations such as AmB colloidal dispersion and lipid formulations. Furthermore, combination therapy has also been considered as it tends to increase treatment efficacy, especially for drug-resistant fungal isolates [11]. Indeed the emergence of resistance is another challenging situation that has increased over the decades due to the use of azole fungicides for agricultural purposes [12]. Therefore, natural antifungal alternatives have emerged as promising agents to overcome resistance concerns and unwanted side-effects. The present review provides an updated compilation of the studies performed in the last 20 years based on a bibliographic search conducted using PubMed, Scopus and Google Scholar databases.

2. Types of Fungal Infections of the Respiratory Tract

Respiratory infections can be caused by the inhalation of spores from fungi that inhabit the environment. Although most people have been exposed to these pathogens, symptoms are rare in healthy individuals and infections are generally not transmitted between humans. However, in immunocompromised individuals these infections are of major concern being *Aspergillus*, *Cryptococcus*, *Pneumocystis*, and endemic fungi the major pulmonary fungal pathogens able to cause life-threatening invasive diseases [5]. Moreover, several risk factors, such as prolonged antibiotic use, hematologic malignancies and other immunocompromised states, worsen infection outcomes. Although the number of known fungi is relatively high—around 70,000—only 100 have been detected in respiratory infections with only a few being consistently considered pathogenic. Several of these fungi present dimorphism, with primary pathogens and *Sporothrix schenckii*, showing a morphological transformation in host tissues from a hyphal form to a yeast-like form (or spherule in the case of *Coccidioides immitis*). Contrarily, *Candida albicans* alters from blastoconidia to germ tubes that further develop into hyphae and *Penicillium marneffei* converts to transversely dividing sausage-shaped cells [13]. Moreover, many of these pathogens are capable of biofilm growth, forming highly organized communities that are resistant to antifungals and are responsible for recurrent infections [14]. Table 1 summarizes the main fungi responsible for relevant respiratory mycoses and points out for each pathogen (divided into endemic or opportunistic) the main source of infection, relevant pathological manifestations, diagnostic methods, and current therapy. The majority of information on some of the pathogens included in the table was based on the online textbook *Microbiology*, an OpenStax resource [15], the information on the remaining fungi being completed with other bibliographic sources such as clinical guidelines [16], centers for disease and health control [17] and professional manuals [18] available online.

Table 1. Causal agents of respiratory mycosis.

Fungi	Disease	Source of Infection	Pathological Manifestations	Diagnosis	Antifungals
Endemic					
Blastomyces dermatitidis	Blastomycosis	Soil	Mild flu-like symptoms; chronic cutaneous disease with subcutaneous lesions on the face and hands	Microscopic observation of sputum samples; urine antigen test; enzyme immunoassay	Amphotericin B, ketoconazole
Coccidioides immitis	Coccidioidomycosis (valley fever)	Soil	Granulomatous lesions on the face and nose; meningitis in severe cases	Serological tests	Amphotericin B
Histoplasma capsulatum	Histoplasmosis	Soils with bird or bat droppings	Fever, headache, and weakness with some chest discomfort	Chest X-ray; cultures grown on fungal selective media; direct fluorescence antibody and Giemsa staining	Amphotericin B, ketoconazole, itraconazole
Paracoccidioides sp.	Paracoccidioidomycosis	Soil near armadillo burrows	Adults: affects lungs and causes lesions in the mouth and throat; Children: swollen lymph nodes and skin lesions	Chest X-ray, biopsy for fungal culture or to be examined under the microscope and blood tests	Itraconazole and amphotericin B; trimetho-prim/sulfamethoxazole
Talaromyces marneffei (formerly *Penicillium marneffei*)	Talaromycosis (formerly Penicilliosis)	Plants and farmed animals	Fever, weight loss, hepatosplenomegaly, lymphadenopathy, skin lesions	Microscopy, histology, and culture	Amphotericin B or voriconazole followed by itraconazole
Sporothrix schenckii	Sporotrichosis (rose gardener's disease)	Soil, *Sphagnum* moss, rose bushes and hay	Cutaneous nodules that spread and break down into abscesses and ulcers, with rare pulmonary involvement	Culture	Itraconazole, amphotericin B
Opportunistic					
Aspergillus fumigatus	Aspergillosis	Soils and organic debris	Asthma-like allergic reactions; shortness of breath, wheezing, coughing, runny nose and headaches	Chest X-ray; microscopic examination of tissue and respiratory fluid samples	Itraconazole, voriconazole
Candida spp.	Candidiasis	Skin and inside the body	Localized or diffuse pneumonia, nodular lesions, abscesses, and empyema	Isolation of the organism from lung tissue samples	Fluconazole (milder cases), amphotericin B deoxycholate, lipid formulations

Table 1. *Cont.*

Fungi	Disease	Source of Infection	Pathological Manifestations	Diagnosis	Antifungals
Cryptococcus neoformans	Cryptococcosis	Soil, pigeon guano and tropical and subtropical trees	Fever, fatigue, and a dry cough; when spreading to the brain causes meningitis (headaches, sensitivity to light, and confusion)	Microscopic examination of lung tissues or cerebrospinal fluids	Amphotericin B combined with flucytosine followed by fluconazole for up to 6 months
Nonpigmented fungi (other than *Aspergillus* and *Penicillium* or *Zygomycetes*) *	Hyalohyphomycosis	Soil, water or on decomposing organic debris	Lesions from local cutaneous, subcutaneous, corneal, or nasal mucosal disease to disseminated disease involving multiple organs	Culture isolation and/or PCR	Surgical removal with or without azole antifungal therapy
Rhizopus spp. and *Mucor* spp.	Mucormycosis (formerly zygomycosis)	Throughout the environment	Fever, cough, chest pain, and shortness of breath	Tissue biopsy specimens	Amphotericin B and surgical debridement removal in superficial infections
Dark melanin-pigmented dematiaceous fungi **	Phaeohyphomycosis	Soil	Sinusitis, subcutaneous nodules or abscesses, keratitis, lung masses, osteomyelitis, mycotic arthritis, endocarditis, brain abscess, and disseminated infection	Examination using Masson-Fontana staining; culture to identify causative species	Surgery and/or itraconazole
Pneumocystis jirovecii	*Pneumocystis* pneumonia (PCP)	Person to person through the air	Fever, cough, and shortness of breath	Microscopic examination of tissue and fluid samples from the lungs	Trimethoprim-sulfamethoxazole combination

Includes: * *Acremonium, Fusarium, Geotrichum, Paecilomyces, Pseudallescheria, Sagenomella, Phialosimplex, Geosmithia, Geomyces*, and *Scedosporium*; ** *Bipolaris, Cladophialophora, Cladosporium, Exophiala, Fonsecaea, Phialophora, Ochronosis, Rhinocladiella*, and *Wangiella*.

3. In Vitro and In Vivo Models to Assess Antifungal Properties

The search for new antifungal agents involves in vitro susceptibility tests. Reference methods, namely the Clinical and Laboratory Standards Institute (CLSI) and the European Committee on Antimicrobial Susceptibility Testing (EUCAST) standard methods, are available enabling more practical, reliable, and reproducible protocols for antifungal susceptibility testing. Both resort to the broth microdilution method and assess the minimal inhibitory concentration (MIC) of an antifungal drug, which indicates the minimal drug concentration that inhibits fungal growth. Although these tests present some slight methodological differences, MICs are comparable [19]. Commercial antifungal susceptibility tests are also available. Examples include broth microdilution methods such as the Sensititre™ or YeastOne™ that use color endpoints based on metabolic dyes such as AlamarBlue, incorporated into the growth media or agar-based methods, such as the Etest® and the automated system VITEK® 2 [20].

In addition to in vitro testing, pre-clinical animal models have enabled us to increase the knowledge on fungal infections and putative therapeutic strategies. In these assays it is important to consider the model species and its immune status, the route of infection, and the fungal strain used, as these aspects can impact experimental outcomes. Overall, inbred strains of laboratory mice are the most common models mainly due to their relatively low cost and the wide availability of immunological reagents. However, other vertebrates such as rats, guinea pigs, rabbits, and zebrafish have gained popularity [21]. Recently, the

Galleria mellonella model has been one of the most used as it is inexpensive, easy to use, and does not require a dedicated infrastructure, the antifungal efficacy of the drug being estimated by fungal burden or mortality rate in infected and treated larvae [22]. To mimic respiratory mycosis, an intranasal or intratracheal injection of a liquid fungal suspension or the inhalation of dry fungal cells can be used. A catheter can be inserted beyond the vocal cords to facilitate intratracheal delivery of fungal cells and their dispersion into the pulmonary parenchyma may be enhanced by mechanical ventilation [23] or by using a microsprayer attached to the syringe tip [24]. Classical readouts include organ fungal burden and histopathology.

4. The Relevance of Essential Oils

Essential oils are mixtures of volatile compounds present in various organs of aromatic plants and obtained from the plant by hydrodistillation, steam distillation or dry distillation or, in the case of *Citrus* fruits, by a suitable mechanical process [25,26]. Essential oils generally present a strong odor and high lipophilicity, being primarily composed of terpenic compounds, such as monoterpenes and sesquiterpenes. In some cases, phenylpropanoids may also occur in high amounts and more rarely nitrogen and sulphur derivatives can be found. An important aspect to consider is their high chemical variability mainly due to environmental factors or genetic variations [27]. The latter may result in the expression of different metabolic pathways with significant variations in essential oils' composition that justify the identification of chemotypes. To assure high quality of the commercialized essential oils, analytical guidelines published by several institutions, such as the European Pharmacopoeia, the International Standard Organization (ISO), and the World Health Organization (WHO), should be followed.

Essential oils play relevant natural roles primarily in plant defense and signaling processes and have been explored by several industries due to particular features such as aroma, taste and bioactive potential. Indeed, they are valuable raw materials in the pharmaceutical, agronomic, food, sanitary, cosmetic, and perfume industries [28]. In what concerns bioactive potential, several studies have highlighted the antifungal and anti-inflammatory potential of these compounds as reviewed elsewhere [29,30]. These properties are quite relevant in the context of respiratory mycosis as fungal infections are associated with acute inflammation, which exacerbates the infection and delays its eradication. Therefore, extracts or compounds that combine both antifungal and anti-inflammatory activities, at concentrations devoid of toxicity, emerge as suitable preventive/therapeutic agents.

4.1. *Antifungal Effect of Essential Oils on Respiratory Mycosis*

Essential oils and their volatile compounds have shown promising effects against fungi involved in infections of the respiratory tract. Overall, *Aspergillus* spp., *Candida albicans*, and *Cryptococcus neoformans* are by far the most assessed strains, with only a few studies being carried out on other relevant pathogens referred in Table 1. Next, a compilation of selected studies is presented, the assortment criteria being defined in each section.

4.1.1. Essential Oils in Aspergillosis

Regarding *Aspergillus* pathogens, *A. fumigatus* is the most ubiquitous strain and the major causal agent of aspergillosis, followed by *A. flavus*, *A. niger*, *A. terreus*, and *A. nidulans* [31]. Besides species diversity, several types of aspergillosis are known, namely pulmonary aspergillosis that generally develops in patients with underlying lung pathology; allergic bronchopulmonary aspergillosis, an allergic reaction that results from hypersensitivity to *Aspergillus* colonization and is generally exclusive to asthma and cystic fibrosis patients and invasive aspergillosis, the most severe type that occurs when the infection travels from the lungs into the bloodstream [32]. As *A. fumigatus* is the main pathogen involved in aspergillosis, a compilation of the studies showing minimal inhibitory (MIC) or minimal fungicidal/lethal concentrations (MFC/MLC) is presented in Table 2. Only essential oils presenting MICs lower than 10 mg/mL were considered, being the plant name, part of the plant used to obtain the essential oils and antifungal effect highlighted, whenever this information was available in the original study. Broader studies, gathering more than three species, are not included in the table, and are discussed next. Furthermore, since only a few studies assessed a possible mechanism of action underlying these effects, this topic is discussed in Section 4.2.

Table 2. Antifungal effects of essential oils against *Aspergillus fumigatus*.

Essential Oil (Family)	Plant Part Used	Main Compounds	Antifungal Effect	Ref
Achillea millefolium (Asteraceae)	flowering aerial parts	Sample A: α-asarone (33.3%), α-pinene (17.2%), β-bisabolene (16.6%); Sample B: *trans*-thujone (29.0%), *trans*-crhysanthenyl acetate (15.8%), β-pinene (11.1%)	Broth macrodilution method: MIC = 1.25 μL/mL and MLC > 20 μL/mL (A); MIC = 2.5–5 μL/mL and MLC > 20 μL/mL (B)	[33]
Acorus calamus (Acoraceae)	*not identified*	β-asarone (80.6%)	Tube-dilution method: MFC = 0.104 ± 0.016 mg/mL	[34]
Allium hookeri (Amaryllidaceae)	rhizomes	di-2-propenyl trisulfide (31.8%), diallyl disulfide (28.4%)	Broth macrodilution method: MIC = 32 μg/mL and MFC = 64 μg/mL	[35]
Apium graveolens (Apiaceae)	flowering aerial parts	Sample A: neophytadiene (34.6%), phytol isomer (11.8%); Sample B: neophytadiene (45.2%), limonene (24.0%)	Broth macrodilution method: MIC = 0.16–0.32 μL/mL and MLC > 125 μL/mL (A); MIC = 0.64 μL/mL and MLC > 20 μL/mL (B)	[36]
Artemisia absinthium (Asteraceae)	leaves	borneol (18.7%), methyl hinokiate (11.9%)	Tube-dilution method: MIC = 91 ± 13 μg/mL	[37]
Artemisia persica (Asteraceae)	aerial parts	laciniata furanone E (17.1%), artedouglasia oxide C (13.2%), *trans*-pinocarveol (10.2%)	Broth macrodilution method: MIC = 2.5 μL/mL and MFC = 10 μL/mL	[38]
Beilschmiedia madang (Lauraceae)	bark	δ-cadinene (20.5%), α-cubebene (15.6%), α-cadinol (10.6%)	Microdilution method: MIC = 62.5 μg/mL	[39]
	leaf	δ-cadinene (17.0%), β-caryophyllene (10.3%), α-cubebene (11.3%)	Microdilution method: MIC = 250 μg/mL	
Carum copticum (Apiaceae)	*not identified*	p-cymene (33.7%), thymol (22.8%), γ-terpinene (21.6%)	Broth macrodilution method: MIC = 144 μg/mL	[40]
Centaurea solstitialis (Asteraceae)	aerial parts	germacrene D (15.3%), hexadecanoic acid (26.1%), α-linolenic acid (17.9%)	Microdilution assay: MIC = 1.9 μg/mL	[41]
Cinnamomum camphora (Lauraceae)	leaf, branch, wood, root, leaf/branch, leaf/branch/wood	leaf: camphor (93.1%); branch: camphor (53.6%); wood: camphor (53.2%) and 1,8-cineole (19.9%); root: safrole (57.6%) and 1.8-cineole (18.1%); leaf/branch: camphor (53.3%); leaf/branch/wood: (59.5%)	Broth microdilution method: MIC = 312.5 μg/mL (leaf, wood and root); MIC = 156.3 μg/mL (branch and leaf/branch); MIC = 78.1 μg/mL (leaf/branch/wood)	[42]

Table 2. *Cont.*

Essential Oil (Family)	Plant Part Used	Main Compounds	Antifungal Effect	Ref
Cinnamomum glanduliferum (Lauraceae)	bark	eucalyptol (65.9%)	Broth microdilution method: MIC = 32.5 μg/mL	[43]
Cuminum cyminum (Apiaceae)	aerial parts	*not assessed*	Broth microdilution method: MIC = 0.5 mg/mL Broth macrodilution method: MIC = 0.25 mg/mL	[44]
Daucus carota subsp. *carota* (Apiaceae)	flowering and ripe umbels	flowering umbels: α-pinene (37.9%), geranyl acetate (15%); ripe umbels—Sample A: α-pinene (13.0%), geranyl acetate (65.0%); Sample B: β-bisabolene (51.0%), (E)-methyl isoeugenol (10.0%)	Broth macrodilution method—Flowering umbels: MIC = 2.5–5 μL/mL and MLC > 20 μL/mL; Ripe umbels: MIC = 0.64–1.25 μL/mL and MLC > 20 μL/mL (A); MIC = 0.64 μL/mL and MLC > 20 (B)	[45]
Gallesia integrifolia (Phytolaccaceae)	fruit	2,8-dithianonane (52.6%), dimethyl trisulfide (15.5%), lenthionine (14.7%)	Modified microdilution method: MFC = 0.02—0.18 mg/mL	[46]
Juniperus communis subsp. *alpina* (Cupressaceae)	needles	sabinene (26.2%), α-pinene (12.9%), limonene (10.4%)	Macrodilution broth method: MIC = 2.5 μL/mL and MFC = 10 μL/mL	[47]
Leptospermum petersonii (Myrtaceae)	*not identified*	*not assessed*	Broth macrodilution method: MIC and MFC = 0.05%	[48]
Lippia alba (Verbenaceae)	stems and leaves	Carvone chemotype: carvone (25.3%), limonene (22.4%), geranial (10.4%); Citral chemotype: geranial (30.5%), neral (23.6%)	Microdilution broth method: MIC > 500 μg/mL (both chemotypes)	[49]
Lavandula luisieri (Lamiaceae)	flowering aerial parts	Sample A: α-*trans*-necrodyl acetate (17.4%); Sample B: 1,8-cineole (33.9%), fenchone (18.2%)	Broth macrodilution method: MIC = 0.64 μL/mL and MLC = 10–20 μL/mL (A); MIC = 1.25 μL/mL and MLC = 10 μL/mL (B)	[50]
Lavandula multifida (Lamiaceae)	flowering aerial parts	carvacrol (42.8%), *cis*-β-ocimene (27.4%);	Broth macrodilution method: MIC = 0.32 μL/mL and MLC = 0.64 μL/mL	[51]
Lavandula pedunculata (Lamiaceae)	aerial parts	Sample A: 1,8-cineole (34.3%); Sample B: camphor (34.0%), 1,8-cineole (25.1%); Sample C: fenchone (44.5%)	Broth macrodilution method: MIC = 2.5 μL/mL and MLC = 10 μL/mL (A); MIC and MLC = 5 μL/mL (B); MIC = 5 μL/mL and MLC = 10 μL/mL (C)	[52]
Lavandula stoechas (Lamiaceae)	aerial parts	fenchone (37.0%), camphor (27.3%)	Broth macrodilution method: MIC = 1.25 μL/mL and MLC ≥ 20μL/mL	[53]
Lavandula viridis (Lamiaceae)	aerial parts	1,8-cineole(34.5%), camphor (13.4%), α-pinene (9.0%), linalool (7.9%)	Broth macrodilution method: MIC = 2.5 μL/mL and MLC = 5–10 μL/mL	[54]

Table 2. *Cont.*

Essential Oil (Family)	Plant Part Used	Main Compounds	Antifungal Effect	Ref
Melaleuca alternifolia (Myrtaceae)	*not identified*	*not assessed*	Broth microdilution method: MIC = 0.06–0.12%	[55]
Myrtus communis (Myrtaceae)	leaves and flowers	Sample A: α-pinene (50.8%), linalool (14.8%), 1,8-cineole (13.3%); Sample B: α-pinene (33.6%), linalool (14.8%), 1,8-cineole (13.3%)	Broth macrodilution method: MIC = 2.5 mg/mL and MLC > 10 mg/mL (both samples)	[56]
Nigella sativa (Ranunculaceae)	aerial parts	*not assessed*	Broth microdilution method: MIC = 2 mg/mL Broth macrodilution method: MIC = 1.5 mg/mL	[44]
Piper flaviflorum (Piperaceae)	leaf	(E)-nerolidol (40.5%), β-caryophyllene (14.6%)	Broth microdilution method: MIC = 256 μg/mL and MLC = 1024 μg/mL	[57]
Ruta angustifolia (Rutaceae)	flowering aerial parts	2-undecanone (82.5%), 2-decanone (10.0%)	Agar dilution method: MIC < 3.5 μg/mL	[58]
Ruta chalepensis (Rutaceae)	flowering aerial parts	2-nonanone (32.8%), 2-undecanone (32.6%), 1-nonene (14.0%)	Agar dilution method: MIC = 6.2–7.4 μg/mL	
Ruta graveolens (Rutaceae)	flowering aerial parts	2-undecanone (55.4%), nonanone (21.6%)	Agar dilution method: MIC < 3.5 μg/mL	
Ruta tuberculata (Rutaceae)	flowering aerial parts	piperitone (13.6%), *trans*-p-menth-2-en-1-ol (13.1%), *cis*-piperitol (12.3%), *cis*-p-menth-2-en-1-ol (11.2%)	Agar dilution method: MIC < 4.5 μg/mL	
Satureja thymbra (Lamiaceae)	aerial parts	thymol (57.3%), γ-terpinene (9.8%), p-cymene (9.8%)	Broth macrodilution method: MIC = 0.32 μL/mL and MLC = 0.64 μL/mL	[59]
Spondias pinnata (Anacardiaceae)	fruit peels	furfural (17.1%), α-terpineol (13.1%)	Broth microdilution method: MIC = 16 μg/mL and MFC = 32 μg/mL	[60]
Thymus villosus subsp. *lusitanicus* (Lamiaceae)	aerial parts	geranyl acetate (25.0%), terpinen-4-ol (13.5%)	Broth macrodilution method: MIC = 0.64–125 μL/mL and MLC = 2.5–5.0 μL/mL	[61]
Thymus vulgaris (Lamiaceae)	*not identified*	thymol (44.7%), γ-terpinene (26.0%), α-cymene (21.2%)	Broth macrodilution method: MIC = 144 μg/mL	[40]
Ziziphora clinopodioides (Lamiaceae)	aerial parts	*not assessed*	Broth microdilution method: MIC = 1 mg/mL Broth macrodilution method: MIC = 0.5 mg/mL	[44]

MIC—Minimal Inhibitory Concentration; MBC—Minimal Bacterial Concentration; MFC—Minimal Fungicidal Concentration.

As previously mentioned, some studies have included several essential oils. For example, in a study involving fifteen species from the Asteraceae family, only *Achyrocline alata* and *Baccharis latifolia* essential oils were effective against *A. fumigatus* (geometric mean MIC value of 78.7 and 157.4 μg/mL, respectively) [62]. Moreover, in some cases both essential oil direct contact or vapor phase are assessed, thus foreseeing different modes of administration and distinct applications. For example, resorting to these two strategies, the essential of *Eugenia caryophyllata*, *Lavandula angustifolia*, *Origanum vulgare*, *Salvia sclarea*, *Thuja plicata* and *Thymus vulgaris* were assessed, being *Origanum vulgare* essential oil the most effective, through direct contact, with a MIC value of 0.025%. However, the volatile vapor of 0.075% of the majority of the oils was also able to exert a fungicidal effect on *A. fumigatus*, proving that these oils can be used, for example, for environmental disinfection [63].

Synergistic studies between essential oils and isolated compounds or essential oils with conventional antifungals are also relevant, thus reducing the effective dose needed and improving therapeutic outcomes [64]. Indeed, a chessboard assay using twenty-five essential oils showed that a combination of *Cymbopogon citratus* and *Thymus serpyllum* oils was the most potent against *A. fumigatus* (fractional inhibitory concentration index -FICI- of 0.1875; with a total synergistic effect considered for values ≤ 0.5) [65]. Additionally, *Thymus vulgaris* essential oils and thymol showed significant levels of synergistic interaction with fluconazole presenting an FICI value of 0.187 [40].

Moreover, mixed microbial infections are known to cause significant health care burdens and are commonly found in patients with chronic infections. Therefore, some studies have also considered this reality by assessing the polymicrobial antibiofilm effect of essential oils. A study carried out by Pekmezovic and colleagues addressed this ability using selected *Citrus* essential oils and showed that 10 mg/L of the oils were able to both inhibit a mixed biofilm formation composed of *Pseudomonas aeruginosa* and *Aspergillus fumigatus* and affect quorum sensing [66].

Finally, in vivo validations are of utmost importance but are quite sparse regarding respiratory mycosis. A pre-clinical study confirmed the potential of *Leptospermum petersonii* essential oil by showing a significant reduction of fungal infection in the lungs of animals that completed the treatment regimen and the effect was strikingly superior to that reported for conventional antifungal drugs [48].

Regarding the potential of isolated compounds, very few studies have been performed. Once again phenolic compounds stand out as very effective, namely eugenol—MIC = 250 ug/mL [49], thymol—MIC = 192 μg/mL [40] and carvacrol—MIC = 0.16 μL/mL [51]. Indeed, this could explain the excellent activity of some of the essential oils pointed out above such as *Thymus vulgaris* (rich in thymol) and *Lavandula multifida* (with high amounts of carvacrol). Besides phenolic compounds, other essential oils constituents such as safrole have shown very effective antifungal abilities against *A. fumigatus*, with a MIC of 39.1 μg/mL [42]. Interestingly, although the essential oil obtained from the roots of *Cinnamomum camphora* had a high concentration of this compound (57.6%), the antifungal potential of the whole extract was not so prominent (MIC = 312.5 μg/mL), thus showing that, in some cases, antagonistic effects may also take place between different compounds present in the mixture. Additionally, 1,8-cineole, a compound widely found in many essential oils, showed promising effects with a MIC value of 156.3 μg/mL [42].

4.1.2. Essential Oils in Respiratory Candidiasis

Candida infections are quite difficult to detect as *Candida* spp. inhabit the normal microflora of the skin, oral cavity, gastrointestinal mucosa, respiratory tract, and genitourinary tract [67]. Moreover, despite covering more than two-thirds of fungal infections, invasive candidiasis (defined as the presence of *Candida* in the blood—candidemia—or firmly established *Candida* infections) rarely manifest as *Candida* pneumonia and antifungal treatments in these cases is a matter of ongoing debate [68]. Nevertheless, *Candida albicans*, *C. glabrata*, *C. tropicalis*, *C. parapsilosis* and *C. krusei* have been pointed out as the most prevalent strains in invasive candidiasis. Since several reviews have largely covered the antifungal potential of essential oils on these pathogens [69–73], we now focus on a once considered rare species, *Candida auris*, that is currently emerging as a major pathogen in certain parts of the world. *C. auris* infections are difficult to treat as this pathogen does not respond to conventional antifungal drugs. This fungus also presents a high risk of transmission as it persists on several equipment and medical devices [74]. Importantly, some essential oils have shown promising antifungal potential against this species. Table 3 compiles this information by pointing out the plant species, part of the plant used to obtain the essential oil and its main compounds, whenever referred to in the original study. The antifungal potential of the essential oil is also highlighted, as well as the mechanism of action underlying the observed effect, if assessed.

Table 3. Antifungal effects of essential oils against *Candida auris*.

<table>
<tr><th>Essential Oil (Family)</th><th>Plant Part Used</th><th>Main Compounds</th><th>Antifungal Effect</th><th>Mechanism of Action</th><th>Ref</th></tr>
<tr><td rowspan="2">Cinnamomum zeylanicum (Lauraceae)</td><td>bark</td><td>trans-cinnamaldehyde (66.4%)</td><td>Disc diffusion method (diameter of inhibition): 77.37 ± 1.72 mm
Broth dilution method: MIC < 0.03% and MFC < 0.03%</td><td rowspan="2">Modulation of cell membrane permeability (nuclei acid and protein release);
Hemolytic activity</td><td rowspan="2">[75]</td></tr>
<tr><td>leaf</td><td>eugenol (62.6%)</td><td>Disc diffusion method (diameter of inhibition): 35.40 ± 1.08 mm
Broth dilution method: MIC = 0.06% and MFC = 0.25%</td></tr>
<tr><td>Eugenia florida (Myrtaceae)</td><td>aerial parts</td><td>seline-3,11-dien-6-α-ol (12.9%), eremoligenol (11.0%), γ-elemene (10.7%)</td><td>Disc diffusion method (diameter of inhibition): 8 mm</td><td></td><td>[76]</td></tr>
<tr><td>Juniperus oxycedrus subsp. macrocarpa (Cupressaceae)</td><td>aerial parts</td><td>α-pinene (56.6 ± 0.2%), limonene (14.6 ± 0.11%), β-pinene (13.4 ± 0.09%)</td><td>Planktonic MIC: 0.02%; Sessile MIC: no biofilm formation</td><td></td><td>[77]</td></tr>
<tr><td>Lavandula angustifolia (Lamiaceae)</td><td>not identified</td><td>linalyl acetate (48.5%), linalool (39.3%)</td><td>Growth reduction (0.01%)</td><td>Modulation of biofilm related genes</td><td>[78]</td></tr>
<tr><td>Lippia sidoides (Verbenaceae)</td><td>leaves</td><td>thymol (68.2%)</td><td>Broth microdilution method: Geometric mean MIC = 0.281 mg/mL</td><td></td><td>[79]</td></tr>
<tr><td>Myrcia multiflora (Myrtaceae)</td><td>aerial parts</td><td>Sample A: α-bulnesene (26.8%), pogostol (21.3%);
Sample B: (E)-nerolidol (44.4%);
Sample C: (E)-nerolidol (92.2%)</td><td>Disc diffusion method (diameter of inhibition): 9 mm (A), 10 mm (B) and 8 mm (C);
Broth dilution method: MIC = 3.12 μL/mL (A) and 5 μL/mL (C); MFC > 12.5 μL/mL (A) and >50 μL/mL (C)
[Sample B not assessed]</td><td></td><td>[76]</td></tr>
<tr><td>Thuja plicata (Verbenaceae)</td><td>not identified</td><td>not assessed</td><td>Inhibition of the intrinsic rate of growth at 0.09% and 0.39%</td><td></td><td>[80]</td></tr>
<tr><td>Thymus mastichina (Lamiaceae)</td><td>flowering cups</td><td>linalool (31.9%), α-terpineol (10.0%)</td><td>Disc diffusion method (diameter of inhibition): 13.60 ± 1.36 mm</td><td rowspan="4">Biofilm inhibition (direct application and vapor phase application)</td><td rowspan="4">[81]</td></tr>
<tr><td>Thymus satureioides (Lamiaceae)</td><td>flowering cups</td><td>borneol (29.3%), α-terpineol (15.9%)</td><td>Disc diffusion method (diameter of inhibition): 20.00 ± 0.63 mm</td></tr>
<tr><td>Thymus vulgaris (Lamiaceae)</td><td>flowering cups</td><td>thymol (63.1%), 1,8-cineole (10.0%)</td><td>Disc diffusion method (diameter of inhibition): 42.33 ± 3.77 mm</td></tr>
<tr><td>Thymus zygis subsp. sylvestris (Lamiaceae)</td><td>flowering aerial parts</td><td>thymol (26.5%), carvacrol (22.7%)</td><td>Disc diffusion method (diameter of inhibition): 28.25 ± 1.0 mm</td></tr>
</table>

MIC—Minimal Inhibitory Concentration; MFC—Minimal Fungicidal Concentration.

Although *C. auris* is an emergent species, several studies have already shown the potential of essential oils as effective antifungal agents, as highlighted in Table 3. Nevertheless, as different susceptibility tests are used and units lack uniformity among studies, it is quite difficult to perform comparisons between studies and understand which essential oil is the most effective. Even so, it seems that essential oils rich in phenolic compounds exert a more potent inhibitory effect. Indeed, the phenolic compound carvacrol assessed alone has already shown promising effects. A median MIC of 125 µg/mL was reported as well as synergistic and additive effects in combination with the conventional antifungals fluconazole, amphotericin B, nystatin and caspofungin [82].

4.1.3. Essential Oils in Cryptococcosis

Pulmonary cryptococcosis, contrarily to cryptococcal meningitis, remains underdiagnosed mainly due to limitations in diagnostic tools. Indeed, the infection presents similar clinical and radiological features to lung cancer, pulmonary tuberculosis, bacterial pneumonia, and other pulmonary mycoses. The genus *Cryptococcus* consists of more than 70 species, the two main human pathogens being *C. neoformans* and *C. gattii* [83]. Several studies have also pointed out the antifungal potential of essential oils, mainly for *C. neoformans*. Table 4 compiles these studies referring to the plant name, part of the plant used to obtain the essential oils, and antifungal effect. Since the mechanisms of action underlying these effects are poorly explored, this topic will be discussed collectively in Section 4.2.

Table 4. Antifungal effects of essential oils against *Cryptococcus neoformans*.

Essential Oil (Family)	Plant Part Used	Main Compounds	Antifungal Effect	Ref
Achillea millefolium (Asteraceae)	flowering aerial parts	Sample A: α-asarone (33.3%), β-bisabolene (16.6%) and α-pinene (17.2%); Sample B: *trans*-thujone (29.0%), *trans*-crhysanthenyl acetate (15.8%) and β-pinene (11.1%)	Broth macrodilution method: MIC = 0.64 µL/mL and MLC = 1.25 µL/mL (A); MIC = 1.25 µL/mL and MLC = 1.25 µL/mL (B)	[33]
Angelica major (Apiaceae)	aerial parts	*cis*-β-ocimene (30.4 %), α-pinene (21.8 %)	Broth macrodilution method: MIC = 0.16 µL/mL and MLC = 0.64 µL/mL	[84]
Apium graveolens (Apiaceae)	flowering aerial parts	Sample A: neophytadiene (34.6%), γ-himachalene (10.3%); Sample B: neocnidilide (45.2%), limonene (24.0%)	Broth macrodilution method: MIC and MLC = 0.16 µL/mL (A); MIC = 0.32 µL/mL and MLC = 0.64 µL/mL (B)	[36]
Aristolochia delavayi (Aristolochiaceae)	aerial parts	(E)-dec-2-enal (52.0%)	Broth microdilution method: MIC = 7.81 µg/mL and MFC = 62.5 µg/mL	[85]
Artemisia herba-alba (Asteraceae)	aerial parts	β-thujone (25.1%), α-thujone (22.9%), 1,8-cineole (20.1%), camphor (10.5%)	Broth macrodilution method: MIC = 0.64 mg/mL and MLC = 0.64–125 mg/mL	[86]
Artemisia judaica (Asteraceae)	aerial parts	piperitone (30.4%), camphor (16.1%), ethyl cinnamate (11.0%)	Broth macrodilution method: MIC = 0.16 µL/mL and MLC = 0.64 µL/mL	[87]
Bupleurum rigidum subsp. *paniculatum* (Apiaceae)	aerial parts	α-pinene (36.0%), β-pinene (26.1%), limonene (10.5%)	Broth macrodilution method: MIC = 72 µg/mL and MLC = 144 µg/mL	[88]

Table 4. *Cont.*

Essential Oil (Family)	Plant Part Used	Main Compounds	Antifungal Effect	Ref
Chamaecyparis obtusa (Cupressaceae)	needles and twigs	bicyclo[2.2.1]heptan-2-ol (18.8%), (+)-2-carene (17.4%), sabinene (12.8%)	Broth microdilution method: MIC > 2.18 mg/mL	[89]
Croton gratissimus (Euphorbiaceae)	leaves	not assessed	Broth microdilution method: MIC = 4 mg/mL	[90]
Cryptomeria japonica (Cupressaceae)	needles and twigs	kaur-16-ene (31.5%), sabinene (11.1%),	Broth microdilution method: MIC > 2.18 mg/mL	[89]
Daucus carota subsp. *carota* (Apiaceae)	flowering and ripe umbels	Flowering umbels—Sample A: α-pinene (37.9%), geranyl acetate (15.0%); Sample B: carotol (25.1%), β-bisabolene (17.6%); Ripe umbels—Sample A: α-pinene (13.0%), geranyl acetate (65.0%); Sample B: β-bisabolene (51.0%), (E)-methyl isoeugenol (10.0%)	Broth macrodilution method—Flowering umbels: MIC = 0.32–0.64 μL/mL and MLC = 1.25–2.5 μL/mL (A); MIC = 0.32 μL/mL and MLC = 0.32–0.64 μL/mL (B); Ripe umbels: MIC and MLC = 0.64 μL/mL (A); MIC and MLC = 0.64–1.25μL/mL (B)	[45]
	ripe umbels with seeds	geranyl acetate (29.0%), α-pinene (27.2%)	Broth macrodilution method: MIC and MLC = 0.16 μL/mL	[91]
Distichoselinum tenuifolium (Apiaceae)	ripe umbels	myrcene (84.6%)	Broth macrodilution method: MIC = 0.32–0.64 μL/mL and MLC = 0.64 μL/mL	[92]
Foeniculum vulgare (Apiaceae)	umbels and fruits	E-anetol (47.0%), α-phellandrene (11.0%), α-pinene (10.1%), fenchone (10.8%)	Broth macrodilution method: MIC = 0.32 μL/mL and MLC = 0.64 μL/mL	[93]
Hirtellina lobelia (Asteraceae)	aerial parts	α-bisabolol (34.5%), fokienol (12.0%)	Broth microdilution method: MIC = 128 μL/mL	[94]
Hyptis crenata (Lamiaceae)	aerial parts	borneol (17.8%), 1,8-cineol (15.6%)	Broth microdilution method: MIC = 62.5 μg/mL and MFC = 125 μg/mL	[95]
Juniperus communis subsp. *alpina* (Cupressaceae)	needles	sabinene (26.2%), α-pinene (12.9%)	Broth macrodilution method: MIC = 1.25 μL/mL and MFC = 1.25 μL/mL	[47]
Lavandula luisieri (Lamiaceae)	flowering aerial parts	Sample A: α-*trans*-necrodyl acetate (17.4%); Sample B: 1,8-cineole (33.9%), fenchone (18.2%)	Broth macrodilution method: MIC and MLC = 0.64 μL/mL (A); MIC = 0.64 μL/mL and MLC = 0.64–1.2 μL/mL (B)	[50]
Lavandula multifida (Lamiaceae)	flowering aerial parts	carvacrol (42.8%), cis-β-ocimene (27.4%)	Broth macrodilution method: MIC = 0.16 μL/mL and MLC = 0.32 μL/mL	[51]
Lavandula pedunculata (Lamiaceae)	aerial parts	Sample A: 1,8-cineole (34.3%); Sample B: camphor (34%), 1,8-cineole (25.1%); Sample C: fenchone (44.5%)	Broth macrodilution method: MIC and MLC = 1.25 μL/mL (A); MIC and MLC = 0.32–0.64 μL/mL (B); MIC = 1.25 μL/mL and MLC = 1.25–2.5μL/mL (C)	[52]
Lavandula viridis (Lamiaceae)	aerial parts	1,8-cineole(34.5%), camphor (13.4%), α-pinene (9.0%), linalool (7.9%)	Broth macrodilution method: MIC = 0.64 μL/mL and MLC = 0.64 μL/mL	[54]

Table 4. *Cont.*

Essential Oil (Family)	Plant Part Used	Main Compounds	Antifungal Effect	Ref
Melaleuca alternifolia (Myrtaceae)	*not identified*	terpinen-4-ol (42.4%), γ-terpinene (20.7%)	Broth microdilution method: MIC = 0.06–0.2%	[96]
Mentha x piperita (Lamiaceae)	leaves	menthol (41.7%), menthone (21.8%)	Broth microdilution method: MIC = 0.06–0.125% and MFC = 0.06–0.125%	[97]
Mentha pulegium (Lamiaceae)	aerial parts	pulegone (86.2%)	Broth macrodilution method: MIC = 0.64 mg/mL and MLC = 1.25 μL/mL	[98]
Mentha spicata (Lamiaceae)	aerial parts	carvone (62.9%)	Broth macrodilution method: MIC = 0.32 μL/mL and MLC = 0.64–1.25 μL/mL	
Mesembryanthemum edule(Aizoaceae)	leaves	tetradecamethylcycloheptasiloxane (13.6%), phytol (12.4%)	Broth microdilution method: MIC = 0.08 mg/mL	[99]
Mitracarpus frigidus (Rubiaceae)	aerial parts	linalool (29.9%), eugenol acetate (15.9%)	Broth microdilution method: MIC = 8 μg/mL	[100]
Myrtus communis (Myrtaceae)	dried leaves and flowers	Sample A: α-pinene (50.8%), 1,8-cineole (21.9%); Sample B: α-pinene (33.6%), linalool (14.8%), 1,8-cineole (13.3%)	Broth macrodilution method: MIC = 0.64 mg/mL and MLC = 0.64–1.25 mg/mL (A); MIC and MLC = 0.64 mg/mL (B)	[56]
Myrtus nivellei (Myrtaceae)	aerial parts (10 samples)	1,8-cineole (37.5%), limonene (25.0%)	Broth macrodilution method: MIC = 0.16 μL/mL and MLC = 0.32 μL/mL	[101]
Oenanthe crocata (Apiaceae)	aerial parts	*trans*-β-ocimene (31.3%), sabinene (29.0%), *cis*-β-ocimene (12.3%)	Broth macrodilution method: MIC = 0.16 μL/mL and MLC = 0.32 μL/mL	[102]
Pinus densiflora (Pinaceae)	needles and twigs	β-phellandrene (16.7%), (-)-α-pinene (14.9%), 1-β-pinene (10.5%), α-fenchyl acetate (10.3%)	Broth microdilution method: MIC = 0.545 mg/mL	[89]
Pistacia x saportae (Anacardiaceae)	aerial parts	α-pinene (30.3%), (Z)-β-ocimene (26.7%), (E)-β-ocimene (11.1%)	Broth macrodilution method: MIC = 0.32 mg/mL and MLC = 0.64–1.25 mg/mL	[103]
Pistacia lentiscus (Anacardiaceae)	aerial parts	terpinen-4-ol (25.2%), α-phellandrene (11.9%), β-phellandrene (10.2%), γ-terpinene (10.1%)	Broth macrodilution method: MIC = 0.32 mg/mL and MLC = 0.64 mg/mL	
Pistacia terebinthus (Anacardiaceae)	aerial parts	terpinolene (35.2%), α-pinene (35%)	Broth macrodilution method: MIC = 1.25 mg/mL and MLC = 2.5 mg/mL	
	leaves	Sample A: α-pinene (62.4%), β-pinene (12.1%); Sample B: terpinolene (35.2%), α-pinene (35.0%)	Broth macrodilution method: MIC = 0.32 μL/mL and MLC = 0.64 μL/mL (A); MIC = 1.25 μL/mL and MLC = 2.5 μL/mL (B)	[104]
Protium amazonicum (Myrtaceae)	oleoresin	δ-3-carene (47.9%)	Broth microdilution method: MIC = 0 156 μg/mL	[105]
Santiria trimera (Burseraceae)	bark	α-pinene (66.6%), β-pinene (20.0%)	Agar dilution method: MIC < 0.71 μL/mL	[106]

Table 4. *Cont.*

Essential Oil (Family)	Plant Part Used	Main Compounds	Antifungal Effect	Ref
Santolina impressa (Asteraceae)	flowering aerial parts	β-pinene (22.5%), 1,8-cineole (10.0%)	Broth macrodilution method: MIC = 0.27 mg/mL	[107]
Santolina insularis (Asteraceae)	aerial parts	β-phellandrene (22.6%), myrcene (11.4%)	Broth macrodilution method: MIC = 0.13 mg/mL	[108]
Satureja thymbra (Lamiaceae)	aerial parts	thymol (57.3%), γ-terpinene (9.8%), β-caryophyllene, p-cymene (9.8%)	Broth macrodilution method: MIC = 0.16 μL/mL and MLC = 0.32 μL/mL	[59]
Smyrnium olusatrum (Apiaceae)	fruiting umbels	Sample A: β-phellandrene (42.7%); Sample B: acetoxyfurano-4(15)-eudesmene (17.6%)	Broth macrodilution method: MIC = 0.32 μL/mL and MLC = 0.64 μL/mL (A); MIC and MLC = 0.64 μL/mL (B)	[109]
Tanacetum vulgare (Asteraceae)	flowering aerial parts	1,8-cineole (18.2%), myrtenol (10.3%)	Broth macrodilution method: MIC = 0.16 μL/mL and MLC = 0.16–0.32 μL/mL	[110]
Teucrium scordium subsp. *scordioides* (Lamiaceae)	aerial parts	germacrene D (25.1%), δ-cadinene (12.9%), alloaromadendrene (11.3%)	Broth macrodilution method: MIC = 0.32 μL/mL and MLC = 0.32 μL/mL	[111]
Thapsia villosa (Apiaceae)	aerial parts	limonene (57.5%), methyleugenol (35.9%)	Broth macrodilution method: MIC and MFC = 0.16 μL/mL	[112]
Thymus camphoratus (Lamiaceae)	flowering aerial parts	1,8-cineole (15.5%), α-pinene (12.7%)	Broth macrodilution method: MIC = 0.14 mg/mL and MLC = 0.28 mg/mL (both oils)	[113]
Thymus carnosus (Lamiaceae)	flowering aerial parts	borneol (29.0%), camphene (19.5%)		
Thymus villosus subsp. *lusitanicus* (Lamiaceae)	aerial parts	terpinen-4-ol (13.5%), geranyl acetate (25%)	Broth macrodilution method: MIC = 0.16 μL/mL and MLC = 0.16–0.32 μL/mL	[61]
Vitex rivularis (Lamiaceae)	leaves and flowers	Sample A: germacrene D (12.6%); Sample B: germacrene D (20.6%)	Broth macrodilution method: MIC = 0.64–1.25 μL/mL and MLC = 2.5–5 μL/mL (A); MIC = 1.25 μL/mL and MLC = 5–10 μL/mL (B)	[114]
Ziziphora tenuior (Lamiaceae)	aerial parts	pulegone (46.8%), p-menth-3-en-8-ol (12.5%)	Broth macrodilution method: MIC = 0.16 μL/mL and MLC = 0.64 μL/mL	[115]

MIC—Minimal Inhibitory Concentration; MFC—Minimal Fungicidal Concentration; MLC—Minimal Lethal Concentration.

Besides the studies referred to in Table 4, others were performed gathering a higher number of species, with bigger data approaches enabling more extensive analysis. For example, using this approach, eighty-two essential oils were analyzed, with fifteen being highlighted as very potent (MIC $\leq$ 100 μg/mL), and from these *Cedrus atlantica* standing out as the most effective, with a MIC value of 20 μg/mL [116]. In another extensive study, the antifungal potential of sixty commercially-available essential oils was also assessed, with *Cinnamomum cassia*, *C. zeylanicum*, *Coriandrum sativum*, *Pogostemon cablin*, *Santalum album*, *S. austrocaledonicum*, *S. paniculatum* and *Vetiveria zizanoides* being very effective (MIC = 20 μg/mL) [117].

Some studies also resort to ethnopharmacological evidence and tend to assess the potential of essential oils based on their traditional uses, thus recovering relevant knowledge that tends to be lost over time. For example, the study carried out by Lawson and colleagues considered a specific group of plants from the Asteraceae family used in Cherokee and other Native American traditional medicines, and showed a very potent effect of *Eupatorium serotinum* essential oil against *Cryptococcus neoformans* with a MIC value of 78 μg/mL [118].

Regarding synergistic studies between essential oils and conventional drugs, interesting results against *Cryptococcus neoformans* have also been reported. For example, Scalas and colleagues, through checkerboard testing and isobolographic analysis, showed synergistic effects (FICI $\leq$ 0.5) between itraconazole and *Origanum vulgare*, *Pinus sylvestris* or *Thymus vulgaris* essential oils. Importantly, a synergistic effect was also observed with itraconazole and *Thymus vulgaris* essential oil (chemotype: thymol 26.52%; carvacrol 7.85%) on an azole not susceptible strain of *Cryptococcus neoformans*, thus confirming the potential of essential oils as cost-effective adjuvants in antifungal therapy [119]. Using similar methodology, the combination of *Mentha* x *piperita* essential oil with itraconazole also exerted a synergistic effect (FICI $\leq$ 0.5), with a decrease in the MIC. Nevertheless, on the azole-resistant strain, the binary combination of itraconazole and the oil yielded additive effects [97]. Furthermore, the combination of *Ocimum basilicum* var. Maria Bonita (a genetically improved cultivar) essential oil with fluconazole enhanced the antifungal activity, especially against the resistant strain of *Cryptococcus neoformans*, with MIC being reduced from 1250 μg/mL to 625 μg/mL [120].

Combinations between essential oils have also shown positive effects on *Cryptococcus neoformans*. For example, combining the essential oils of *Boswellia rivae*, *B. neglecta* and *B. papyrifera* with *Commiphora guidotti* or *C. myrrha* oils displayed synergistic, additive and noninteractive properties with MICs ranging from 0.5–5.3 mg/mL [121].

Regarding isolated compounds, several studies have been performed, showing that the most potent compounds against *Cryptococcus neoformans* were (E)-Dec-2-enal with a MIC of 25.65 ug/mL [85], α-pinene with a MIC value of 0.07 mg/mL [113], and (R)-(+)-limonene with a fungicidal effect at 0.08μL/mL [112]. These effects may explain the potential of *Aristolochia delavayi* essential oil, with high amounts of (E)-dec-2-enal [85], and *Bupleurum rigidum* subsp. *paniculatum* and *Thapsia villosa* rich in α-pinene [88,112]. Geraniol also showed promising effects with a study pointing out a MIC of 76 μg/mL [120]; nevertheless in other studies, the MICs were quite distinct (MIC_{80} = 128 μg/mL [122] and MIC = 0.32 μL/mL [61]). Variability between MIC values can occur mainly due to variations between laboratories and strains [123]. Other compounds showed some antifungal potential but to a less extent, namely linalool (MIC = 5 μL/mL) and terpinen-4-ol (MIC = 1.25 μL/mL) [61], sabinene (MIC = 0.32μL/mL) and *cis*-*β*-ocimene (MIC = 0.16–0.32μL/mL) [102], methyleugenol (MIC= 0.32 μL/mL) [112], and carvacrol (MIC= 0.16 μL/mL) [51].

4.1.4. Essential Oils on Other Respiratory Infections

Concerning other fungal strains involved in respiratory mycosis, namely those referred to in Table 1, in vitro studies have been performed, although to a much lesser extent. Table 5 compiles these studies and points out their main findings.

Table 5. In vitro antifungal effects of volatile compounds against respiratory fungi.

Fungal Strain Tested	Essential Oil (Family)	Plant Part Used	MainCompounds	Antifungal Effect	Ref
Rhizopus oryzae	*Thymus vulgaris* (Lamiaceae)	*not identified*	*not assessed*	Disc diffusion method (diameter of inhibition): 32 mm; Broth microdilution method: MIC = 256–512 μg/mL; MFC = 512–1024 μg/mL (several strains tested)	[124]
Paracoccidioides brasiliensis	*Schinus molle* (Anacardiaceae)	leaves	β-pinene (25.2%), epi-α-cadinol (21.3%), α-pinene (18.7%), myrcene (11.5%)	Broth microdilution method: MIC and MLC = 39.06 μg/mL	[125]

Table 5. *Cont.*

Fungal Strain Tested	Essential Oil (Family)	Plant Part Used	MainCompounds	Antifungal Effect	Ref
Sporothrix schenckii	*Origanum majorana* (Lamiaceae)	*not identified*	1.8-cineole (20.9%), terpeninen-4-ol (20.4%)	Broth microdilution method: MIC and MLC ≤ 2.25 mg/mL	[126]

Regarding isolated compounds, several studies have been performed. Overall, farnesol, a sesquiterpenic compound, and terpinene-4-ol, a terpene alcohol, were the most assessed and showed very promising effects. For example, farnesol was tested against *Coccidioides posadasii*—MIC = 0.002–0.01 mg/L [127], *Histoplasma capsulatum* var. *capsulatum*—MIC = 0.008–0.003 µM [128] and *Sporothrix schenckii*—MIC = 0.003 to 0.222 µg/mL [129]. Importantly, in the first two pathogens, synergistic effects with amphotericin B, itraconazole, voriconazole and caspofungin or itraconazole, respectively were observed [127,128], thus confirming its potential as an adjuvant in fungal infections. On the other hand, terpinene-4-ol was tested against *Sporothrix schenckii* (MIC = 4–32 mg/L) and was able to decrease its cellular ergosterol content. In combination with itraconazole or terbinafine, it also exerted a synergistic effect [130]. Furthermore, this compound was also effective against *Coccidioides posadasii* (MIC = 350–1420 µg/mL) and *Histoplasma capsulatum* [MIC = 20–1420 µg/mL (filamentous phase) and 40–350 µg/mL (yeast phase)], although to a lesser extent [131].

Other compounds have also been tested, such as the monoterpene p-cymene (MIC = 1024 µg/mL) and the phenolic compound thymol (MIC = 128–256 µg/mL), against *Rhizopus oryzae* [124]. Furthermore, the sesquiterpene dialdehyde polygodial was also very effective against *Penicillium marneffei* with a MIC value of 3.3 µg/mL [132]. These studies, although still limited in number, show that other fungal pathogens are being considered for antifungal susceptibility testing, besides the gold standard strains, thus opening new avenues for the development of new preventive/therapeutic strategies.

4.2. Mechanism of Action Underlying Essential Oils Antifungal Effects

The majority of the in vitro studies performed tend to identify the MIC and MLC/MFC of the volatile extract/compound and in some cases, the putative mechanism of action underlying the antifungal effect. Indeed, several methods have been proposed to elucidate the target site or the mechanism of action of the essential oils, in fungal cells [133]. Generally, mechanistic studies are performed on *Candida albicans* and the main effects considered are biofilm disruption, cell morphology and plasma membrane integrity. In less extent mitochondrial enzymes, reactive oxygen species (ROS) and gene expression are also assessed. Next, studies performed on other pathogens rather than *Candida albicans* are highlighted, namely for biofilm and cell wall/membrane integrity.

Biofilms are important virulence factors for pathogenic fungi naturally formed when fungi change from a planktonic to a sessile state and attach to surfaces and to each other, being involved and protected by a polymeric extracellular matrix. The fungi also secrete quorum-sensing molecules that play a relevant role in fungal resistance and pathogenicity [134] and, therefore, constitute a very promising therapeutic target.

Kumari and colleagues carried out a study using several compounds present in essential oils and confirmed their anti-biofilm effect on *Cryptococcus laurentii* and *Cryptococcus neoformans* in the following order: thymol > carvacrol > citral > eugenol = cinnamaldehyde > menthol. Indeed, for the most effective compounds, a potent effect on biofilm morphology was confirmed by scanning electron microscopy and confocal laser scanning microscopy that showed the absence of extracellular polymeric matrix, reduction in cellular density and alteration in the surface morphology of biofilm cells [122].

Other relevant therapeutic targets that are highly assessed are the structural elements of fungi cell walls and membranes, as their inhibition can affect cell wall maturation, septum formation, and bud ring formation, by damaging cell division and cell growth. Essential oils are able to disrupt the cell wall, leading to cytoplasm leakage and compromise fungi membrane permeability and fluidity by altering its properties and compromising membrane-associated functions. Indeed, morphological and ultrastructural alterations were observed in *Aspergillus fumigatus* exposed to *Cuminum cyminum*, *Nigella sativa* and *Ziziphora clinopodioides* essential oil that interfered with the enzymes involved in cell wall synthesis, caused high vacuolation of the cytoplasm, detachment of the fibrillar layer of cell wall, and plasma membrane disruption. Additionally, disorganization of nuclear and mitochondrial structures was observed [44]. Similarly, *Leptospermum petersonii* essential oil disturbed *Aspergillus fumigatus* cell membrane with alterations observed in hyphal morphology, susceptibility of spheroplasts and uptake of propidium iodide following exposure to the oil [48]. Regarding isolated compounds, eugenol reduced the cell diameter and capsule size of *Cryptococcus gattii* and *C. neoformans*. The compound was also able to increase the levels of ROS, leading to increased lipid peroxidation, mitochondrial membrane depolarization and reduction of lysosomal integrity in these fungi [135]. Another relevant target for essential oils is ergosterol, a compound present in the fungal cell membrane whose biosynthesis can be altered by disturbing sterol biosynthetic pathways. Importantly, the absence or reduced presence of ergosterol in fungal membranes results in osmotic and metabolic instability of the fungal cell, compromising reproduction and infectious activity [136]. Indeed, it was shown that *Thymus vulgaris* essential oil and its main compound, thymol, were effective against *Rhizopus oryzae* due to their interaction with ergosterol, thus supporting their use in the management of mucormycosis [124].

5. Translation to the Clinic: Limitations and Future Perspectives

Studies comparing different treatments for respiratory mycosis, namely cryptococcal meningitis, candidemia, endemic mycoses, and invasive aspergillosis, are among the most cited clinical trials and are very relevant for the development of treatment guidelines for these infections [137]. Nevertheless, regarding essential oils as anti-infective agents, clinical trials are lacking. Moreover, essential oils present several features such as hydrophobicity, instability, high volatility, and possible toxicity that compromise their use. Therefore, to overcome these limitations, encapsulation resorting to delivery systems, namely lipid-based carriers, have been developed to stabilize these compounds, improve the shelf-life of the formulated products and prolong the biological effect of the active molecules [138]. For example, *Lavandula angustifolia* essential oil encapsulated in liposomes was effective against persistent biofilms of *Candida auris* [78] while *Lippia sidoides* essential oil encapsulated in nanostructured lipid carriers showed anti-*Candida auris* potential and low toxicity, suggesting a new strategy to overcome multidrug-resistant pathogens [79]. On the other hand, due to their volatility, essential oils can easily reach the upper and lower parts of the respiratory tract via active or passive inhalation. In the first case, an inhalation device is needed for the patients to directly inhale the volatile compounds whereas in the latter, heating, vaporization or ventilation is used to deliver these compounds to the environment [139]. Interestingly, over the last years, several patents on portable inhalation devices have been registered and have been shown to be suitable delivery systems for these volatile compounds. Another possible form of administration resorts to patches, the volatiles being released to the skin and/or inhaled by the patient [140].

6. Final Remarks

The present review provides, for the first time, an updated compilation of relevant information on the antifungal potential of essential oils and their volatile compounds on respiratory mycosis. Overall, our bibliographic search showed that the majority of the studies are performed on strains involved in opportunistic infections, namely aspergillosis, candidiasis and cryptococcosis, with the present review focusing on strains involved in life-threatening invasive diseases such as *Aspergillus fumigatus* and *Cryptococcus neoformans* and relevant emergent strains such as *Candida auris*. Moreover, it is quite evident that essential oils rich in phenolic compounds, namely thymol and carvacrol, are very effective, and therapeutic improvements can be achieved by combining essential oils and/or their volatile compounds with conventional antifungal drugs. Furthermore, several administration strategies and devices have been designed to effectively deliver these volatile compounds but clinical translation still requires in vivo validations and human trials to confirm the efficacy and tolerability of these extracts/compounds in respiratory mycosis.

Author Contributions: Conceptualization, M.Z. and L.S.; methodology, M.Z. and L.S.; validation, M.Z. and L.S.; writing—original draft preparation, M.Z.; writing—review and editing, L.S.; funding acquisition, M.Z. All authors have read and agreed to the published version of the manuscript.

Funding: This research was funded by FCT and Fundação La Caixa under the project: PLANTS4AGEING.

Institutional Review Board Statement: Not applicable.

Informed Consent Statement: Not applicable.

Conflicts of Interest: The authors declare no conflict of interest.

References

1. Chowdhary, A.; Agarwal, K.; Meis, J.F. Filamentous fungi in respiratory infections. What lies beyond aspergillosis and mucormycosis? *PLoS Pathog.* **2016**, *12*, e1005491. [CrossRef] [PubMed]
2. Brown, G.D.; Denning, D.W.; Gow, N.A.R.; Levitz, S.M.; Netea, M.G.; White, T.C. Hidden Killers: Human fungal infections. *Sci. Transl. Med.* **2012**, *4*, 165rv13. [CrossRef] [PubMed]
3. Ravikumar, S.; Win, M.S.; Chai, L.Y.A. Optimizing outcomes in immunocompromised hosts: Understanding the role of immunotherapy in invasive fungal diseases. *Front. Microbiol.* **2015**, *6*, 1322. [CrossRef] [PubMed]
4. Di Mango, A.L.; Zanetti, G.; Penha, D.; Menna Barreto, M.; Marchiori, E. Endemic pulmonary fungal diseases in immunocompetent patients: An emphasis on thoracic imaging. *Expert. Rev. Respir. Med.* **2019**, *13*, 263–277. [CrossRef]
5. Li, Z.; Lu, G.; Meng, G. Pathogenic fungal infection in the lung. *Front. Immunol.* **2019**, *10*, 1524. [CrossRef]
6. Miller, A.S.; Wilmott, R.W. The Pulmonary Mycoses. In *Kendig's Disorders of the Respiratory Tract in Children*; Elsevier: Amsterdam, The Netherlands, 2019; pp. 507–527.e3. [CrossRef]
7. Boucher, H.W.; Groll, A.H.; Chiou, C.C.; Walsh, T.J. Newer systemic antifungal agents. *Drugs* **2004**, *64*, 1997–2020. [CrossRef]
8. Limper, A.H.; Knox, K.S.; Sarosi, G.A.; Ampel, N.M.; Bennett, J.E.; Catanzaro, A.; Davies, S.F.; Dismukes, W.E.; Hage, C.A.; Marr, K.A.; et al. An official American thoracic society statement: Treatment of fungal infections in adult pulmonary and critical care patients. *Am. J. Respir. Crit. Care Med.* **2011**, *183*, 96–128. [CrossRef]
9. Cook, S.; Confer, J. Assessment and treatment of fungal lung infections. *US Pharm.* **2011**, *36*, HS-17-HS-24.
10. Moen, M.D.; Lyseng-Williamson, K.A.; Scott, L.J. Liposomal amphotericin B. *Drugs* **2009**, *69*, 361–392. [CrossRef]
11. Chang, Y.-L.; Yu, S.-J.; Heitman, J.; Wellington, M.; Chen, Y.-L. New facets of antifungal therapy. *Virulence* **2017**, *8*, 222–236. [CrossRef]
12. Chowdhary, A.; Kathuria, S.; Xu, J.; Meis, J.F. Emergence of azole-resistant *Aspergillus fumigatus* strains due to agricultural azole use creates an increasing threat to human health. *PLoS Pathog.* **2013**, *9*, e1003633. [CrossRef]
13. Walsh, T.J.; Dixon, D.M. Spectrum of mycoses. *Med. Microbiol.* **1996**, *75*, 919–925. Available online: http://www.ncbi.nlm.nih.gov/pubmed/21413276 (accessed on 1 June 2022).
14. Desai, J.V.; Mitchell, A.P.; Andes, D.R. Fungal biofilms, drug resistance, and recurrent infection. *Cold Spring Harb. Perspect Med.* **2014**, *4*, a019729. [CrossRef] [PubMed]
15. Parker, N.; Schneegurt, M.; Thi Tu, A.-H.; Lister, P.; Forster, B.M. Microbiology. 2016. Available online: https://openstax.org/books/microbiology/pages/22-4-respiratory-mycoses (accessed on 1 June 2022).
16. Clinical info HIV gov. Guidelines for the Prevention and Treatment of Opportunistic Infections in Adults and Adolescents with HIV. Available online: https://clinicalinfo.hiv.gov/en/guidelines/adult-and-adolescent-opportunistic-infection/talaromycosis-formerly-penicilliosis (accessed on 1 June 2022).
17. Centers for Disease Control and Prevention. Available online: https://www.cdc.gov/ (accessed on 1 June 2022).
18. MSD Manual for the Professional. Available online: https://www.msdmanuals.com/en-pt/professional) (accessed on 1 June 2022).
19. Pfaller, M.A.; Espinel-Ingroff, A.; Boyken, L.; Hollis, R.J.; Kroeger, J.; Messer, S.A.; Tendolkar, S.; Diekema, D.J. Comparison of the broth microdilution (BMD) method of the European Committee on Antimicrobial Susceptibility Testing with the 24-Hour CLSI BMD method for testing susceptibility of *Candida* species to fluconazole, posaconazole, and voriconazole by use of epidemiological cut-off values. *J. Clin. Microbiol.* **2011**, *49*, 845–850. [CrossRef]
20. Sanguinetti, M.; Posteraro, B. Susceptibility testing of fungi to antifungal drugs. *J. Fungi* **2018**, *4*, 110. [CrossRef]
21. Hohl, T.M. Overview of vertebrate animal models of fungal infection. *J. Immunol. Methods* **2014**, *410*, 100–112. [CrossRef]
22. Jemel, S.; Guillot, J.; Kallel, K.; Botterel, F.; Dannaoui, E. *Galleria mellonella* for the evaluation of antifungal efficacy against medically important fungi, a narrative review. *Microorganisms* **2020**, *8*, 390. [CrossRef]
23. Hasenberg, M.; Köhler, A.; Bonifatius, S.; Jeron, A.; Gunzer, M. Direct observation of phagocytosis and NET-formation by neutrophils in infected lungs using 2-photon microscopy. *J. Vis. Exp.* **2011**, *52*, e2659. [CrossRef]

24. Kelly, M.M.; McNagny, K.; Williams, D.L.; van Rooijen, N.; Maxwell, L.; Gwozd, C.; Mody, C.H.; Kubes, P. The lung responds to zymosan in a unique manner independent of Toll-like receptors, complement, and dectin-1. *Am. J. Respir. Cell Mol. Biol.* **2008**, *38*, 227–238. [CrossRef]
25. *ISO 9235*. Aromatic Natural Raw Materials—Vocabulary. ISO: Geneva, Switzerland, 2013.
26. Council of Europe. *Directorate for the Quality of Medicines & HealthCare of the Council of Europe*, 7th ed.; European Pharmacopoeia: Strasbourg, France, 2010.
27. Figueiredo, A.C.; Barroso, J.G.; Pedro, L.G.; Johannes, S.J.C. Factors affecting secondary metabolite production in plants: Volatile components and essential oils. *Flavour Fragr. J.* **2008**, *23*, 213–226. [CrossRef]
28. Irshad, M.; Ali Subhani, M.; Ali, S.; Hussain, A. Biological importance of essential oils. In *Essential Oils-Oils of Nature*; IntechOpen: London, UK, 2020. [CrossRef]
29. Wińska, K.; Mączka, W.; Łyczko, J.; Grabarczyk, M.; Czubaszek, A.; Szumny, A. Essential oils as antimicrobial agents—Myth or real alternative? *Molecules* **2019**, *24*, 2130. [CrossRef] [PubMed]
30. Lucca, L.G.; Romão, P.R.T.; Vignoli-Silva, M.; da Veiga-Junior, V.F.; Koester, L.S. In vivo acute anti-inflammatory activity of essential oils: A review. *Mini-Rev. Med. Chem.* **2021**, *21*, 1495–1515. [CrossRef] [PubMed]
31. Sugui, J.A.; Kwon-Chung, K.J.; Juvvadi, P.R.; Latge, J.-P.; Steinbach, W.J. *Aspergillus fumigatus* and related species. *Cold Spring Harb. Perspect. Med.* **2015**, *5*, a019786. [CrossRef] [PubMed]
32. El-Baba, F.; Gao, Y.; Soubani, A.O. Pulmonary aspergillosis: What the generalist needs to know. *Am. J. Med.* **2020**, *133*, 668–674. [CrossRef] [PubMed]
33. Falconieri, D.; Piras, A.; Porcedda, S.; Marongiu, B.; Gonçalves, M.J.; Cabral, C.; Cavaleiro, C.; Salgueiro, L. Chemical composition and biological activity of the volatile extracts of *Achillea millefolium*. *Nat. Prod. Commun.* **2011**, *6*, 1527–1530. [CrossRef]
34. Joshi, R.K. *Acorus calamus* Linn.: Phytoconstituents and bactericidal property. *World J. Microbiol. Biotechnol.* **2016**, *32*, 164. [CrossRef]
35. Li, R.; Wang, Y.-F.; Sun, Q.; Hu, H.-B. Chemical composition and antimicrobial activity of the essential oil from *Allium hookeri* consumed in Xishuangbanna, southwest China. *Nat. Prod. Commun.* **2014**, *9*, 863–864. [CrossRef]
36. Marongiu, B.; Piras, A.; Porcedda, S.; Falconieri, D.; Maxia, A.; Frau, M.A.; Gonçalves, M.J.; Cavaleiro, C.; Salgueiro, L. Isolation of the volatile fraction from *Apium graveolens* L. (Apiaceae) by supercritical carbon dioxide extraction and hydrodistillation: Chemical composition and antifungal activity. *Nat. Prod. Res.* **2013**, *27*, 1521–1527. [CrossRef]
37. Joshi, R.K. Volatile composition and antimicrobial activity of the essential oil of Artemisia absinthium growing in Western Ghats region of North West Karnataka. *India Pharm. Biol.* **2013**, *51*, 888–892. [CrossRef]
38. Dehghani Bidgoli, R. Chemical composition of essential oil and antifungal activity of Artemisia persica Boiss. *Iran. J. Food Sci. Technol.* **2021**, *58*, 1313–1318. [CrossRef]
39. Salleh, W.M.N.H.W.; Ahmad, F.; Yen, K.H. Chemical compositions and biological activities of the essential oils of *Beilschmiedia madang* Blume (Lauraceae). *Arch. Pharm. Res.* **2015**, *38*, 485–493. [CrossRef] [PubMed]
40. Khan, M.S.A.; Ahmad, I.; Cameotra, S.S. Carum copticum and Thymus vulgaris oils inhibit virulence in *Trichophyton rubrum* and *Aspergillus* spp. *Braz. J. Microbiol.* **2014**, *45*, 523–531. [CrossRef] [PubMed]
41. Carev, I.; Ruščić, M.; Skočibušić, M.; Maravić, A.; Siljak-Yakovlev, S.; Politeo, O. Phytochemical and cytogenetic characterization of *Centaurea solstitialis* L. (Asteraceae) from Croatia. *Chem. Biodivers.* **2017**, *14*, e1600213. [CrossRef] [PubMed]
42. Poudel, D.K.; Rokaya, A.; Ojha, P.K.; Timsina, S.; Satyal, R.; Dosoky, N.S.; Satyal, P.; Setzer, W.N. The Chemical profiling of essential oils from different tissues of *Cinnamomum camphora* L. and their antimicrobial activities. *Molecules* **2021**, *26*, 5132. [CrossRef]
43. Taha, A.M.; Eldahshan, O.A. Chemical characteristics, antimicrobial, and cytotoxic activities of the essential oil of Egyptian *Cinnamomum glanduliferum* Bark. *Chem. Biodivers.* **2017**, *14*, e1600443. [CrossRef]
44. Khosravi, A.R.; Minooeianhaghighi, M.H.; Shokri, H.; Emami, S.A.S.M.A.; Asili, J. The potential inhibitory effect of *Cuminum cyminum*, *Ziziphora clinopodioides* and *Nigella sativa* essential oils on the growth of *Aspergillus fumigatus* and *Aspergillus*. *Braz. J. Microbiol.* **2011**, *42*, 216–224. [CrossRef]
45. Maxia, A.; Marongiu, B.; Piras, A.; Porcedda, S.; Tuveri, E.; Gonçalves, M.J.; Cavaleiro, C.; Salgueiro, L. Chemical characterization and biological activity of essential oils from *Daucus carota* L. subsp. *carota* growing wild on the Mediterranean coast and on the Atlantic coast. *Fitoterapia* **2009**, *80*, 57–61. [CrossRef]
46. Raimundo, K.F.; Bortolucci, W.d.C.; Glamočlija, J.; Soković, M.; Gonçalves, J.E.; Linde, G.A.; Gazim, Z.C. Antifungal activity of *Gallesia integrifolia* fruit essential oil. *Braz. J. Microbiol.* **2018**, *49*, 229–235. [CrossRef]
47. Cabral, C.; Francisco, V.; Cavaleiro, C.; Gonçalves, M.J.; Cruz, M.T.; Sales, F.; Batista, M.T.; Salgueiro, L. Essential oil of *Juniperus communis* subsp. *alpina* (Suter) Čelak needles: Chemical composition, antifungal activity and cytotoxicity. *Phyther. Res.* **2012**, *26*, 1352–1357. [CrossRef]
48. Hood, J.R.; Burton, D.M.; Wilkinson, J.M.; Cavanagh, H.M.A. The effect of *Leptospermum petersonii* essential oil on *Candida albicans* and *Aspergillus fumigatus*. *Med. Mycol.* **2010**, *48*, 922–931. [CrossRef]
49. Mesa-Arango, A.C.; Montiel-Ramos, J.; Zapata, B.; Durán, C.; Betancur-Galvis, L.; Stashenko, E. Citral and carvone chemotypes from the essential oils of Colombian Lippia alba (Mill.) N.E. Brown: Composition, cytotoxicity and antifungal activity. *Mem. Inst. Oswaldo Cruz.* **2009**, *104*, 878–884. [CrossRef] [PubMed]

50. Zuzarte, M.; Gonçalves, M.J.; Cruz, M.T.; Cavaleiro, C.; Canhoto, J.; Vaz, S.; Pinto, E.; Salgueiro, L. Lavandula luisieri essential oil as a source of antifungal drugs. *Food Chem.* **2012**, *135*, 1505–1510. [CrossRef] [PubMed]
51. Zuzarte, M.; Vale-Silva, L.; Gonçalves, M.J.; Cavaleiro, C.; Vaz, S.; Canhoto, J.; Pinto, E.; Salgueiro, L. Antifungal activity of phenolic-rich *Lavandula multifida* L. essential oil. *Eur. J. Clin. Microbiol. Infect. Dis.* **2012**, *31*, 1359–1366. [CrossRef]
52. Zuzarte, M.; Gonçalves, M.J.; Cavaleiro, C.; Dinis, A.M.; Canhoto, J.M.; Salgueiro, L.R. Chemical composition and antifungal activity of the essential oils of *Lavandula pedunculata* (Miller) Cav. *Chem. Biodivers.* **2009**, *6*, 1283–1292. [CrossRef] [PubMed]
53. Zuzarte, M.; Gonçalves, M.J.; Cavaleiro, C.; Cruz, M.T.; Benzarti, A.; Marongiu, B.; Maxia, A.; Piras, A.; Salgueiro, L. Antifungal and anti-inflammatory potential of *Lavandula stoechas* and *Thymus herba-barona* essential oils. *Ind. Crops Prod.* **2013**, *44*, 97–103. [CrossRef]
54. Zuzarte, M.; Gonçalves, M.J.; Cavaleiro, C.; Canhoto, J.; Vale-Silva, L.; Silva, M.J.; Pinto, E.; Salgueiro, L. Chemical composition and antifungal activity of the essential oils of Lavandula viridis L'Hér. *J. Med. Microbiol.* **2011**, *60*, 612–618. [CrossRef]
55. Hammer, K.A. In vitro activity of *Melaleuca alternifolia* (tea tree) oil against dermatophytes and other filamentous fungi. *J. Antimicrob. Chemother.* **2002**, *50*, 195–199. [CrossRef]
56. Bouzabata, A.; Cabral, C.; Gonçalves, M.J.; Cruz, M.T.; Bighelli, A.; Cavaleiro, C.; Casanova, J.; Tomi, F.; Salgueiro, L. *Myrtus communis* L. as source of a bioactive and safe essential oil. *Food Chem. Toxicol.* **2015**, *75*, 166–172. [CrossRef]
57. Li, R.; Yang, J.-J.; Wang, Y.-F.; Sun, Q.; Hu, H.-B. Chemical composition, antioxidant, antimicrobial and anti-inflammatory activities of the stem and leaf essential oils from Piper flaviflorum from Xishuangbanna, SW China. *Nat. Prod. Commun.* **2014**, *9*, 1011–1014. [CrossRef]
58. Haddouchi, F.; Chaouche, T.M.; Zaouali, Y.; Ksouri, R.; Attou, A.; Benmansour, A. Chemical composition and antimicrobial activity of the essential oils from four Ruta species growing in Algeria. *Food Chem.* **2013**, *141*, 253–258. [CrossRef]
59. Piras, A.; Cocco, V.; Falconieri, D.; Porcedda, S.; Marongiu, B.; Maxia, A.; Frau, M.A.; Gonçalves, M.J.; Cavaleiro, C.; Salgueiro, L. Isolation of the volatile oil from *Satureja thymbra* by supercritical carbon dioxide extraction: Chemical composition and biological activity. *Nat. Prod. Commun.* **2011**, *6*, 1523–1526. [CrossRef] [PubMed]
60. Li, R.; Yang, J.-J.; Song, X.-Z.; Wang, Y.-F.; Corlett, R.; Xu, Y.-K.; Hu, H.B. Chemical composition and the cytotoxic, antimicrobial, and anti-inflammatory activities of the fruit peel essential oil from *Spondias pinnata* (Anacardiaceae) in Xishuangbanna, southwest China. *Molecules* **2020**, *25*, 343. [CrossRef] [PubMed]
61. Pinto, E.; Gonçalves, M.J.; Hrimpeng, K.; Pinto, J.; Vaz, S.; Vale-Silva, L.A. Antifungal activity of the essential oil of *Thymus villosus* subsp. *lusitanicus* against *Candida*, *Cryptococcus*, *Aspergillus* and dermatophyte species. *Ind Crops Prod.* **2013**, *51*, 93–99. [CrossRef]
62. Zapata, B.; Durán, C.; Stashenko, E.; Betancur-Galvis, L.; Mesa-Arango, A.C. Actividad antimicótica y citotóxica de aceites esenciales de plantas de la familia Asteraceae. *Rev. Iberoam. Micol.* **2010**, *27*, 101–103. [CrossRef] [PubMed]
63. Puškárová, A.; Bučková, M.; Kraková, L.; Pangallo, D.; Kozics, K. The antibacterial and antifungal activity of six essential oils and their cyto/genotoxicity to human HEL 12469 cells. *Sci. Rep.* **2017**, *7*, 8211. [CrossRef]
64. Aljaafari, M.N.; AlAli, A.O.; Baqais, L.; Alqubaisy, M.; AlAli, M.; Molouki, A.; Ong-Abdullah, J.; Abushelaibi, A.; Lai, K.S.; Lim, S.H. An overview of the potential therapeutic applications of essential oils. *Molecules* **2021**, *26*, 628. [CrossRef]
65. Hlebová, M.; Hleba, L.; Medo, J.; Kováčik, A.; Čuboň, J.; Ivana, C.; Uzsáková, V.; Božik, M.; Klouček, P. Antifungal and synergistic activities of some selected essential oils on the growth of significant indoor fungi of the genus *Aspergillus*. *J. Environ. Sci. Health Part A* **2021**, *56*, 1335–1346. [CrossRef]
66. Pekmezovic, M.; Aleksic, I.; Barac, A.; Arsic-Arsenijevic, V.; Vasiljevic, B.; Nikodinovic-Runic, J.; Senerovic, L. Prevention of polymicrobial biofilms composed of *Pseudomonas aeruginosa* and pathogenic fungi by essential oils from selected *Citrus* species. *Pathog. Dis.* **2016**, *74*, 102. [CrossRef]
67. Kumamoto, C.A.; Vinces, M.D. Alternative *Candida albicans* lifestyles growth on surfaces. *Annu. Rev. Microbiol.* **2005**, *59*, 113–133. [CrossRef]
68. Meena, D.S.; Kumar, D. *Candida* Pneumonia: An innocent bystander or a silent killer? *Med. Princ. Pract.* **2022**, *31*, 98–102. [CrossRef]
69. Palmeira-de-Oliveira, A.; Salgueiro, L.; Palmeira-de-Oliveira, R.; Martinez-de-Oliveira, J.; Pina-Vaz, C.; Queiroz, J.; Queiroz, J.A.; Rodrigues, A.G. Anti-*Candida* activity of essential oils. *Mini-Rev. Med. Chem.* **2009**, *9*, 1292–1305. [CrossRef]
70. Karpiński, T.M. Essential oils of Lamiaceae family plants as antifungals. *Biomolecules* **2020**, *10*, 103. [CrossRef] [PubMed]
71. Guimarães, R.; Milho, C.; Liberal, Â.; Silva, J.; Fonseca, C.; Barbosa, A.; Ferreira, I.C.; Alves, M.J.; Barros, L. Antibiofilm potential of medicinal plants against *Candida* spp. oral biofilms: A review. *Antibiotics* **2021**, *10*, 1142. [CrossRef] [PubMed]
72. Karpiński, T.M.; Ożarowski, M.; Seremak-Mrozikiewicz, A.; Wolski, H.; Adamczak, A. Plant preparations and compounds with activities against biofilms formed by *Candida* spp. *J. Fungi* **2021**, *7*, 360. [CrossRef] [PubMed]
73. Shala, A.; Singh, S.; Hameed, S.; Khurana, S.M.P. Essential oils as alternative promising anti-candidal agents: Progress and prospects. *Curr. Pharm. Des.* **2022**, *28*, 58–70. [CrossRef]
74. Horton, M.V.; Johnson, C.J.; Kernien, J.F.; Patel, T.D.; Lam, B.C.; Cheong, J.Z.A.; Meudt, J.J.; Shanmuganayagam, D.; Kalan, L.R.; Nett, J.E. *Candida auris* forms high-burden biofilms in skin niche conditions and on porcine skin. *mSphere* **2020**, *5*, e00910-19. [CrossRef]
75. Tran, H.N.H.; Graham, L.; Adukwu, E.C. In vitro antifungal activity of Cinnamomum zeylanicum bark and leaf essential oils against *Candida albicans* and *Candida auris*. *Appl. Microbiol. Biotechnol.* **2020**, *104*, 8911–8924. [CrossRef]

76. Ferreira, O.O.; da Silva, S.H.M.; de Oliveira, M.S.; Andrade, E.H.d.A. Chemical composition and antifungal activity of *Myrcia multiflora* and *Eugenia florida* essential oils. *Molecules* **2021**, *26*, 7259. [CrossRef]
77. Spengler, G.; Gajdács, M.; Donadu, M.G.; Usai, M.; Marchetti, M.; Ferrari, M.; Mazzarello, V.; Zanetti, S.; Nagy, F.; Kovács, R. Evaluation of the antimicrobial and antivirulent potential of essential oils isolated from *Juniperus oxycedrus* L. ssp. *macrocarpa* aerial parts. *Microorganisms* **2022**, *10*, 758. [CrossRef]
78. de Alteriis, E.; Maione, A.; Falanga, A.; Bellavita, R.; Galdiero, S.; Albarano, L.; Salvatore, M.M.; Galdiero, E.; Guida, M. Activity of free and liposome-encapsulated essential oil from *Lavandula angustifolia* against persister-derived biofilm of *Candida auris*. *Antibiotics* **2021**, *11*, 26. [CrossRef]
79. Baldim, I.; Paziani, M.H.; Grizante Barião, P.H.; Kress, M.R.v.Z.; Oliveira, W.P. Nanostructured lipid carriers loaded with *Lippia sidoides* essential oil as a strategy to combat the multidrug-resistant *Candida auris*. *Pharmaceutics* **2022**, *14*, 180. [CrossRef]
80. McGregor, R.C.; Parker, K.A.; Hornby, J.M.; Latta, L.C. Microbial population dynamics under microdoses of the essential oil arborvitae. *BMC Complement. Altern. Med.* **2019**, *19*, 247. [CrossRef] [PubMed]
81. Ribeiro, R.; Fernandes, L.; Costa, R.; Cavaleiro, C.; Salgueiro, L.; Henriques, M.; Rodrigues, M.E. Comparing the effect of Thymus spp. essential oils on *Candida auris*. *Ind. Crops Prod.* **2022**, *178*, 114667. [CrossRef]
82. Shaban, S.; Patel, M.; Ahmad, A. Improved efficacy of antifungal drugs in combination with monoterpene phenols against Candida auris. *Sci. Rep.* **2020**, *10*, 1162. [CrossRef] [PubMed]
83. Setianingrum, F.; Rautemaa-Richardson, R.; Denning, D.W. Pulmonary cryptococcosis: A review of pathobiology and clinical aspects. *Med. Mycol.* **2019**, *57*, 133–150. [CrossRef]
84. Cavaleiro, C.; Salgueiro, L.; Gonçalves, M.-J.; Hrimpeng, K.; Pinto, J.; Pinto, E. Antifungal activity of the essential oil of *Angelica major* against *Candida, Cryptococcus, Aspergillus* and dermatophyte species. *J. Nat. Med.* **2015**, *69*, 241–248. [CrossRef]
85. Li, Z.-J.; Njateng, G.S.S.; He, W.-J.; Zhang, H.-X.; Gu, J.-L.; Chen, S.-N.; Du, Z.Z. Chemical composition and antimicrobial activity of the essential oil from the edible aromatic plant *Aristolochia delavayi*. *Chem. Biodivers.* **2013**, *10*, 2032–2041. [CrossRef]
86. Abu-Darwish, M.S.; Cabral, C.; Gonçalves, M.J.; Cavaleiro, C.; Cruz, M.T.; Efferth, T.; Salgueiro, L. Artemisia herba-alba essential oil from Buseirah (South Jordan): Chemical characterization and assessment of safe antifungal and anti-inflammatory doses. *J. Ethnopharmacol.* **2015**, *174*, 153–160. [CrossRef]
87. Abu-Darwish, M.S.; Cabral, C.; Gonçalves, M.J.; Cavaleiro, C.; Cruz, M.T.; Zulfiqar, A.; Khan, I.A.; Efferth, T.; Salgueiro, L. Chemical composition and biological activities of *Artemisia judaica* essential oil from southern desert of Jordan. *J. Ethnopharmacol.* **2016**, *191*, 161–168. [CrossRef]
88. Zuzarte, M.; Correia, P.M.P.; Alves-Silva, J.M.; Gonçalves, M.J.; Cavaleiro, C.; Cruz, T.; Salgueiro, L. Antifungal and anti-inflammatory potential of *Bupleurum rigidum* subsp. *paniculatum* (Brot.) H.Wolff essential oil. *Antibiotics* **2021**, *10*, 592. [CrossRef]
89. Lee, J.-H. Comparison of chemical compositions and antimicrobial activities of essential oils from three conifer trees; *Pinus densiflora, Cryptomeria japonica,* and *Chamaecyparis obtusa*. *J. Microbiol. Biotechnol.* **2009**, *19*, 391–396. [CrossRef]
90. van Vuuren, S.F.; Viljoen, A.M. In vitro evidence of phyto-synergy for plant part combinations of Croton gratissimus (Euphorbiaceae) used in African traditional healing. *J. Ethnopharmacol.* **2008**, *119*, 700–704. [CrossRef] [PubMed]
91. Alves-Silva, J.M.; Zuzarte, M.; Gonçalves, M.J.; Cavaleiro, C.; Cruz, M.T.; Cardoso, S.M.; Salgueiro, L. New claims for wild carrot (*Daucus carota* subsp. *carota*) essential oil. *Evid.-Based Complement. Altern. Med.* **2016**, *2016*, 9045196. [CrossRef] [PubMed]
92. Tavares, A.C.; Gonçalves, M.J.; Cruz, M.T.; Cavaleiro, C.; Lopes, M.C.; Canhoto, J.; Salgueiro, L.R. Essential oils from Distichoselinum tenuifolium: Chemical composition, cytotoxicity, antifungal and anti-inflammatory properties. *J. Ethnopharmacol.* **2010**, *130*, 593–598. [CrossRef] [PubMed]
93. Cabral, C.; Miranda, M.; Gonçalves, M.J.; Cavaleiro, C.; Cruz, M.T.; Salgueiro, L. Assessment of safe bioactive doses of *Foeniculum vulgare* Mill. essential oil from Portugal. *Nat. Prod. Res.* **2017**, *31*, 2654–2659. [CrossRef]
94. Khoury, M.; El Beyrouthy, M.; Ouaini, N.; Eparvier, V.; Stien, D.; Hirtellina Lobelii, D.C. Essential oil, its constituents, its combination with antimicrobial drugs and its mode of action. *Fitoterapia* **2019**, *133*, 130–136. [CrossRef]
95. Violante, I.M.P.; Garcez, W.S.; Barbosa, C.d.S.; Garcez, F.R. Chemical composition and biological activities of essential oil from Hyptis crenata growing in the Brazilian cerrado. *Nat. Prod. Commun.* **2012**, *7*, 1387–1389. [CrossRef]
96. Mondello, F. In vitro and in vivo activity of tea tree oil against azole-susceptible and -resistant human pathogenic yeasts. *J. Antimicrob. Chemother.* **2003**, *51*, 1223–1229. [CrossRef]
97. Tullio, V.; Roana, J.; Scalas, D.; Mandras, N. Evaluation of the antifungal activity of *Mentha* x *piperita* (Lamiaceae) of Pancalieri (Turin, Italy) essential oil and its synergistic interaction with azoles. *Molecules* **2019**, *24*, 3148. [CrossRef]
98. Piras, A.; Porcedda, S.; Falconieri, D.; Maxia, A.; Gonçalves, M.; Cavaleiro, C.; Salgueiro, L. Antifungal activity of essential oil from *Mentha spicata* L. and *Mentha pulegium* L. growing wild in Sardinia Island (Italy). *Nat. Prod. Res.* **2021**, *35*, 993–999. [CrossRef]
99. Omoruyi, B.; Afolayan, A.; Bradley, G. Chemical composition profiling and antifungal activity of the essential oil and plant extracts of *Mesembryanthemum edule* (L.) bolus leaves. *Afr. J. Tradit. Complement Altern. Med.* **2014**, *11*, 19. [CrossRef]
100. Fabri, R.L.; Coimbra, E.S.; Almeida, A.C.; Siqueira, E.P.; Alves, T.M.A.; Zani, C.L.; Scio, E. Essential oil of Mitracarpus frigidus as a potent source of bioactive compounds. *An. Acad. Bras. Cienc.* **2012**, *84*, 1073–1080. [CrossRef] [PubMed]
101. Bouzabata, A.; Bazzali, O.; Cabral, C.; Gonçalves, M.J.; Cruz, M.T.; Bighelli, A.; Tomi, F. New compounds, chemical composition, antifungal activity and cytotoxicity of the essential oil from *Myrtus nivellei* Batt. & Trab., an endemic species of Central Sahara. *J. Ethnopharmacol.* **2013**, *149*, 613–620. [CrossRef] [PubMed]

102. Valente, J.; Zuzarte, M.; Gonçalves, M.J.; Lopes, M.C.; Cavaleiro, C.; Salgueiro, L.; Cruz, M.T. Antifungal, antioxidant and anti-inflammatory activities of Oenanthe crocata L. essential oil. *Food Chem. Toxicol.* **2013**, *62*, 349–354. [CrossRef] [PubMed]
103. Marengo, A.; Piras, A.; Falconieri, D.; Porcedda, S.; Caboni, P.; Cortis, P.; Maxia, A. Chemical and biomolecular analyses to discriminate three taxa of *Pistacia genus* from Sardinia Island (Italy) and their antifungal activity. *Nat. Prod. Res.* **2018**, *32*, 2766–2774. [CrossRef] [PubMed]
104. Piras, A.; Marzouki, H.; Maxia, A.; Marengo, A.; Porcedda, S.; Falconieri, D.; Gonçalves, M.J.; Cavaleiro, C.; Salgueiro, L. Chemical characterisation and biological activity of leaf essential oils obtained from *Pistacia terebinthus* growing wild in Tunisia and Sardinia Island. *Nat. Prod. Res.* **2017**, *31*, 2684–2689. [CrossRef]
105. Satyal, P.; Powers, C.; Parducci, V.R.; McFeeters, R.; Setzer, W. Chemical composition, enantiomeric distribution, and antifungal activity of the oleoresin essential oil of *Protium amazonicum* from Ecuador. *Medicines* **2017**, *4*, 70. [CrossRef]
106. Martins, A.; Salgueiro, L.; Gonçalves, M.; Proença da Cunha, A.; Vila, R.; Cañigueral, S. Essential oil composition and antimicrobial activity of *Santiria trimera* Bark. *Planta Med.* **2003**, *69*, 77–79. [CrossRef]
107. Alves-Silva, J.M.; Zuzarte, M.; Gonçalves, M.J.; Cruz, M.T.; Cavaleiro, C.; Salgueiro, L. Unveiling the bioactive potential of the essential oil of a Portuguese endemism, *Santolina impressa*. *J. Ethnopharmacol.* **2019**, *244*, 112120. [CrossRef]
108. Alves-Silva, J.M.; Piras, A.; Porcedda, S.; Falconieri, D.; Maxia, A.; Gonçalves, M.J.; Cruz, M.T.; Salgueiro, L. Chemical characterization and bioactivity of the essential oil from *Santolina insularis*, a Sardinian endemism. *Nat. Prod. Res.* **2022**, *36*, 445–449. [CrossRef]
109. Marongiu, B.; Piras, A.; Porcedda, S.; Falconieri, D.; Frau, M.A.; Maxia, A.; Gonçalves, M.J.; Cavaleiro, C.; Salgueiro, L. Antifungal activity and chemical composition of essential oils from *Smyrnium olusatrum* L. (Apiaceae) from Italy and Portugal. *Nat. Prod. Res.* **2012**, *26*, 993–1003. [CrossRef]
110. Piras, A.; Falconieri, D.; Bagdonaite, E.; Maxia, A.; Gonçalves, M.J.; Cavaleiro, C.; Salgueiro, L.; Porcedda, S. Chemical composition and antifungal activity of supercritical extract and essential oil of *Tanacetum vulgare* growing wild in Lithuania. *Nat. Prod. Res.* **2014**, *28*, 1906–1909. [CrossRef] [PubMed]
111. Piras, A.; Maccioni, A.; Falconieri, D.; Porcedda, S.; Gonçalves, M.J.; Alves-Silva, J.M.; Silva, A.; Cruz, M.T.; Salgueiro, L.; Maxia, A. Chemical composition and biological activity of essential oil of *Teucrium scordium* L. subsp. scordioides (Schreb.) Arcang. (Lamiaceae) from Sardinia Island (Italy). *Nat. Prod. Res.* **2021**, 1–8. [CrossRef] [PubMed]
112. Pinto, E.; Gonçalves, M.-J.; Cavaleiro, C.; Salgueiro, L. Antifungal activity of *Thapsia villosa* essential oil against *Candida, Cryptococcus, Malassezia, Aspergillus* and dermatophyte species. *Molecules* **2017**, *22*, 1595. [CrossRef]
113. Alves, M.; Gonçalves, M.J.; Zuzarte, M.; Alves-Silva, J.M.; Cavaleiro, C.; Cruz, M.T.; Salgueiro, L. Unveiling the antifungal potential of two Iberian *Thyme* essential oils: Effect on *C. albicans* germ tube and preformed biofilms. *Front. Pharm.* **2019**, *10*, 446. [CrossRef]
114. Cabral, C.; Gonçalves, M.J.; Cavaleiro, C.; Sales, F.; Boyom, F.; Salgueiro, L. Composition and anti-fungal activity of the essential oil from Cameroonian Vitex rivularis Gürke. *Nat. Prod. Res.* **2009**, *23*, 1478–1484. [CrossRef]
115. Abu-Darwish, M.S.; Cabral, C.; Gonçalves, M.J.; Cavaleiro, C.; Cruz, M.T.; Paoli MTomi, F.; Efferth, T.; Salgueiro, L. *Ziziphora tenuior*, L. essential oil from Dana Biosphere Reserve (Southern Jordan); Chemical characterization and assessment of biological activities. *J. Ethnopharmacol.* **2016**, *194*, 963–970. [CrossRef]
116. Powers, C.N.; Satyal, P.; Mayo, J.A.; McFeeters, H.; McFeeters, R.L. Bigger data approach to analysis of essential oils and their antifungal activity against *Aspergillus niger*, *Candida albicans*, and *Cryptococcus neoformans*. *Molecules* **2019**, *24*, 2868. [CrossRef]
117. Powers, C.; Osier, J.; McFeeters, R.; Brazell, C.; Olsen, E.; Moriarity DSatyal, P.; Setzer, W.N. Antifungal and cytotoxic activities of sixty commercially-available essential oils. *Molecules* **2018**, *23*, 1549. [CrossRef]
118. Lawson, S.K.; Sharp, L.G.; Powers, C.N.; McFeeters, R.L.; Satyal, P.; Setzer, W.N. Volatile compositions and antifungal activities of native American medicinal plants: Focus on the Asteraceae. *Plants* **2020**, *9*, 126. [CrossRef]
119. Scalas, D.; Mandras, N.; Roana, J.; Tardugno, R.; Cuffini, A.M.; Ghisetti, V.; Benvenuti, S.; Tullio, V. Use of *Pinus sylvestris* L. (Pinaceae), *Origanum vulgare* L. (Lamiaceae), and *Thymus vulgaris* L. (Lamiaceae) essential oils and their main components to enhance itraconazole activity against azole susceptible/not-susceptible Cryptococcus neoformans strains. *BMC Complement Altern. Med.* **2018**, *18*, 143. [CrossRef] [PubMed]
120. Cardoso, N.N.R.; Alviano, C.S.; Blank, A.F.; Romanos, M.T.V.; Fonseca, B.B.; Rozental, S.; Rodrigues, I.A.; Alviano, D.S. Synergism effect of the essential oil from *Ocimum basilicum* var. Maria Bonita and its major components with fluconazole and its influence on ergosterol biosynthesis. *Evid.-Based Complement. Altern. Med.* **2016**, *2016*, 5647182. [CrossRef] [PubMed]
121. de Rapper, S.; Van Vuuren, S.F.; Kamatou, G.P.P.; Viljoen, A.M.; Dagne, E. The additive and synergistic antimicrobial effects of select frankincense and myrrh oils—A combination from the pharaonic pharmacopoeia. *Lett. Appl. Microbiol.* **2012**, *54*, 352–358. [CrossRef] [PubMed]
122. Kumari, P.; Mishra, R.; Arora, N.; Chatrath, A.; Gangwar, R.; Roy, P.; Prasad, R. Antifungal and anti-biofilm activity of essential oil active components against *Cryptococcus neoformans* and *Cryptococcus laurentii*. *Front. Microbiol.* **2017**, *8*, 2161. [CrossRef]
123. Mouton, J.W.; Meletiadis, J.; Voss, A.; Turnidge, J. Variation of MIC measurements: The contribution of strain and laboratory variability to measurement precision. *J. Antimicrob. Chemother.* **2018**, *73*, 2374–2379. [CrossRef]
124. de Lira Mota, K.; de Oliveira Pereira, F.; de Oliveira, W.; Lima, I.; de Oliveira Lima, E. Antifungal activity of *Thymus vulgaris* L. essential oil and its constituent phytochemicals against *Rhizopus oryzae*: Interaction with ergosterol. *Molecules* **2012**, *17*, 14418–14433. [CrossRef]

125. do Prado, A.C.; Garces, H.G.; Bagagli, E.; Rall, V.L.M.; Furlanetto, A.; Junior, A.F.; Furtado, F.B. *Schinus molle* essential oil as a potential source of bioactive compounds: Antifungal and antibacterial properties. *J. Appl. Microbiol.* **2019**, *126*, 516–522. [CrossRef]
126. Waller, S.B.; Madrid, I.M.; Ferraz, V.; Picoli, T.; Cleff, M.B.; de Faria, R.O.; Meireles, M.C.; de Mello, J.R. Cytotoxicity and anti-Sporothrix brasiliensis activity of the *Origanum majorana* Linn. oil. *Braz. J. Microbiol.* **2016**, *47*, 896–901. [CrossRef]
127. Brilhante, R.S.N.; de Lima, R.A.C.; Caetano, E.P.; Leite, J.J.G.; Castelo-Branco D de, S.C.M.; Ribeiro, J.F.; Bandeira, T.D.; Cordeiro, R.D.; Monteiro, A.J.; Sidrim,, J.C.C.; et al. Effect of farnesol on growth, ergosterol biosynthesis, and cell permeability in *Coccidioides posadasii*. *Antimicrob. Agents Chemother.* **2013**, *57*, 2167–2170. [CrossRef]
128. Brilhante, R.S.N.; de Lima, R.A.C.; Marques, F.J.d.F.; Silva, N.F.; Caetano, É.P.; Castelo, D.D.; Bandeira, T.D.; Moreira, J.L.; de Aguiar Cordeiro, R.; Monteiro, A.J.; et al. Histoplasma capsulatum in planktonic and biofilm forms: In vitro susceptibility to amphotericin B, itraconazole and farnesol. *J. Med. Microbiol.* **2015**, *64*, 394–399. [CrossRef] [PubMed]
129. Brilhante, R.S.; Silva, N.F.; Marques, F.J.; Castelo-Branco, D.D.; Lima, R.A.; Malaquias, A.D.; Caetano, E.P.; Barbosa, G.R.; Camargo, Z.P.; Rodrigues, A.M.; et al. In vitro inhibitory activity of terpenic derivatives against clinical and environmental strains of the Sporothrix schenkii complex. *Med. Mycol.* **2015**, *53*, 93–98. [CrossRef]
130. Brilhante, R.S.; Pereira, V.S.; Oliveira, J.S.; Rodrigues, A.M.; de Camargo, Z.P.; Pereira-Neto, W.A.; Nascimento, N.R.; Castelo-Branco, D.S.; Cordeiro, R.A.; Sidrim, J.J.; et al. Terpinen-4-ol inhibits the growth of Sporothrix schenckii complex and exhibits synergism with antifungal agents. *Future Microbiol.* **2019**, *14*, 1221–1233. [CrossRef] [PubMed]
131. Brilhante, R.S.; Caetano, É.P.; Lima, R.A.; Marques, F.J.; Castelo-Branco, D.D.; Melo, C.V.; Guedes, G.M.; Oliveira, J.S.; Camargo, Z.P.; Moreira, J.L.; et al. Terpinen-4-ol, tyrosol, and β-lapachone as potential antifungals against dimorphic fungi. *Braz. J. Microbiol.* **2016**, *47*, 917–924. [CrossRef] [PubMed]
132. Lee, S.H.; Lee, J.R.; Lunde, C.S.; Kubo, I. In vitro antifungal susceptibilities of *Candida albicans* and other fungal pathogens to polygodial, a sesquiterpene dialdehyde. *Planta Med.* **1999**, *65*, 204–208. [CrossRef]
133. Mani-López, E.; Cortés-Zavaleta, O.; López-Malo, A. A review of the methods used to determine the target site or the mechanism of action of essential oils and their components against fungi. *SN Appl. Sci.* **2021**, *3*, 44. [CrossRef]
134. Costa-Orlandi, C.; Sardi, J.; Pitangui, N.; de Oliveira, H.; Scorzoni, L.; Galeane, M.C.; Medina-Alarcón, K.P.; Melo, W.C.; Marcelino, M.Y.; Braz, J.D.; et al. Fungal biofilms and polymicrobial diseases. *J. Fungi* **2017**, *3*, 22. [CrossRef]
135. Alves, J.C.O.; Ferreira, G.F.; Santos, J.R.; Silva, L.C.N.; Rodrigues, J.F.S.; Neto, W.R.; Farah, E.I.; Santos, Á.R.; Mendes, B.S.; Sousa, L.V.; et al. Eugenol induces phenotypic alterations and increases the oxidative burst in *Cryptococcus*. *Front. Microbiol.* **2017**, *8*, 2419. [CrossRef]
136. Rajput, S.B.; Karuppayil, S.M. Small molecules inhibit growth, viability and ergosterol biosynthesis in *Candida albicans*. *SpringerPlus* **2013**, *2*, 26. [CrossRef]
137. Pappas, P.G. Antifungal clinical trials and guidelines: What we know and do not know. *Cold Spring Harb. Perspect. Med.* **2014**, *4*, a019745. [CrossRef]
138. Cimino, C.; Maurel, O.M.; Musumeci, T.; Bonaccorso, A.; Drago, F.; Souto, E.M.; Pignatello, R.; Carbone, C. Essential oils: Pharmaceutical applications and encapsulation strategies into lipid-based delivery systems. *Pharmaceutics* **2021**, *13*, 327. [CrossRef]
139. Kamyar, M. Essential oil diffusion. Patent US20090169487 A1, 2 July 2009.
140. Horváth, G.; Ács, K. Essential oils in the treatment of respiratory tract diseases highlighting their role in bacterial infections and their anti-inflammatory action: A review. *Flavour Fragr. J.* **2015**, *30*, 331–341. [CrossRef] [PubMed]

molecules

MDPI

Review

Pharmacological Profile, Bioactivities, and Safety of Turmeric Oil

Adriana Monserrath Orellana-Paucar [1,2,*] and María Gabriela Machado-Orellana [2,3]

1 Nutrition and Dietetics School, Faculty of Medical Sciences, University of Cuenca, Cuenca 010204, Ecuador
2 Pharmacology and Nutritional Sciences Interdisciplinary Research Group, Faculty of Medical Sciences, University of Cuenca, Cuenca 010204, Ecuador
3 Medicine and Surgery School, Faculty of Medical Sciences, University of Cuenca, Cuenca 010204, Ecuador
* Correspondence: adriana.orellanap@ucuenca.edu.ec

Abstract: The pharmacological attributes of turmeric have been extensively described and frequently related to the action of curcuminoids. However, there is also scientific evidence of the contribution of turmeric oil. Since the oil does not contain curcuminoids in its composition, it is crucial to better understand the therapeutic role of other constituents in turmeric. The present review discusses the pharmacokinetics of turmeric oil, pointing to the potential application of its active molecules as therapeutic compounds. In addition, the bioactivities of turmeric oil and its safety in preclinical and clinical studies were revised. This literature-based research intends to provide an updated overview to promote further research on turmeric oil and its constituents.

Keywords: *Curcuma longa*; turmeric; curcuma oil; turmeric oil; pharmacological profile; pharmacological activity; safety; toxicity

Citation: Orellana-Paucar, A.M.; Machado-Orellana, M.G. Pharmacological Profile, Bioactivities, and Safety of Turmeric Oil. *Molecules* **2022**, *27*, 5055. https://doi.org/10.3390/molecules27165055

Academic Editors: Ana Paula Duarte, Ângelo Luís and Eugenia Gallardo

Received: 11 July 2022
Accepted: 2 August 2022
Published: 9 August 2022

Publisher's Note: MDPI stays neutral with regard to jurisdictional claims in published maps and institutional affiliations.

1. Introduction

Curcuma longa L. (syn. *Curcuma domestica*), commonly known as turmeric, is a perennial herb native to Asia. After curing, drying, and milling, turmeric rhizomes are usually employed as a dye, cosmetic, and food seasoning. In traditional medicine, turmeric is used for treating hepatic and gastrointestinal disorders, arthritis, rheumatism, skin diseases, fever, inflammation, amenorrhea, sepsis, and as an anthelmintic and laxative [1–3]. Some of these properties are supported by scientific evidence, and other novel activities have also been uncovered.

Most pharmacological activities of turmeric have been explained by the properties of curcumin, mainly because turmeric oil has not been as extensively studied as curcuminoids. Turmeric rhizome oil (TO) is responsible for this spice's characteristic taste and smell. Dried rhizomes contain about 3–6% essential oil [4]. The oil is extracted from powdered turmeric rhizomes through steam distillation. The major TO constituents are sesquiterpenes: bisabolanes, guaianes, germacranes, caranes, elemanes, spironolactones, selinanes, santalanes, and caryophyllanes [5,6]. Ar-turmerone, α-turmerone, and β-turmerone are the principal bisabolane sesquiterpenes [6,7]. Other notable TO compounds with reported bioactivity are α-atlantone, ar-curcumene, γ-curcumene, curlone, p-cymene, z-citral, eucalyptol, β-(Z)-farnesene, germacrone, β-sesquiphellandrene, α-santalene, α-zingiberene, and l-zingiberene [1,5] (Figure 1).

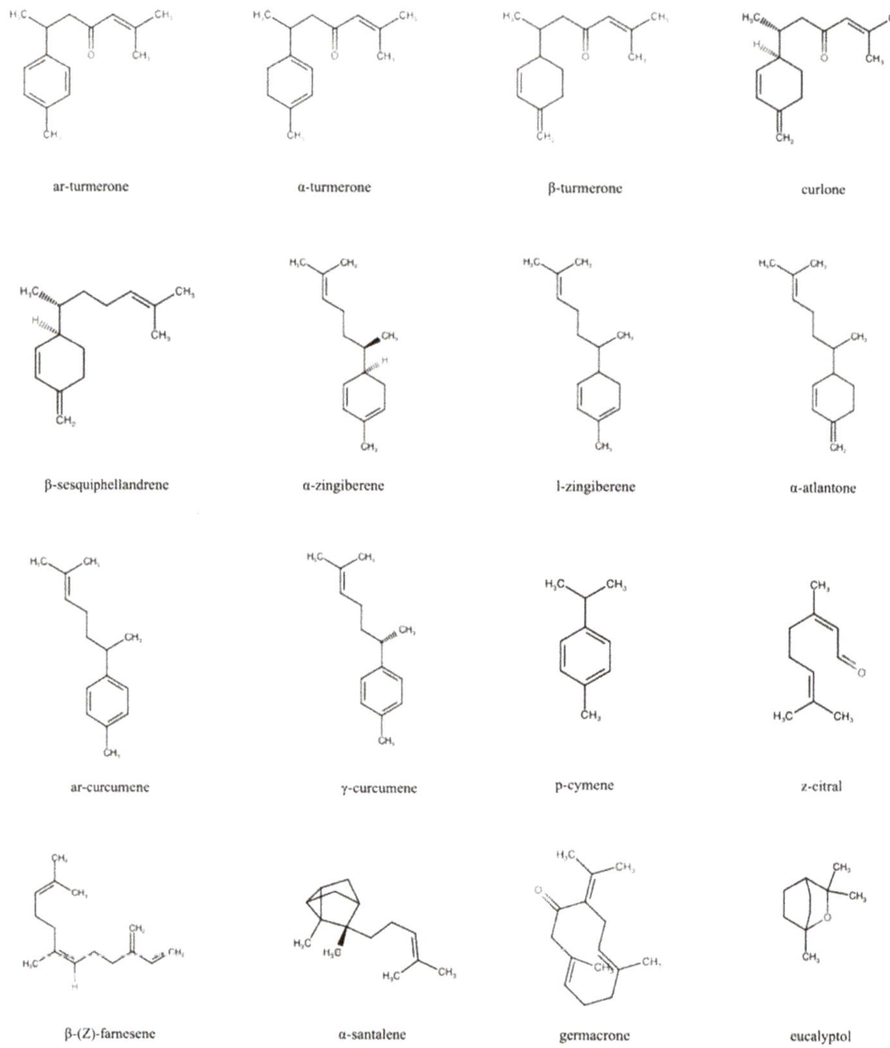

Figure 1. Chemical structure of bioactive turmeric oil constituents.

2. Pharmacological Profile

The murine pharmacokinetics profile of TO (500 mg/kg; p.o.) revealed the oil capability to be absorbed after oral administration, the high bioavailability, and the extended residence time for ar-turmerone, α, β-turmerone, and curlone [8]. TO displayed a peak plasma level 2 h after administration. Plasma concentrations of ar-turmerone and α-, β-turmerone remained uniform (100–135 ng/mL) from 8 to 18 h. Ar-turmerone showed a bioavailability of 13%, α,β-turmerone 11%, and curlone 7%. The mean residence time was 13.2, 11.6, and 14 h, respectively [8]. The high bioavailability of ar-turmerone in mice brains suggests the feasibility of orally administering the oil (or its components) through prolonged dosage periods since the plasma concentration remains stable for a considerable time lapse [9].

Regarding excretion, ar-turmerone was detected intact in 24 h urine of healthy adult volunteers after dry curcuma extract consumption. In addition, two prostaglandin-D2 metabolites were identified. This finding aligns with the anti-inflammatory effect attributed to this plant [10].

Additionally, the role of TO as a bioavailability enhancer is of current interest. Amyloid-β peptide accumulation and increased β-secretase activity have been associated with Alzheimer's dementia pathogenesis [11,12]. The lignans, sesamin and sesamolin, inhibit β-secretase. Nevertheless, both showed low bioavailability in murine models. TO acted as an efficient carrier, enhancing brain permeation of sesamin and sesamolin [13]. Likewise, the molecules responsible for the preventive effect of *Ginkgo biloba* in cases of dementia and Alzheimer's disease are flavonoids and terpene lactones with a well-known poor brain bioavailability. This limitation was surpassed in mice by joint administration with TO and sesame extract [14]. When curcumin is co-administered with ar-turmerone, curcumin has a significant permeation increase in the Caco-2 cell monolayer [15]. The high brain bioavailability of TO is relevant not only for its role as an enhancer but also for its intrinsic therapeutic properties. For instance, the antitumor properties attributed to TO are interesting in neurology due to its capability to cross the blood–brain barrier [16].

3. Bioactivity

3.1. Antioxidant

Nitric oxide, superoxide, and peroxynitrite levels increase in response to infections or inflammation. The radical scavenging assay and the ferric reducing antioxidant power test displayed significant antioxidant activity for TO [17,18]. TO inhibited superoxide generation triggered by phorbol-12-myristate-13-acetate in mice after i.p. administration. Additionally, oral administration of TO for 30 days prompted a relevant increase in glutathione and antioxidant enzyme concentrations of superoxide dismutase and glutathione reductase in plasma ($p < 0.001$). In the liver, glutathione-S-transferase and superoxide dismutase concentrations were increased by TO's influence ($p < 0.01$) [19]. TO decreased nitric oxide (NO) synthase expression and displayed immune-modulatory properties since it restricted neutrophil infiltration in the ischemic area of a murine cerebral focal ischemia model. Altogether, NO, peroxynitrite, and reactive oxygen neuronal concentration diminished and the number of apoptotic cells [20,21].

Moreover, carrageenan-induced inflammation is characterized by two phases: (a) rise in serotonin and histamine levels and (b) augment of proteases, prostaglandins, and superoxide concentrations [22]. The TO response in this model suggests a powerful influence as an antioxidant and critical modulator of inflammation mediators. In addition, it has been suggested that the capability of TO to scavenge free radicals and activate antioxidant enzymes may be related to its antimutagenic action [21].

3.2. Anti-Inflammatory

The models of acute inflammation stimulated by carrageenan and dextran and the chronic model induced by formalin were used to assess the anti-inflammatory properties of TO following i.p. administration in mice. In all cases, the oil reduced paw thickness. In the chronic model, the effect observed with TO was comparable to that obtained with diclofenac [19]. Accordingly, TO administration in rats significantly decreased paw edema in the carrageenan-induced inflammation model (76%) compared to aspirin. This reduction was only 62% when TO was administered with fish oil [23]. Thus, TO sesquiterpenes seems to act as antagonists or inhibitors of EPA and DHA action.

Furthermore, when employed as a pre- and post-treatment for myocardial ischemia/reperfusion (MI/RP), endothelial cell-mediated inflammation was lessened by TO in rats. TO reduced the ischemic area, adhesion of inflammatory cells to endothelial cells, and expression of pro-inflammatory genes and adhesion factors such as E-selective and intercellular adhesion molecule (ICAM) [24]. The oil's anti-inflammatory activity appears to be coupled with its ability to affect adhesion factors. Thus, spontaneous healing was promoted by avoiding the risk of ventricular rupture.

Lipophilic compounds (turmerones) were isolated from TO through hexane extraction and evaluated in a streptococcal cell wall (SCW)-induced rheumatoid arthritis murine model. Turmeric hexane extract (>28 mg/kg/day; i.p.) displayed a potent anti-inflammatory response accompanied by a high murine hepatotoxicity and mortality (56 mg/kg/day). Conversely, no toxicity signs or mortality were observed with the oral administration of this extract, employing a superior dose (560 mg/kg). This study's authors admitted hexane extract contamination with curcuminoids [25]. Based on the preclinical and clinical evidence, the toxicity observed with TO i.p. administration may be associated with an incomplete solvent elimination or a potential synergistic effect between curcuminoids and bisabolane-type sesquiterpenes. Since it is well known that curcumin exhibits a significant oral absorption limitation [26], the absence of toxicity signs (including mortality) in using this route supports this hypothesis. In addition, extraction of TO via hydro-distillation is strongly recommended. The use of organic solvents could lead to confounding factors related to toxicity triggered by an incomplete solvent removal or unintentional collection of compounds other than those described in the regular composition of turmeric rhizome oil.

TO constitutes an exciting drug candidate for preventing, delaying, or treating cardiovascular, pulmonary, neurodegenerative, and metabolic diseases and cancer, based on the correlation of their pathogenesis to inflammation.

3.3. Antidiabetic

A murine model of insulin resistance evidenced the capability of TO for controlling diabetic dyslipidemia, impaired glucose tolerance, insulin resistance, and insulin sensitivity indices, and altered plasma glucose and insulin levels. The oil reduced plasma glucose ($p < 0.05$), triacylglycerides ($p < 0.01$), total cholesterol ($p < 0.001$), LDL ($p < 0.001$), malondialdehyde ($p < 0.01$); INF-δ, IL-6, and C-reactive protein levels ($p < 0.001$); and the hepatic expression of lipogenic genes (SREBP-1c, PGC-1α, and PGC-1β). Moreover, TO increased the HDL concentration ($p < 0.001$) and restored the vasorelaxation response to acetylcholine [27].

Oils obtained from fresh and dried turmeric rhizomes showed a higher glucosidase inhibitory activity than acarbose ($p < 0.05$). Ar-turmerone reduced the expression of α-glucosidase and α-amylase [28]. These results support the hypoglycemic effect of TO hexane extract containing ar-turmerone as its primary component. It was stated that ar-turmerone exerted this action through PPAR-δ activation ($p < 0.05$) [29]. Since the effect of TO from dried rhizomes was 3.5 times more than the oil from fresh samples to control α-glucosidase [28], this finding could suggest an increased concentration of ar-turmerone in dried turmeric samples.

TO's capability for regulating SREBPAc, PGC1-α, and PGC1-β expression implies the preventive action of the oil for insulin resistance, type II diabetes, and diabetic dyslipidemia since PGC1-α promotes hepatic gluconeogenesis, PGC1-β stimulates SREBPAc expression, and SREBPAc endorses de novo hepatic lipogenesis [30,31].

3.4. Anticancer

The aryl hydrocarbon receptor (AhR) is a transcription factor involved in chemically induced toxicity and carcinogenesis due to its capability to generate free radicals and carcinogens. AhR has been associated with cancer, immunotoxicity, diabetes, atherosclerosis, liver fibrosis, and chronic kidney disease. A study to identify AhR ligands from the diet revealed an antagonist response of TO to human AhR probably related to the action of p-cymene [32].

Concerning oral submucous fibrosis, a precancerous oral lesion, TO prevented in vitro micronuclei formation in lymphocytes from healthy subjects. In addition, a combination of TO and turmeric extract reduced micro-nucleated cells in circulating lymphocytes and oral mucosal cells [33]. TO prevented in vitro mutagenicity caused by tobacco extract and inhibited microsomal activation of mutagens. The topical application of TO displayed anti-carcinogenic activity in a murine model of skin papilloma induced by 7,12–dimethylbenz

[a] anthracene and croton oil. In vitro, TO significantly hindered cytochrome p450 enzymes involved in carcinogens activation (CYP1A, A2, 2B, 2A, 2D, and 3A) ($p < 0.001$) [34]. In addition, TO prevented A431 human skin cancer cell proliferation ($p < 0.05$) and stimulated apoptosis in vitro. It is implied that these properties are related to increased caspase-3 and caspase-9 expression [35]. Activation of caspase-3, -8, and -9 mediated by TO was observed in a murine model of benign prostatic hyperplasia (BPH) [36]. Apoptosis appears to be triggered by upregulation of Bax, caspase-3, and caspase-9 and inhibition of Bcl-2 and COX-2 expression in rat tissues. TO suppressed NF-κB, a transcription factor responsible for regulating transcriptional activation of apoptosis, inflammation, cell proliferation, angiogenesis, cellular adhesion, cell invasion, and metastasis [36]. Therefore, a relevant role of NF-κB in oncogenesis promotion and cancer therapy resistance has been suggested [37]. These findings uncovered the role of NF-κB and inflammatory factors in BPH progression and turmeric oil's contribution to treating benign hyperplasia and cancer.

Regarding cervical cancer, oral TO pre-treatment in mice with tumor xenograft implants displayed a chemopreventive effect by decreasing tumor size ($p = 0.163$). In vitro, no cytotoxic effect was observed in three cancer cell lines (HeLa, SiHa, and ME180) with a maximum concentration of 80 μg/mL [38]. Conversely, another study reported TO's cytotoxicity against HeLa cells at higher concentrations (2100 μg/mL). TO triggered morphological changes in cancer cells and death by apoptosis [39]. TO's chemopreventive properties may be linked to its antimutagenic action and its ability to inhibit cytochrome p450 enzymes involved in carcinogen activation [34,40]. Since ROS could act as initiators and promoters of mutagenesis and carcinogenesis, the antioxidant properties of TO may contribute to its chemopreventive action [33,41]. Nevertheless, the exact contribution of antioxidants supplementation as adjuvants in radiotherapy and certain types of chemotherapy remains unclear to date [42]. Since TO's capability to affect the viability and morphological changes in cervical cancer cells and control tumor size corresponds to a dose-dependent relationship, it is advisable to determine the therapeutical index to better understand its antitumor activity [38,39].

A clinical study on liver cancer compared the effectiveness of hepatic arterial infusion with embolized TO vs. transcatheter artery chemoembolization (TACE). No differences were observed in the number of complete and partial remission cases, total effective rate, and incidence of post-embolism syndrome between both groups. Nevertheless, hepatic arterial infusion with TO promoted a longer survival time in liver cancer patients ($p < 0.05$) and minor myelosuppression occurrence ($p < 0.01$) than TACE treatment [43]. Additionally, TO inhibited in vitro growth of two human colon cancer cells (HT-29 and HCT-116). A murine model of HCT-116 xenograft confirmed a synergistic effect of TO with curcumin and vitamin E to inhibit cell growth in vitro and in vivo. This additive action may be attributable to the influence of TO on curcumin bioavailability, improving its absorption and distribution to the target organ [44]. The antioxidant effects of vitamin E and turmeric oil could also play an essential role in the antitumor outcome.

Interestingly, the oral administration of TO and curcumin in mice also shifted the fecal microbial composition. The *Bacteroidaceae*, *Ruminococcaceae*, *Clostridiales*, *Firmicutes*, and *Parabacteroids* families were markedly reduced, and the concentration of anti-inflammatory *Clostridium* XIVa was augmented. TO and curcumin also increased the probiotic concentration of *Lactobacillaceae* (20-fold) and *Bifidobacteriaceae* (6-fold) [45]. Additional studies are required to determine the precise association between anticancer properties and the gut microbiome composition.

3.5. Analgesic and Antinociceptive

An antinociceptive evaluation in mice exhibited a significant writhing reduction ($p < 0.001$) after treatment with TO. This effect was comparable to the response triggered by aspirin [19]. Accordingly, the tail-flick model evidenced the analgesic properties of TO in rats [23]. In line with these findings, the analgesic and antinociceptive properties of TO were confirmed through the hot plate and the acetic acid writhing tests. TO was capable of substantially increasing the pain threshold ($p < 0.050$) [46].

The analgesic and antinociceptive properties of TO deserve a detailed exploration. Pharmacological characterization of TO's active ingredients is also required to further evaluate their mechanism of action and prevent possible inhibitory activities with other analgesics or related drugs. The appealing combination of analgesic and anti-inflammatory properties confers TO the status of a promissory therapeutic alternative.

3.6. Cardiovascular

Hemorheology assay in mice evidenced TO's capability to decrease blood viscosity and the erythrocyte aggregation index [46]. These findings correspond to the antithrombotic activity of TO identified through the rat model of myocardial reperfusion injury. In a dose-dependent response, TO suppressed ADP-, collagen-, and thrombin-induced platelet aggregation in the presence and absence of plasma ($p < 0.001$). Thus, TO antithrombotic activity appears independent of plasma activators [8]. TO mitigated protein tyrosine phosphorylation in active platelets, augmented the total time to occlusion in arterial thrombosis induced by ferric chloride, and reduced thrombus weight ($p < 0.001$). No TO influence was identified on coagulation parameters such as prothrombin time (PT) or activated partial thromboplastin time (aPTT). The minor impact of the oil on the bleeding time and TO's capability to diminish tyrosine phosphorylation demonstrates its potential for modulating specific pathways during platelet activation [8,27]. A thrombus-specific mechanism of action of TO with no influence on normal hemostasis is supported by the significant effect of the oil in the rat model of arterial thrombosis induced by ferric chloride, which involves platelet-rich thrombus formation. Noteworthy, TO did not exhibit cardioprotective properties in this myocardial reperfusion injury murine model. TO triggered no noticeable variation in infarct size, myeloperoxidase activity, or CK-MB serum concentrations [8]. Paradoxically to what occurs in cerebral ischemia, iNOS induction prevents cardiac injury in rats. As aforementioned, TO exerts its neuroprotective activity by reducing NOS expression, NO-mediated peroxynitrite synthesis, oxidative stress, and neuronal apoptosis. The inhibitory action of TO on NO explains the absenteeism of cardioprotective action.

Additionally, the disease-modifying capability of TO was assessed in hyperlipidemic hamsters [27,47]. TO reduced the plasma total cholesterol, LDL-cholesterol, and TAG and increased HDL-cholesterol levels. In the liver, TO decreased cholesterol synthesis and oxidative stress. TO augmented the hepatic and intestinal expression of lipid metabolism and transport genes, such as PPARa, LXRa, CYP7A1, ABCA1, ABCG5, ABCG8, and LPL. Moreover, the oil suppressed the hepatic expression of SREBP-2, HMGCR, and intestinal NPC1L1 expression.

Furthermore, TO restored eNOS mRNA expression, improved vascular function, and controlled platelet activation and oxidative stress [47]. Since lipid metabolism pathways are similar in hamsters and humans, it is suggested that the antihyperlipidemic and antiatherogenic properties of TO entangle PPARa and LXRa activation and associated enterohepatic genes involved in cholesterol absorption, transport, and metabolism [48]. PPARa, LXRa, and the target genes are responsible for cholesterol homeostasis. Lipoprotein lipase (LPL), associated with PPARa, is also activated by TO. LPL catalyzes triacylglycerol hydrolysis, thus producing antiatherogenic and hypolipidemic effects. TO also triggered LXRa and CYP7A1 activation. LXRa is responsible for the fecal excretion of cholesterol, and CYP7A1 is involved in converting cholesterol into bile acids. Accordingly, TO downregulated NPC1L1 and upregulated enterohepatic expression of ABCG5 and ABCG8 transporters, leading to a prominent decrease in cholesterol intestinal absorption [49]. ABC transporters favor the biliary excretion of cholesterol [50]. TO upregulated ABCA1 expression, thus encouraging HDL synthesis. Altogether, these mechanisms of action explain dyslipidemia improvement mediated by TO. Therefore, the cardiovascular protection exerted by TO may be a consequence of the sum of additive mechanisms of action responsible for the antioxidant, anti-inflammatory, antiplatelet, and hypolipidemic effects.

3.7. Neuroprotective

The pathogenesis of Parkinson's and Alzheimer's diseases and multiple sclerosis involves neuroinflammation. This inflammatory process is triggered by the activation of IL-6, IL-1 beta, and TNF-α. TO controlled the expression of these inflammatory cytokines in rat brains after Cd-induced neurotoxicity. In addition, the oil inhibited the activity of acetylcholinesterase ($p < 0.01$) and adenosine deaminase ($p < 0.05$). Both are key regulatory enzymes of neurodegeneration [51]. β-secretase is involved in the plaque development in the hippocampus, cerebral cortex, and amygdaloid body associated with Alzheimer's disease pathogenesis [52]. TO inhibited in vitro β-secretase activity (53.4%), thus suggesting its potential to prevent dementia [11]. Noteworthy, turmerones, TO compounds with proven neurological activity, possess remarkable lipophilicity and brain bioavailability without showing neurotoxicity [9,11,53].

TO controlled seizure onset in zebrafish and mice models of seizures. The anticonvulsant properties of the oil and its isolated constituents were confirmed through electrophysiological analysis in zebrafish [53]. TO administration significantly decreased nitrosative stress, caspase-3 activation, and apoptosis in a murine model of cerebral infarct and corrected the mitochondrial membrane potential [54]. Findings from studies carried out in zebrafish and murine models of seizures, ischemic attack, and neuroinflammation suggest that, along with modulation of sodium ion channels, neurotransmitters signaling (γ-aminobutyric acid), and upregulation of brain-derived neurotrophic factor (BDNF), the neuroprotective properties of the oil are also related to its antioxidant and anti-inflammatory activity. The neuroprotection exerted by TO and its main components seems to be associated with its ability to up- or downregulate strategic pathways. Therefore, the antioxidant properties of turmeric oil are supported by its capability to scavenge free radicals, attenuating upregulation of Bax and Bcl-2, controlling the release of cytochrome c, inhibiting the formation of ROS and NO, and preventing lipid peroxidation and oxidative DNA damage. Additionally, TO controls neuroinflammation through the inhibition of adenosine deaminase and acetylcholinesterase. These properties guarantee the protection of tissue unaffected by neuropathogenic processes and promote a total removal of injured cells [51,53–55].

3.8. Nephroprotective

TO triggered protective activity in Cd-induced nephrotoxicity in rats and prevented modifications in renal function biomarkers (creatinine, urea, BUN), inflammatory cytokines (IL-6 and TNF-α), and renal adenosine deaminase (ADA) activity [56]. Since Cd induces a significant decrease in IL-10, an interleukin involved in cytokines synthesis inhibition, the action of TO infers a crucial role of the inflammatory response in kidney diseases.

The therapeutic application extent of TO regarding its anti-inflammatory properties must be further understood considering the crucial role of inflammation in the pathogenesis of cancer, neurological, cardiovascular, metabolic, renal, and respiratory diseases.

3.9. Antibacterial

TO inhibited *Porphyromonas gingivalis*, a pathogen responsible for periodontitis, showing a significant inhibition zone in vitro [57]. Likewise, TO inhibited in vitro growth, acid production, adherence to saliva-coated hydroxyapatite beads, and biofilm formation of *Streptococcus mutans*, a cariogenic bacterium [58]. *S. mutans* must adhere to tooth surfaces to metabolize dietary sugars, transforming them into lactic and formic acids. These acids lower oral pH, demineralize tooth enamel, and cause dental caries. Biofilm formation is a bacterial defense mechanism against host and antibacterial action [59]. Therefore, it is implied that TO may be helpful for the prevention and treatment of dental caries [60].

Additionally, TO restrained the growth of Gram-positive bacteria (*Bacillus cereus*, *B. coagulans*, *B. subtilis*, and *Staphylococcus aureus*) and Gram-negative bacteria (*Escherichia coli*, *Klebsiella pneumoniae*, and *Pseudomonas aeruginosa*) [61–63]. There is evidence that turmeric extract exerts its antibacterial activity through cell wall degradation, cytoplasmic membrane disruption, leakage of cellular components, alterations in DNA and RNA synthesis, electron transport, and nutrient uptake [64]. TO antibacterial properties are especially relevant in the current context, where antibiotic resistance is a global health problem.

3.10. Antifungal

Drug resistance against dermatophytes is also a growing health concern. TO showed in vitro antifungal activity against *Candida tropicalis*, *Penicillium notatum*, *Aspergillus fumigatus*, *A. niger*, *A. flavus*, *Trichophyton rubrum*, *T. violceum*, *T. mentagrophytes*, *Epidermophyton floccosum*, *Microsporum gypseum*, and *Sporothrix schenckii* [65–67], and a synergistic effect with the commercially available antifungal drugs: clotrimazole, fluconazole, ketoconazole, and terbinafine [65]. These findings are in line with the antifungal activity described in guinea pigs. Lesions improved after 2–5 days and completely disappeared after 6–7 days of topical TO application [66]. Ar-turmerone displayed higher anti-dermatophytic activity than ketoconazole [67]. Fungistatic and fungicidal mechanisms of the oil include structural modifications in fungal cells, functional changes, and inactivation of enzymes, proteins, and nuclear material [68]. Since no adverse skin reactions were observed in guinea pigs after topical application, TO appears to be an interesting add-on treatment for dermatological infections. Further studies focused on the underlying mechanisms of action are warranted.

On the other hand, fungal infections of cereal crops constitute a relevant health threat and economic loss factor. TO hindered mycotoxin (deoxynivalenol and zearalenone) production from *Fusarium gramineum* [69]. *F. gramineum* infection of wheat and barley causes *Fusarium* head blight, a harmful plant pathogen and mycotoxin producer. Likewise, *Aspergillus flavus* is well known for its capability to infect cereal crops and produce toxic and hepatocarcinogenic compounds. TO prevented in vitro and in vivo *A. flavus* mycelial growth, spore germination, sporulation, and aflatoxin production. Additionally, TO damaged hyphae membranes and conidiophores, controlled mycotoxin gene expression in maize, and inhibited ergosterol synthesis in fungal cells [70,71]. TO displayed in vitro antifungal activity against *Aspergillus parasiticus*, *Fusarium moniliforme*, and *Penicillium digitatum* [72]. *Fusarium verticillioides* is one of the most common producers of mycotoxins (fumonisins) in stored grains. TO moderately inhibits in vitro *F. verticillioides* growth and conidial production. A noticeable color change on *F. verticillioides* mycelium triggered by TO implies its interference in pathways related to fungal cell synthesis, affecting cell wall integrity and permeability [73]. Thus, TO appears to be a potential eco-friendly substitute for controlling fungal contamination in food since synthetic chemical fungicides produce harmful effects on crops and people.

3.11. Antiparasitic

TO prevented the growth of *Leishmania amazonensis* promastigotes in mice. The oil diminished the number of intracellular amastigotes and infected cells without cytotoxicity in vitro. Macrophage activation in infected mice suggests a contribution of additional mechanisms, excepting NO or ROS production [63].

Likewise, TO exhibited in vitro anti-trypanosomal activity and high selectivity indices in cytotoxicity evaluation [74]. The antipromastigote activity of TO comprises a direct action on the parasite, including structural modifications, changes in mitochondrial physiology, and cell death apoptosis. These results agree with previous reports on turmeric extract and *L. donovani* [75].

Additionally, TO displayed scolicidal properties. The oil was capable of exterminating isolated protoscoleces (in vitro) and protoscoleces into hydatid cysts collected from sheep liver (ex vivo) [76]. The major limitation in preventing protoscolex spillage during hydatid cyst surgery is the adverse effects of protoscolicide drugs (i.e., chemically induced sclerosing cholangitis) [77]. It is implied that TO exerts this activity by disrupting the protoscolex cyst wall and interfering in vital intracellular pathways. In addition, the safety evaluation of TO performed in mice showed no significant biochemical or hematological modification [76].

In this context, TO appears to be a safe and efficient antiparasitic alternative. Nevertheless, additional studies should be performed to understand this property better.

3.12. Insecticidal

TO displayed larvicidal activity against *Aedes albopictus*, *Aedes aegypti*, and *Culex pipiens* [78]. The oil could be used as a repellent, eco-friendly larvicide, and pupacide in breeding places. Further research on this potential application is needed in response to the current interest in substituting synthetic pesticides with natural products exhibiting improved action and minor adverse effects.

Information of published studies on the antioxidant, anti-inflammatory, antidiabetic, anticancer, analgesic, antinociceptive, cardiovascular, neuroprotective, nephroprotective, antibacterial, antifungal, antiparasitic, and insecticidal properties of TO is depicted in Table 1.

Table 1. A summary of studies focused on the bioactivities of turmeric oil.

Bioactivity	Main Compounds of Turmeric Oil	Model	Concentration/Dose; Administration Route	Source
Antioxidant	ar-turmerone, a-turmerone, β-turmerone	In vitro	0.025 g/3 mL	[17]
	ar-turmerone, β-turmerone, ar-curcumene	In vitro	80% ethanol	[18]
	ar-turmerone, curlone, ar-curcumene	In vitro	200 mg/mL	[19]
		In vivo (mouse)	250 mg/kg; i.p.	[19]
	ar-turmerone, a-turmerone, β-turmerone, curlone	In vivo (rat)	250 mg/kg; i.p.	[20]
Anti-inflammatory	ar-turmerone, curlone, ar-curcumene	In vivo (mouse)	500 mg/kg; i.p.	[19]
	ar-turmerone, a-turmerone, β-turmerone	In vivo (rat)	100 mg/kg; p.o.	[23]
	ar-turmerone, turmerone, and curlone	In vivo (rat)	250 mg/kg; p.o.	[24]
	ar-turmerone, α-turmerone, β-turmerone	In vivo (rat)	>28 mg/kg/day; i.p.	[25]
		In vivo (rat)	560 mg/kg; p.o.	[25]
Antidiabetic	ar-turmerone, a-turmerone, β-turmerone, curlone	In vivo (hamster)	300 mg/kg/day; p.o.	[27]
		In vivo (rat)	300 mg/kg/day; p.o.	[27]
	ar-turmerone	In vitro	0.38 mg/mL	[28]
	ar-turmerone	In vivo (mouse)	0.5 g/100 g of diet; p.o.	[29]
Anticancer	p-cymene	In vitro	100 μg/mL	[32]
	not reported	In vivo (human)	600 mg TO + 3 g turmeric extract; p.o.	[33]
	ar-turmerone	In vitro	1 mg/plate	[34]
		In vivo (mice)	50%; topical	[34]
		In vitro	200 mg/mL	[34]
	not reported	In vitro	80 mg/L	[35]
	not reported	In vitro	40 mg/mL	[36]
		In vivo (rat)	7.2 mg/kg; i.g.	[36]
	ar-turmerone, a-turmerone, β-turmerone	In vitro	2100 mg/mL	[39]
	ar-turmerone, curlone	In vitro	3000 mg/plate	[40]
		In vivo (rat)	1 g/kg; p.o.	[40]
	not reported	In vivo (human)	3 mL (embolized)	[43]
	ar-turmerone, a-turmerone, β-turmerone, a-santalene, ar-curcumene	In vivo (mice)	5 mg (TO + curcumin)/kg; p.o.	[45]

Table 1. *Cont.*

Bioactivity	Main Compounds of Turmeric Oil	Model	Concentration/Dose; Administration Route	Source
Analgesic and antinociceptive	ar-turmerone, curlone, ar-curcumene	In vivo (mouse)	100 mg/kg; i.p.	[19]
	ar-turmerone, a-turmerone, β-turmerone	In vivo (rat)	100 mg/kg; p.o.	[23]
	ar-turmerone, curlone, turmerone	In vivo (mouse)	9.75 mL/kg; i.p.	[46]
Cardiovascular	ar-turmerone, a-turmerone, β-turmerone, curlone	In vivo (rat)	1 g/kg; p.o.	[8]
		In vivo (mouse)	1 g/kg; p.o.	[8]
	ar-turmerone, a-turmerone, β-turmerone, curlone	In vivo (hamster)	300 mg/kg/day; p.o.	[27]
		In vivo (rat)	300 mg/kg/day; p.o.	[27]
	ar-turmerone, curlone, turmerone	In vivo (mouse)	9.75 mL/kg; i.p.	[46]
	ar-turmerone, a-turmerone, β-turmerone, curlone	In vivo (hamster)	300 mg/kg; p.o.	[47]
Neuroprotective	ar-turmerone, a-turmerone, β-turmerone, l-zingiberene, β-sesquiphellandrene	In vitro	250 mg/mL	[11]
	eucalyptol	In vivo (rat)	50 mg/kg; p.o.	[51]
	ar-turmerone, a-turmerone, β-turmerone, a-atlantone	In vivo (zebrafish)	10 mg/mL	[53]
		In vivo (mouse)	100 mg/kg; i.v.	[53]
	ar-turmerone, a-turmerone, β-turmerone, curlone	In vivo (rat)	50 mg/kg, p.o.	[54]
Nephroprotective	eucalyptol	In vivo (rat)	50 mg/kg; p.o.	[56]
Antibacterial	not reported	In vitro	100%	[57]
	a-turmerone, germacrone	In vitro	>0.5 mg/mL	[58]
	not reported	In vitro	1000 ppm	[61]
	ar-turmerone, turmerone, curlone	In vitro	100 ppm	[62]
	turmerone, b-turmerone, γ-curcumene	In vitro	75 mL	[63]
Antifungal	z-citral	In vitro	10 mg/mL	[65]
	not reported	In vitro	114.9 mg/mL	[66]
		In vivo (guinea pig)	topical	[66]
	ar-turmerone	In vitro	6% *w/w*; topical	[67]
	not reported	In vitro	11,580 mg/mL	[69]
	ar-turmerone, turmerone, b-sesquiphellandrene, curcumene	In vitro	4 mL/mL	[70]
	ar-turmerone, a-turmerone, β-turmerone	In vitro	0.5% *v/v*	[71]
	ar-turmerone, a-zingiberene, b-(Z)-farnesene, ar-curcumene	In vitro	6 mg/mL	[72]
	ar-turmerone, a-turmerone, β-turmerone	In vitro	1000 ppm	[73]

Table 1. *Cont.*

Bioactivity	Main Compounds of Turmeric Oil	Model	Concentration/Dose; Administration Route	Source
Antiparasitic	turmerone, β-turmerone, γ-curcumene	In vitro	500 mg/mL	[63]
	a-zingiberene, b-sesquiphellandrene, ar-turmerone, curlone	In vitro	3.17 nL/mL	[74]
	a-turmerone, b-turmerone	In vitro	200 mg/mL	[76]
Insecticidal	turmerone, curcumene	In vivo (mosquito larvae)	0.2 mg/mL	[78]

4. Safety

The proportion of turmeric oil constituents could vary depending on the crop's location. Nevertheless, no toxicity warning has been reported for any of the components of the oil extracted from turmeric. Methyl eugenol, a genotoxic carcinogen, is the sole constituent of turmeric leaf oil (~3%). It is not present in turmeric rhizome oil [7,79]. Moreover, possible heavy metal contamination is negligible if the oil is extracted through steam distillation (excluding mercury) [4,80].

Reported cases of turmeric toxicity are frequently related to curcumin, a turmeric extract component that is not present in the oil [81–84]. A 13-week oral administration of TO did not cause mortality in rats or adverse effects. TO doses up to 500 mg/kg/day did not modify the hepatic and renal biochemical profiles. In addition, TO genotoxicity analysis reported no mutagenicity to Salmonella typhimurium TA98, 100, 102, and 1535. Furthermore, TO administration (1 g/kg; p.o.) for 14 days did not trigger chromosomal aberration or micronucleus formation in rat bone marrow cells [40].

Accordingly, human subjects did not show relevant adverse reactions associated with TO oral administration. Daily doses of TO (1 mL) were given to healthy volunteers for 3 months. Of nine volunteers, only two subjects discontinued the treatment: one person on the third day due to skin rash and another one on the seventh day due to intercurrent fever demanding antibiotic treatment. TO administration in the remaining seven subjects did not trigger side effects of clinical importance. Only one case of a reversible change in serum lipids was reported [85].

The evaluation of pharmacological properties and toxicity is relevant in searching for novel, efficient, and safe therapeutic alternatives. In the case of TO, no undesired effects have been described in animals or humans. In addition, carcinogens are not present in the oil composition [40,43,85]. Consistent with these findings, even computational models for carcinogenicity prediction pointed to ar-turmerone, a major TO constituent, as a non-mutagenic, non-carcinogenic, and non-hepatotoxic agent with negligible side effects [84]. These characteristics support the classification of TO as 'Generally Recognized As Safe (GRAS)' conferred by the FDA [40].

5. Conclusions

Based on the available information, the bioactivities of the oil are not modified by oral, intravenous, or intraperitoneal administration. The dose–response relationship observed with the oil and its role as a bioavailability enhancer suggest an appropriate absorption and distribution. Furthermore, the high brain bioavailability of the oil supports its therapeutic application for various illnesses, including neurological diseases, due to its ability to overcome the blood–brain barrier.

The therapeutic potential of TO deserves scientific attention due to the vast diversity of possible pharmacological targets. These include antioxidant, anti-inflammatory, analgesic, antinociceptive, neuroprotective, cardiovascular, antidiabetic, nephroprotective, anticancer, antibacterial, antifungal, antiparasitic, and insecticidal properties. Most research studies described in this review were carried out at the preclinical level, reporting interesting pharmacological effects without associated toxicity. Since TO's safety was confirmed in healthy volunteers, the development of clinical research on TO's active compounds remains a pending matter.

TO is a rich source of bioactive molecules. Major chemical constituents are often pointed out as the responsible compounds for the oil's pharmacological properties. Nevertheless, the proportion of each component in the oil can be affected by the crops' geographical location, plant nutritional status, or maturity stage. Thus, the reported bioactivities of TO should be considered a baseline for the development of further research on the isolated compounds since the oil's activity will not always depend on its primary constituents. Due to its potency, a minor component could be accountable for a specific response. Moreover, the oil's pharmacological effect could result from the joint action of two or more compounds leading to an enhanced, synergistic, or inhibitory outcome.

Further studies are required to assess the potential clinical application of TO's active constituents. It remains crucial to determine the pharmacological profile of the isolated TO active compounds and their bioavailability, efficacy, and safety to maximize their therapeutical benefits according to the target organ.

Author Contributions: A.M.O.-P.: conceptualization, writing of original draft preparation; A.M.O.-P. and M.G.M.-O.: writing the final version of the manuscript and editing. All authors have read and agreed to the published version of the manuscript.

Funding: This research was funded by the Research Vice-Chancellorship of the University of Cuenca, Ecuador.

Institutional Review Board Statement: Not applicable.

Informed Consent Statement: Not applicable.

Data Availability Statement: Not applicable.

Conflicts of Interest: A.M.O.-P. holds patents on turmeric oil's applications for treating neurodegenerative disorders. M.G.M.-O. declares not to have conflict of interest.

Abbreviations

ABC	adenosine triphosphate-binding cassette
ADA	adenosine deaminase
AhR	aryl hydrocarbon receptor
aPTT	activated partial thromboplastin time
Bax	B-cell leukaemia/lymphoma 2-associated X Protein
Bcl	B-cell lymphoma 2
BDNF	brain-derived neurotrophic factor
BPH	benign prostatic hyperplasia
CK-MB	creatine kinase MB
COX	cyclooxygenase
CYP	cytochrome P-450
DHA	docosahexanoic acid
eNOS	endothelial NOS
EPA	eicosapentaenoic acid
GRAS	Generally Recognized As Safe
HMGCR	hydroxymethylglutaryl coenzyme A reductase
ICAM	intercellular adhesion molecule
i.g.	intragastric
IL	interleukin
iNOS	inducible nitric oxide synthase
i.p.	intraperitoneal
i.v.	intravenous
LXR	liver X receptor
LPL	lipoprotein lipase

MI/RP	myocardial ischemia/reperfusion
NF	nuclear factor
NO	nitric oxide
NOS	nitric oxide synthase
NPC1L1	Niemann-Pick C1-Like 1
PGC	peroxisome proliferator-activated receptor gamma coactivator
p.o.	per os (oral)
PPAR	peroxisome proliferator-activated receptors
PT	prothrombin time
ROS	reactive oxygen species
SCW	streptococcal cell wall
SREBP	sterol regulatory element binding protein
TACE	transcatheter artery chemoembolization
TAG	triacylglycerides
TNF	tumor necrosis factor
TO	turmeric oil

References

1. Dosoky, N.S.; Setzer, W.N. Chemical Composition and Biological Activities of Essential Oils of Curcuma Species. *Nutrients* **2018**, *10*, 1196. [CrossRef] [PubMed]
2. World Health Organization. *WHO Monographs on Selected Medicinal Plants*; World Health Organization: Geneva, Switzerland, 1999; Volume 1, pp. 115–119.
3. Villegas, I.; Sánchez-Fidalgo, S.; de La Lastra, C.A. New mechanisms and therapeutic potential of curcumin for colorectal cancer. *Mol. Nutr. Food Res.* **2008**, *52*, 1040–1061. [CrossRef]
4. Bampidis, V.; Azimonti, G.; de Lourdes Bastos, M.; Christensen, H.; Durjava, K.; Kouba, M.; López-Alonso, M.; Puente, S.L.; Marcon, F.; Mayo, B.; et al. Safety and efficacy of turmeric extract, turmeric oil, turmeric oleoresin and turmeric tincture from *Curcuma longa* L. rhizome when used as sensory additives in feed for all animal species EFSA Panel on Additives and Products or Substances used in Animal Feed (FEEDAP), Panel members. *EFSA J.* **2020**, *18*, e06146. [PubMed]
5. Aggarwal, B.B.; Yuan, W.; Li, S.; Gupta, S.C. Curcumin-free turmeric exhibits anti-inflammatory and anticancer activities: Identification of novel components of turmeric. *Mol. Nutr. Food Res.* **2013**, *57*, 1529–1542. [CrossRef] [PubMed]
6. Zhang, H.A.; Kitts, D.D. Turmeric and its bioactive constituents trigger cell signaling mechanisms that protect against diabetes and cardiovascular diseases. *Mol. Cell. Biochem.* **2021**, *476*, 3785–3814. [CrossRef] [PubMed]
7. Dixit, S.; Awasthi, P. Chemical Composition of *Curcuma Longa* Leaves and Rhizome Oil from the Plains of Northern India. *J. Young Pharm.* **2009**, *1*, 312. [CrossRef]
8. Prakash, P.; Misra, A.; Surin, W.R.; Jain, M.; Bhatta, R.S.; Pal, R. Antiplatelet effects of Curcuma oil in experimental models of myocardial ischemia-reperfusion and thrombosis. *Thromb. Res.* **2011**, *127*, 111–118. [CrossRef] [PubMed]
9. Orellana-Paucar, A.M.; Afrikanova, T.; Thomas, J.; Aibuldinov, Y.K.; Dehaen, W.; de Witte, P.A.; Esguerra, C.V. Insights from Zebrafish and Mouse Models on the Activity and Safety of Ar-Turmerone as a Potential Drug Candidate for the Treatment of Epilepsy. *PLoS ONE* **2013**, *8*, e81634. [CrossRef]
10. Peron, G.; Sut, S.; Dal Ben, S.; Voinovich, D.; Dall'Acqua, S. Untargeted UPLC-MS metabolomics reveals multiple changes of urine composition in healthy adult volunteers after consumption of *curcuma longa* L. extract. *Food Res. Int.* **2020**, *127*, 108730. [CrossRef]
11. Matsumura, S.; Murata, K.; Zaima, N.; Yoshioka, Y.; Morimoto, M.; Kugo, H.; Yamamoto, A.; Moriyama, T.; Matsuda, H. Inhibitory Activities of Essential Oil Obtained from Turmeric and Its Constituents against β-Secretase. *Nat. Prod. Commun.* **2016**, *11*, 1785–1788. [CrossRef]
12. Li, R.; Lindholm, K.; Yang, L.B.; Yue, X.; Citron, M.; Yan, R.; Beach, T.; Sue, L.; Sabbagh, M.; Cai, H. Amyloid β peptide load is correlated with increased β-secretase activity in sporadic Alzheimer's disease patients. *Proc. Natl. Acad. Sci. USA* **2004**, *101*, 3632–3637. [CrossRef] [PubMed]
13. Iwamoto, K.; Matsumura, S.; Yoshioka, Y.; Yamamoto, A.; Makino, S.; Moriyama, T.; Zaima, N. Using Turmeric Oil as a Solvent Improves the Distribution of Sesamin-Sesamolin in the Serum and Brain of Mice. *Lipids* **2019**, *54*, 311–320. [CrossRef] [PubMed]
14. Iwamoto, K.; Kawamoto, H.; Takeshita, F.; Matsumura, S.; Ayaki, I.; Moriyama, T.; Zaima, N. Mixing Ginkgo biloba Extract with Sesame Extract and Turmeric Oil Increases Bioavailability of Ginkgolide a in Mice Brain. *J. Oleo Sci.* **2019**, *68*, 923–930. [CrossRef] [PubMed]
15. Yue, G.G.; Cheng, S.W.; Yu, H.; Xu, Z.S.; Lee, J.K.; Hon, P.M.; Lee, M.Y.H.; Kennelly, E.J.; Deng, G.; Yeung, S.K.; et al. The role of turmerones on curcumin transportation and P-glycoprotein activities in intestinal Caco-2 cells. *J. Med. Food* **2012**, *15*, 242–252. [CrossRef] [PubMed]
16. Wu, X.S.; Xie, T.; Lin, J.; Fan, H.Z.; Huang-Fu, H.J.; Ni, L.F. An investigation of the ability of elemene to pass through the blood-brain barrier and its effect on brain carcinomas. *J. Pharm. Pharmacol.* **2009**, *61*, 1653–1656. [CrossRef] [PubMed]

17. Fernández-Marín, R.; Fernandes, S.C.M.; Andrés, M.A.; Labidi, J. Microwave-assisted extraction of *curcuma longa* l. Oil: Optimization, chemical structure and composition, antioxidant activity and comparison with conventional soxhlet extraction. *Molecules* **2021**, *26*, 1516. [CrossRef] [PubMed]
18. Ivanović, M.; Makoter, K.; Razboršek, M.I. Comparative study of chemical composition and antioxidant activity of essential oils and crude extracts of four characteristic zingiberaceae herbs. *Plants* **2021**, *10*, 501. [CrossRef]
19. Liju, V.B.; Jeena, K.; Kuttan, R. An evaluation of antioxidant, anti-inflammatory, and antinociceptive activities of essential oil from *Curcuma longa*. L. *Indian J. Pharmacol.* **2011**, *43*, 526.
20. Dohare, P.; Varma, S.; Ray, M. Curcuma oil modulates the nitric oxide system response to cerebral ischemia/reperfusion injury. *Nitric Oxide* **2008**, *19*, 1–11. [CrossRef]
21. Jayaprakasha, G.K.; Jena, B.S.; Negi, P.S.; Sakariah, K.K. Evaluation of antioxidant activities and antimutagenicity of turmeric oil: A byproduct from curcumin production. *Z. Nat. Sect. C J. Biosci.* **2002**, *57*, 828–835. [CrossRef]
22. Di Rosa, M.; Giroud, J.P.; Willoughby, D.A. Studies on the mediators of the acute inflammatory response induced in rats in different sites by carrageenan and turpentine. *J. Pathol.* **1971**, *104*, 15–29. [CrossRef] [PubMed]
23. Jacob, J.N.; Badyal, D.K. Biological studies of turmeric oil, part 3: Anti-inflammatory and analgesic properties of turmeric oil and fish oil in comparison with aspirin. *Nat. Prod. Commun.* **2014**, *9*, 225–228. [CrossRef] [PubMed]
24. Manhas, A.; Khanna, V.; Prakash, P.; Goyal, D.; Malasoni, R.; Naqvi, A. Curcuma oil reduces endothelial cell-mediated inflammation in postmyocardial ischemia/reperfusion in rats. *J. Cardiovasc. Pharmacol.* **2014**, *64*, 228–236. [CrossRef] [PubMed]
25. Funk, J.L.; Frye, J.B.; Oyarzo, J.N.; Zhang, H.; Timmermann, B.N. Anti-Arthritic Effects and Toxicity of the Essential Oils of Turmeric (*Curcuma longa* L.). *J. Agric. Food Chem.* **2010**, *58*, 842. [CrossRef]
26. Sabet, S.; Rashidinejad, A.; Melton, L.D.; McGillivray, D.J. Recent advances to improve curcumin oral bioavailability. *Trends Food Sci. Technol.* **2021**, *110*, 253–266. [CrossRef]
27. Singh, V.; Jain, M.; Misra, A.; Khanna, V.; Prakash, P.; Malasoni, R. Curcuma oil ameliorates insulin resistance & associated thrombotic complications in hamster & rat. *Indian J. Med. Res.* **2015**, *141*, 823.
28. Lekshmi, P.C.; Arimboor, R.; Indulekha, P.S.; Nirmala Menon, A. Turmeric (*Curcuma longa* L.) volatile oil inhibits key enzymes linked to type 2 diabetes. *Int. J. Food Sci. Nutr.* **2012**, *63*, 832–834. [CrossRef]
29. Nishiyama, T.; Mae, T.; Kishida, H.; Tsukagawa, M.; Mimaki, Y.; Kuroda, M. Curcuminoids and Sesquiterpenoids in Turmeric (*Curcuma longa* L.) Suppress an Increase in Blood Glucose Level in Type 2 Diabetic KK-Ay Mice. *J. Agric. Food Chem.* **2005**, *53*, 959–963. [CrossRef]
30. Herzig, S.; Long, F.; Jhala, U.S.; Hedrick, S.; Quinn, R.; Bauer, A. CREB regulates hepatic gluconeogenesis through the coactivator PGC-1. *Nature* **2001**, *413*, 179–183. [CrossRef]
31. Dekker, M.J.; Su, Q.; Baker, C.; Rutledge, A.C.; Adeli, K. Fructose: A highly lipogenic nutrient implicated in insulin resistance, hepatic steatosis, and the metabolic syndrome. *Am. J. Physiol. Endocrinol. Metab.* **2010**, *299*, E685–E694. [CrossRef]
32. Bartoňková, I.; Dvořák, Z. Essential oils of culinary herbs and spices display agonist and antagonist activities at human aryl hydrocarbon receptor AhR. *Food Chem. Toxicol.* **2018**, *111*, 374–384. [CrossRef] [PubMed]
33. Hastak, K.; Lubri, N.; Jakhi, S.D.; More, C.; John, A.; Ghaisas, S.D. Effect of turmeric oil and turmeric oleoresin on cytogenetic damage in patients suffering from oral submucous fibrosis. *Cancer Lett.* **1997**, *116*, 265–269. [CrossRef]
34. Liju, V.B.; Jeena, K.; Kuttan, R. Chemopreventive activity of turmeric essential oil and possible mechanisms of action. *Asian Pac. J. Cancer Prev.* **2014**, *15*, 6575–6580. [CrossRef]
35. Zan, X.J.; Rong, D.Y.; Tu, Y.H.; Xue, Y.C.; Ye, Z.Y.; Kang, Y.Q.; Zhou, Y.; Cao, Y. Effect of turmeric volatile oil on proliferation and apoptosis of human skin SCC A431 cells. *Zhongguo Zhong Yao Za Zhi* **2016**, *41*, 2883–2887. [PubMed]
36. Wang, S.; Li, Y.; Li, W.; Zhang, K.; Yuan, Z.; Cai, Y. Curcuma oil ameliorates benign prostatic hyperplasia through suppression of the nuclear factor-kappa B signaling pathway in rats. *J. Ethnopharmacol.* **2021**, *279*, 113703. [CrossRef] [PubMed]
37. Orlowski, R.Z.; Baldwin, A.S. NF-κB as a therapeutic target in cancer. *Trends Mol. Med.* **2002**, *8*, 385–389. [CrossRef]
38. Paradkar, P.H.; Juvekar, A.S.; Barkume, M.S.; Amonkar, A.J.; Joshi, J.V.; Soman, G.; Vaidya, A.D.B. In vitro and in vivo evaluation of a standardized *Curcuma longa* Linn formulation in cervical cancer. *J. Ayurveda Integr. Med.* **2021**, *12*, 616–622. [CrossRef]
39. Santos, P.A.S.R.; Avanço, G.B.; Nerilo, S.B.; Marcelino, R.I.A.; Janeiro, V.; Valadares, M.C. Assessment of Cytotoxic Activity of Rosemary (*Rosmarinus officinalis* L.), Turmeric (*Curcuma longa* L.), and Ginger (*Zingiber officinale* R.) Essential Oils in Cervical Cancer Cells (HeLa). *Sci. World J.* **2016**, *2016*, 9273078. [CrossRef]
40. Liju, V.B.; Jeena, K.; Kuttan, R. Acute and subchronic toxicity as well as mutagenic evaluation of essential oil from turmeric (*Curcuma longa* L). *Food Chem. Toxicol.* **2013**, *53*, 52–61. [CrossRef]
41. Clark, S.F. The biochemistry of antioxidants revisited. *Nutr. Clin. Pract.* **2002**, *17*, 5–17. [CrossRef]
42. Yasueda, A.; Urushima, H.; Ito, T. Efficacy and interaction of antioxidant supplements as adjuvant therapy in cancer treatment: A systematic review. *Integr. Cancer Ther.* **2016**, *15*, 17–39. [CrossRef] [PubMed]
43. Cheng, J.H.; Chang, G.; Wu, W.Y. A controlled clinical study between hepatic arterial infusion with embolized curcuma aromatic oil and chemical drugs in treating primary liver cancer. *Zhongguo Zhong Xi Yi Jie He Za Zhi* **2001**, *21*, 165–167. [CrossRef] [PubMed]
44. Benny Antony. Composition to Enhance the Bioavailability of Curcumin. U.S. Patent US8895087B2, 25 November 2014.
45. Farhana, L.; Sarkar, S.; Nangia-Makker, P.; Yu, Y.; Khosla, P.; Levi, E.; Azmi, A.; Majumdar, A.P.N. Natural agents inhibit colon cancer cell proliferation and alter microbial diversity in mice. *PLoS ONE.* **2020**, *15*, e0229823. [CrossRef] [PubMed]

46. Chen, Z.; Quan, L.; Zhou, H.; Zhao, Y.; Chen, P.; Hu, L. Screening of active fractions from *Curcuma Longa* Radix isolated by HPLC and GC-MS for promotion of blood circulation and relief of pain. *J. Ethnopharmacol.* **2019**, *234*, 68–75. [CrossRef] [PubMed]
47. Singh, V.; Jain, M.; Misra, A.; Khanna, V.; Rana, M.; Prakash, P. Curcuma oil ameliorates hyperlipidaemia and associated deleterious effects in golden Syrian hamsters. *Br. J. Nutr.* **2013**, *110*, 437–446. [CrossRef] [PubMed]
48. Srivastava, R.A.K. Evaluation of anti-atherosclerotic activities of PPAR-α, PPAR-γ, and LXR agonists in hyperlipidemic atherosclerosis-susceptible F(1)B hamsters. *Atherosclerosis* **2011**, *214*, 86–93. [CrossRef]
49. Valasek, M.A.; Repa, J.J.; Quan, G.; Dietschy, J.M.; Turley, S.D. Inhibiting intestinal NPC1L1 activity prevents diet-induced increase in biliary cholesterol in Golden Syrian hamsters. *Am. J. Physiol. Gastrointest Liver Physiol.* **2008**, *295*, G813–G822. [CrossRef]
50. Yu, L.; Li-Hawkins, J.; Hammer, R.E.; Berge, K.E.; Horton, J.D.; Cohen, J.C. Overexpression of ABCG5 and ABCG8 promotes biliary cholesterol secretion and reduces fractional absorption of dietary cholesterol. *J. Clin. Investg.* **2002**, *110*, 671–680. [CrossRef]
51. Akinyemi, A.J.; Adeniyi, P.A. Effect of Essential Oils from Ginger (*Zingiber officinale*) and Turmeric (*Curcuma longa*) Rhizomes on Some Inflammatory Biomarkers in Cadmium Induced Neurotoxicity in Rats. *J. Toxicol.* **2018**, *2018*, 4109491. [CrossRef]
52. Matsumura, S.; Murata, K.; Yoshioka, Y.; Matsuda, H. Search for β-Secretase Inhibitors from Natural Spices. *Nat. Prod. Commun.* **2016**, *11*, 507–510. [CrossRef]
53. Orellana-Paucar, A.M.; Serruys, A.S.K.; Afrikanova, T.; Maes, J.; de Borggraeve, W.; Alen, J. Anticonvulsant activity of bisabolene sesquiterpenoids of *Curcuma longa* in zebrafish and mouse seizure models. *Epilepsy Behav.* **2012**, *24*, 14–22. [CrossRef] [PubMed]
54. Dohare, P.; Garg, P.; Sharma, U.; Jagannathan, N.R.; Ray, M. Neuroprotective efficacy and therapeutic window of curcuma oil: In rat embolic stroke model. *BMC Complement. Altern. Med.* **2008**, *8*, 55. [CrossRef] [PubMed]
55. Choo, B.K.M.; Shaikh, M.F. Mechanism of *Curcuma longa* and Its Neuroactive Components for the Management of Epileptic Seizures: A Systematic Review. *Curr. Neuropharmacol.* **2021**, *19*, 1496. [CrossRef] [PubMed]
56. Akinyemi, A.J.; Faboya, O.L.; Paul, A.A.; Olayide, I.; Faboya, O.A.; Oluwasola, T.A. Nephroprotective Effect of Essential Oils from Ginger (Zingiber officinale) and Turmeric (*Curcuma longa*) Rhizomes against Cadmium-induced Nephrotoxicity in Rats. *J. Oleo Sci.* **2018**, *67*, 1339–1345. [CrossRef]
57. Hans, V.M.; Grover, H.S.; Deswal, H.; Agarwal, P. Antimicrobial Efficacy of Various Essential Oils at Varying Concentrations against Periopathogen Porphyromonas gingivalis. *J. Clin. Diagn. Res.* **2016**, *10*, ZC16. [CrossRef]
58. Lee, K.H.; Kim, B.S.; Keum, K.S.; Yu, H.H.; Kim, Y.H.; Chang, B.S.; Ra, J.-Y.; Moon, H.-D.; Seo, B.-R.; Choi, N.-Y.; et al. Essential Oil of *Curcuma longa* Inhibits Streptococcus mutans Biofilm Formation. *J. Food Sci.* **2011**, *76*, H226–H230. [CrossRef]
59. Köhler, B.; Birkhed, D.; Olsson, S. Acid production by human strains of Streptococcus mutans and Streptococcus sobrinus. *Caries Res.* **1995**, *29*, 402–406. [CrossRef]
60. Nagpal, M.; Sood, S. Role of curcumin in systemic and oral health: An overview. *J. Nat. Sci. Biol. Med.* **2013**, *4*, 3–7.
61. Álvarez, N.M.; Ortíz, A.A.; Martínez, O.C. Actividad antibacteriana in vitro de *Curcuma longa* (Zingiberaceae) frente a bacterias nosocomiales en Montería, Colombia. *Rev. Biol. Trop.* **2016**, *64*, 1201–1208. [CrossRef]
62. Negi, P.S.; Jayaprakasha, G.K.; Rao, L.J.M.; Sakariah, K.K. Antibacterial activity of turmeric oil: A byproduct from curcumin manufacture. *J. Agric. Food Chem.* **1999**, *47*, 4297–4300. [CrossRef]
63. Teles, A.M.; Rosa, T.D.D.S.; Mouchrek, A.N.; Abreu-Silva, A.L.; da Silva Calabrese, K.; Almeida-Souza, F. Cinnamomum zeylanicum, Origanum vulgare, and *Curcuma longa* Essential Oils: Chemical Composition, Antimicrobial and Antileishmanial Activity. *Evid. Based Complement. Altern. Med.* **2019**, *2019*, 2421695. [CrossRef] [PubMed]
64. Gupta, A.; Mahajan, S.; Sharma, R. Evaluation of antimicrobial activity of *Curcuma longa* rhizome extract against Staphylococcus aureus. *Biotechnol. Rep.* **2015**, *6*, 51. [CrossRef] [PubMed]
65. Ogidi, C.O.; Ojo, A.E.; Ajayi-Moses, O.B.; Aladejana, O.M.; Thonda, O.A.; Akinyele, B.J. Synergistic antifungal evaluation of over-the-counter antifungal creams with turmeric essential oil or Aloe vera gel against pathogenic fungi. *BMC Complement. Med. Ther.* **2021**, *21*, 47. [CrossRef] [PubMed]
66. Apisariyakul, A.; Vanittanakom, N.; Buddhasukh, D. Antifungal activity of turmeric oil extracted from *Curcuma longa* (Zingiberaceae). *J. Ethnopharmacol.* **1995**, *49*, 163–169. [CrossRef]
67. Jankasem, M.; Wuthi-udomlert, M.; Gritsanapan, W. Antidermatophytic Properties of Ar-Turmerone, Turmeric Oil, and *Curcuma longa* Preparations. *ISRN Dermatol.* **2013**, *2013*, 250597. [CrossRef] [PubMed]
68. Nazzaro, F.; Fratianni, F.; Coppola, R.; de Feo, V. Essential Oils and Antifungal Activity. *Pharmaceuticals* **2017**, *10*, 86. [CrossRef]
69. Romoli, J.C.Z.; Silva, M.V.; Pante, G.C.; Hoeltgebaum, D.; Castro, J.C.; Oliveira da Rocha, G.H. Anti-mycotoxigenic and antifungal activity of ginger, turmeric, thyme and rosemary essential oils in deoxynivalenol (DON) and zearalenone (ZEA) producing Fusarium graminearum. *Food Addit. Contam. Part A* **2022**, *39*, 362–372. [CrossRef]
70. Hu, Y.; Zhang, J.; Kong, W.; Zhao, G.; Yang, M. Mechanisms of antifungal and anti-aflatoxigenic properties of essential oil derived from turmeric (*Curcuma longa* L.) on Aspergillus flavus. *Food Chem.* **2017**, *220*, 1–8. [CrossRef]
71. Ferreira, F.D.; Mossini, S.A.G.; Ferreira, F.M.D.; Arrotéia, C.C.; da Costa, C.L.; Nakamura, C.V. The inhibitory effects of *Curcuma longa* L. essential oil and curcumin on Aspergillus flavus link growth and morphology. *Sci. World J.* **2013**, *2013*, 343804. [CrossRef]
72. Jayaprakasha, G.K.; Negi, P.S.; Anandharamakrishnan, C.; Sakariah, K.K. Chemical composition of turmeric oil -a byproduct from turmeric oleoresin industry and its inhibitory activity against different fungi. *Z. Nat. C J. Biosci.* **2001**, *56*, 40–44. [CrossRef]

73. Achimón, F.; Brito, V.D.; Pizzolitto, R.P.; Ramirez Sanchez, A.; Gómez, E.A.; Zygadlo, J.A. Chemical composition and antifungal properties of commercial essential oils against the maize phytopathogenic fungus Fusarium verticillioides. *Rev. Argent. Microbiol.* **2021**, *53*, 292–303. [CrossRef] [PubMed]
74. Le, T.B.; Beaufay, C.; Nghiem, D.T.; Pham, T.A.; Mingeot-Leclercq, M.P.; Quetin-Leclercq, J. Evaluation of the Anti-Trypanosomal Activity of Vietnamese Essential Oils, with Emphasis on *Curcuma longa* L. and Its Components. *Molecules* **2019**, *24*, 1158. [CrossRef] [PubMed]
75. Chauhan, I.S.; Rao, G.S.; Shankar, J.; Chauhan, L.K.S.; Kapadia, G.J.; Singh, N. Chemoprevention of Leishmaniasis: In-vitro antiparasitic activity of dibenzalacetone, a synthetic curcumin analog leads to apoptotic cell death in *Leishmania donovani*. *Parasitol. Int.* **2018**, *67*, 627–636. [CrossRef] [PubMed]
76. Mahmoudvand, H.; Pakravanan, M.; Aflatoonian, M.R.; Khalaf, A.K.; Niazi, M.; Mirbadie, S.R. Efficacy and safety of *Curcuma longa* essential oil to inactivate hydatid cyst protoscoleces. *BMC Complement. Altern. Med.* **2019**, *19*, 187. [CrossRef] [PubMed]
77. Brunetti, E.; Kern, P.; Vuitton, D.A. Expert consensus for the diagnosis and treatment of cystic and alveolar echinococcosis in humans. *Acta Trop.* **2010**, *114*, 1–16. [CrossRef]
78. Zhu, J.; Zeng, X.; O'Neal, M.; Schultz, G.; Tucker, B.; Coats, J. Mosquito larvicidal activity of botanical-based mosquito repellents. *J. Am. Mosq. Control Assoc.* **2008**, *24*, 161–168. [CrossRef]
79. Groh, I.A.M.; Rudakovski, O.; Gründken, M.; Schroeter, A.; Marko, D.; Esselen, M. Methyleugenol and oxidative metabolites induce DNA damage and interact with human topoisomerases. *Arch. Toxicol.* **2016**, *90*, 2809–2823. [CrossRef]
80. Tascone, O.; Roy, C.; Filippi, J.J.; Meierhenrich, U.J. Use, analysis, and regulation of pesticides in natural extracts, essential oils, concretes, and absolutes. *Anal. Bioanal. Chem.* **2014**, *406*, 971–980. [CrossRef]
81. Donelli, D.; Antonelli, M.; Firenzuoli, F. Considerations about turmeric-associated hepatotoxicity following a series of cases occurred in Italy: Is turmeric really a new hepatotoxic substance? *Intern. Emerg. Med.* **2020**, *15*, 725–726. [CrossRef]
82. Suhail, F.K.; Masood, U.; Sharma, A.; John, S.; Dhamoon, A. Turmeric supplement induced hepatotoxicity: A rare complication of a poorly regulated substance. *Clin. Toxicol.* **2020**, *58*, 216–217. [CrossRef]
83. Lukefahr, A.L.; McEvoy, S.; Alfafara, C.; Funk, J.L. Drug-induced autoimmune hepatitis associated with turmeric dietary supplement use. *BMJ Case Rep.* **2018**, *2018*, bcr-2018. [CrossRef] [PubMed]
84. Balaji, S.; Chempakam, B. Toxicity prediction of compounds from turmeric (*Curcuma longa* L). *Food Chem. Toxicol.* **2010**, *48*, 2951–2959. [CrossRef] [PubMed]
85. Joshi, J.; Ghaisas, S.; Vaidya, A.; Vaidya, R.; Kamat, D.V.; Bhagwat, A.N.; Bhide, S. Early human safety study of turmeric oil (*Curcuma longa* oil) administered orally in healthy volunteers. *J. Assoc. Physicians India* **2003**, *51*, 1055–1060. [PubMed]

MDPI
St. Alban-Anlage 66
4052 Basel
Switzerland
Tel. +41 61 683 77 34
Fax +41 61 302 89 18
www.mdpi.com

Molecules Editorial Office
E-mail: molecules@mdpi.com
www.mdpi.com/journal/molecules

www.ingramcontent.com/pod-product-compliance
Lightning Source LLC
Chambersburg PA
CBHW041946260926
51850CB00018B/1653
* 9 7 8 3 0 3 6 5 5 6 4 1 3 *